Vulnerability and Adaptation to Drought

Energy, Ecology, and the Environment Series

ISSN 1919-7144 (Print) ISSN 1925-2935 (Online)

This series explores how we live and work with each other on the planet, how we use its resources, and the issues and events that shape our thinking on energy, ecology, and the environment. The Alberta experience in a global arena is showcased.

No. 1 · *Places: Linking Nature, Culture and Planning*
J. Gordon Nelson and Patrick L. Lawrence

No. 2 · *A New Era for Wolves and People: Wolf Recovery, Human Attitudes, and Policy*
Edited by Marco Musiani, Luigi Boitani, and Paul Paquet

No. 3 · *The World of Wolves: New Perspectives on Ecology, Behaviour and Management*
Edited by Marco Musiani, Luigi Boitani, and Paul Paquet

No. 4 · *Parks, Peace, and Partnership: Global Initiatives in Transboundary Conservation*
Edited by Michael S. Quinn, Len Broberg, and Wayne Freimund

No. 5 · *Wilderness and Waterpower: How Banff National Park Became a Hydroelectric Storage Reservoir*
Christopher Armstrong and H. V. Nelles

No. 6 · *L'Alberta Autophage: Identités, mythes et discours du pétrole dans l'Ouest canadien*
Dominique Perron

No. 7 · *Greening the Maple: Canadian Ecocriticism in Context* Edited by Ella Soper and Nicholas Bradley

No. 8 · *Petropolitics: Petroleum Development, Markets and Regulations, Alberta as an Illustrative History*
Alan J. MacFadyen and G. Campbell Watkins

No. 9 · *Vulnerability and Adaptation to Drought: The Canadian Prairies and South America*
Edited by Harry Diaz, Margot Hurlbert, and Jim Warren

Vulnerability and Adaptation to Drought

The Canadian Prairies and South America

Edited by
HARRY DIAZ, MARGOT HURLBERT, AND JIM WARREN

Energy, Ecology, and the Environment Series
ISSN 1919-7144 (Print) ISSN 1925-2935 (Online)

University of Calgary Press
2500 University Drive NW
Calgary, Alberta
Canada T2N 1N4
press.ucalgary.ca

LIBRARY AND ARCHIVES CANADA CATALOGUING IN PUBLICATION

Vulnerability and adaptation to drought : the Canadian prairies and South America / edited by Harry Diaz, Margot Hurlbert, and Jim Warren.

(Energy, ecology, and the environment series, ISSN 1919-7144 ; no. 9)
Includes bibliographical references and index.
Issued in print and electronic formats.
ISBN 978-1-55238-819-8 (paperback).–ISBN 978-1-55238-821-1 (pdf).–
ISBN 978-1-55238-820-4 (open access pdf).–ISBN 978-1-55238-822-8 (epub).–
ISBN 978-1-55238-823-5 (mobi)

1. Droughts–Prairie Provinces. 2. Droughts–South America. 3. Drought management–Prairie Provinces. 4. Drought management–South America. I. Diaz, Harry P., author, editor II. Hurlbert, Margot Ann, 1964-, author, editor III. Warren, Jim W. (Jim William), author, editor IV. Series: Energy, ecology, and the environment series ; no. 9

QC929.28.C32P72 2016 363.34'929709712 C2016-901975-6
C2016-901976-4

The University of Calgary Press acknowledges the support of the Government of Alberta through the Alberta Media Fund for our publications. We acknowledge the financial support of the Government of Canada through the Canada Book Fund for our publishing activities. We acknowledge the financial support of the Canada Council for the Arts for our publishing program.

Canada Council for the Arts
Conseil des Arts du Canada

This book was completed with the generous support of the Social Sciences and Humanities Research Council of Canada.

Printed and bound in Canada by Marquis
This book is printed on Enviro paper

Cover image: #4650692 (colourbox.com)
Copyediting by Kelley Kissner
Cover design, page design, and typesetting by Melina Cusano

CONTENTS

INTRODUCTION

Harry Diaz, Margot Hurlbert, and Jim Warren

Scope and Purpose

Climate change is perhaps one of the most prominent indicators of global environmental change as well as an important source of increased human vulnerability. An unprecedented concentration of greenhouse gases in the atmosphere is linked to an overall warming of the planet, which has been affecting climate and weather patterns. The World Meteorological Organization has estimated that more than 370,000 lives were lost between 2001 and 2010 as a result of extreme climate conditions, including heat waves, cold spells, droughts, storms, and floods, marking a 20% increase in deaths compared to 1991–2000 (WMO 2013). The potential impacts of climate change are of significant concern for those regions of the world subject to drought. Should future droughts exceed previous experience in terms of frequency, duration, and severity, the threats to human lives, livelihoods, and ecosystems could be substantial.

Drought is the one of the most significant natural hazards affecting social and economic systems in many areas of the world. It is particularly hazardous for agricultural communities, where livelihoods depend on natural systems. This is especially true for the Canadian Prairies, where droughts have been one of the most serious recurring natural hazards. Indeed, even when viewed from a national perspective, droughts are among the most economically devastating natural disasters experienced by Canadians over the past century. The most recent widespread severe Canadian drought in 2001–2 produced a $5.8 billion drop in gross domestic product (GDP) and was responsible for an estimated 41,000 lost jobs (Wheaton et al. 2010: 280). Saskatchewan and Alberta were the hardest-hit provinces. Moreover, the magnitude and frequency of Prairie droughts are projected to increase under climate change, potentially increasing people's vulnerabilities and associated risks (Sauchyn et al. 2010).

Drought is arguably one of the most problematic disasters, not only because of the magnitude of its damages but also because "it is one of the most underrated and least understood of disasters" (Sheffield and Wood 2011: xi). The main goal of this book is to contribute to a better understanding of the complexities of drought and its impacts on people's livelihoods from a perspective that emphasizes both vulnerabilities and adaptive capacity in the context of an increasingly complex relationship between nature and society.

The book is the product of a decade of international collaborative interdisciplinary research effort by a network of Canadian and Latin American researchers to understand rural people's vulnerabilities to climate in arid areas. Most of the studies done by the network, both in Canada and Latin America, have adopted the same conceptual and methodological approaches discussed in the first chapter of this book. In most chapters, we have been particularly interested in the ways in which rural people formulate their responses to past, current, and forecasted climate risks and the limits that their social, economic, and political conditions impose on these responses. Most chapters in the book are related to studies on the vulnerabilities and adaptive capacities of Canadian Prairie communities, but we have also included two chapters based on *regional* drought studies in Argentina and Chile from our collaborators in these two countries. Today, more than ever, the global nature of the environmental transformations associated with climate change imposes a need for an explicitly comparative perspective. This perspective allows us to study the complexities and challenges of the processes that emerge in the coupled social and natural systems, which is required to find solutions and alternatives.

A considerable portion of the research supporting many of the chapters in this book was developed in the context of several research projects. Two of these projects were the "Rural Community Adaptation to Drought" (RCAD) project, which focused on the exposure and adaptive practices to drought impacts for rural communities in Saskatchewan (RCAD 2012), and the "Water Governance and Climate Change: The Engagement of Civil Society" project (Hurlbert et al. 2015). The other projects were multinational research efforts involving scholars and drought-management practitioners from Canada and Latin America. These projects were the "Institutional Adaptation to Climate Change" project (IACC 2009), the "Vulnerability and Adaptation to Climate Extremes in the Americas" project, and the "Coming Down the Mountain" project, which allowed for a study of drought vulnerabilities in Argentina, Chile, and Bolivia.[1] The international composition of these projects constituted efforts to share insights from the drought-related experiences of a variety of communities with different levels of exposure and adaptive assets. In these terms, the two Latin American cases presented in this book, Chile and Argentina, are important because they provide insights from regions characterized by long histories of water scarcity and use of irrigation to reduce local vulnerabilities, and also by an unequal distribution of vulnerability among producers.

The book also emphasizes the need to integrate both the natural and social sciences in understanding vulnerabilities. Droughts are "normal" events in the history of the regions covered in this book, but they can become hazards and disasters in the context of the prevailing social and economic conditions that exist in social systems during the drought. The magnitude and severity of the drought are relevant to understanding its impacts, but no less relevant are the social circumstances that shape the capacity of people and their livelihoods to cope with those impacts. In these terms, we have made a special effort to integrate chapters that emphasize both the social and natural scientific perspectives.

This book has been organized into six sections. The first section includes a single chapter that discusses the theoretical and methodological perspectives applied by most of the following chapters. Chapter 1, by Wandel, Diaz, Warren, Hadarits, Hurlbert, and Pittman, frames the discussion in terms of the vulnerability approach. In simple terms, this approach contends that the resilience of a social system—which could be a community, farm, family, or any other social entity—exposed to a natural hazard is a function of the characteristics of the hazard and of the balance between the relative sensitivity of the system to the exposure and its adaptive capacity. More precisely, the chapter defines vulnerability as the degree to which systems, such as a farm or a community, are susceptible to the adverse impacts of climate variability and extremes (such as drought), as well as to other types of stressors and change (Kiparsky et al. 2012; Smit and Wandel 2006; Wisner et al. 2003). The literature indicates that the adaptive capacity or resilience of a community is affected by its access to certain biophysical, social, and economic resources. These resources are referred to as the "determinants of adaptive capacity" and alternatively as the "assets" or "capitals" required to support resilience. Access to the appropriate mix of these assets enhances the capacity of a community to adapt to adverse conditions (IPCC 2001).

Most social systems are capable of adjusting to climate conditions that vary within the parameters of average long-term experience. Indeed, agriculture on the Canadian Prairies, as well as in the agricultural regions of Argentina and Chile discussed in this book, would appear to be relatively resilient to drought conditions that occur below a certain threshold of intensity and duration. However, when climate variability exceeds previous experience and thresholds for resilience, the level of adaptive capacity resident in the community may or may not prove sufficient. When confronted by climate forecasts that predict exposures will far exceed previous experience, one would reasonably expect prudent actors to assess their current levels of vulnerability and adaptive capacity, and endeavour to enhance those facets of adaptive resources that might be lacking.

Access to and control of these resources are important in reducing vulnerabilities, but it is the capabilities of actors to organize them into adaptive activities that define the balance between sensitivity (determined by the lack of or limited resources) and adaptation (defined by the existence of resources that could be mobilized to reduce sensitivity). The first chapter also emphasizes the argument that local vulnerabilities are the product of the multiple interactions between several processes that affect the locality, what Leichenko and O'Brien called "pathways of double exposures" (2008: 5). Thus, issues such as local sensitivities, adaptive capacity, and resiliency acquire a challenging complexity.

The second section of the book deals with the dimension of drought as a hazard to which Canadian rural people are exposed. The section, composed of two chapters, discusses drought from the perspective of climatology, focusing on the past and future features of drought in the Canadian Prairies. In these terms, the section contributes lessons from climate sciences, which should allow for a better understanding of the following three sections.

Chapter 2, by Sauchyn and Kerr, provides an overview of paleoclimatic research for the Prairies region. Their work suggests that if the long-range climate history of the Prairies is a meaningful clue to what we might expect in the short- and long-term future, current drought management practices may not be sufficient to sustain agriculture and Prairie communities in their current form. They show that severe and protracted drought has been a recurrent phenomenon on the Prairies for at least 1,000 years; thus, if past climate is any indication of what the future might bring, residents of the Canadian Prairies can expect to encounter multi-year periods of severe drought in the decades ahead. The adaptations that have sustained Prairie agriculture over the past century were made in response to droughts which were less extreme than many of those which occurred in preceding centuries. Some of those droughts far exceeded the thresholds of severity and duration in which current adaptive strategies were developed.

Sauchyn and Kerr's insight into the Prairie climate of the past is followed by a chapter by Wheaton, Sauchyn, and Bonsal (Chapter 3), which provides a regional assessment of the latest climate science for the Prairies and provides insight into the potential intensity and frequency of future droughts in the region. They explain that the variability of the region's already highly variable climate conditions will increase. Severe weather, including severe drought along with occasional extreme rainfall events, is expected to become more common over the course of the twenty-first century. This suggests the need for major departures from agricultural practices and water management strategies that were developed in response to droughts over the past century. Taken together, these two chapters show that similar warnings arise whether we look into the climate past or future. Communities on the Canadian Prairies could benefit by preparing to adapt to a climate future and associated hazards that present new challenges which threaten to exceed past levels of drought resilience.

The third section contains five chapters, all focused on drought crises and the adaptive responses of Prairie agricultural producers to these crises. It is in the context of the insights presented by climate sciences in the previous section that we endeavour to understand drought and its effects on people and their social systems. The chapters in this section explore the effects of past and recent droughts on the Canadian Prairies and the ways people have adapted to drought conditions. We assume that learning what has and has not worked to enhance people's drought resilience is valuable, in terms of understanding how to better deal with drought both today and in the decades ahead.

Chapter 4, by Kulshreshtha, Wheaton, and Wittrock, discusses the vulnerability of several rural communities in the Canadian Prairies using the 2001–2 drought as a point of reference. This was one of the most serious severe droughts experienced in recent times in the Canadian Prairies, where dry conditions were accompanied by high temperatures, which increased the severity of the drought because of higher evapotranspiration. The chapter provides us with a comprehensive view of the economic and social impacts of drought on different economic sectors and on a group of rural communities in southern Alberta and southwestern Saskatchewan, including the adaptation measures undertaken by agricultural producers.

Chapter 5, by J. Warren, focuses on specific processes of adaptation to drought implemented by Prairie farmers. The chapter deals mainly with technological innovations—new farming practices and the use of appropriate machinery—designed to reduce exposure to drought. Warren shows how community innovation and adaptation processes are well-understood and valued processes that have emerged as a result of increasing human capital in the region and have been integrated into the cultural material of the rural communities. The chapter demonstrates the existence of an adaptive capacity that is linked to creativity, flexibility, and adaptability as important local values.

Chapter 6, also by Warren, discusses "the other side of the coin," showing us that conventional measures of adaptation, such as irrigation, do not always ensure drought resiliency. The chapter is focused on irrigation infrastructure in southwest Saskatchewan, and it demonstrates how the agronomic impacts of drought can be exacerbated by social and economic conditions. The case discussed by Warren shows how the coincidence of drought with depressed farm commodities prices, rising input costs, and institutional weakness can reduce the effectiveness of irrigation, contributing to a heightened state of local vulnerability.

The last chapter in this section, Chapter 7 by Fletcher and Knuttila, provides important insight into the socially constructed and experienced elements of drought on the Prairies. Following the argument that vulnerabilities and adaptive capacity are unequally distributed in society, the chapter examines the gendered characteristics of the impacts of drought and vulnerability reduction in the context of the farm economy. The authors propose that gender vulnerabilities to climate must be linked to an understanding of the processes of industrialization, corporatization, and rapid farm expansion, and accordingly, adaptive policies must consider these processes to increase the resiliency of women.

Section 4 focuses on governance, which is a very specific aspect of vulnerability and adaptive capacity. The first chapter in this volume, which provides the theoretical framework for the book, indicates that institutional capital is an important determinant of adaptive capacity. In most countries, the most important expressions of this form of capital are the programs and policies developed and implemented by governments in co-operation with the institutions of civil society. The availability of resources related to these programs and policies to local people contribute significantly to their adaptive capacity.

The impact of past droughts, particularly the dry decade of the 1930s, has been seared into the socio-economic and political fabric of Prairie communities. And while drought-induced crop failure and water shortages have been the cause of great hardship and adversity, and of the disappearance of many farms and ranches, they have also encouraged the development of a variety of coping strategies. A range of adaptations involving the creation and adoption of new institutional frameworks emerged in response to drought conditions on the Prairies.

Chapter 8, by G. Marchildon, is focused on these institutional interventions developed as a response to the extreme weather conditions of the earlier decades of the past century. Focusing on two case studies, the Special Areas Board and the Prairie Farm Rehabilitation Administration, Marchildon shows how these new institutional arrangements contribute to reduce individual and community vulnerabilities in the most drought-affected areas of the Prairies. An important insight from this chapter is the need for more robust policy interventions in the context of future droughts.

The following two chapters, by M. Hurlbert, assess the present institutional framework, which has evolved to manage the challenges that drought presents to communities. In Chapter 9, Hurlbert assesses the Canadian government agencies and programs that currently deal with drought. She reviews several programs at the federal level, as well as those existing within the provinces of Alberta and Saskatchewan, in terms of their capacity to assist local agricultural producers in drought situations. She argues that these programs have existed for some time, but they have not been reinvigorated to respond to drought periods lasting more than two years. Hurlbert's argument certainly supports the insight developed by Warren in his chapter on irrigation in southwestern Saskatchewan.

In Chapter 10, Hurlbert reviews a set of policies and programs in the context of water governance in the Prairies. Given the essential role of water availability in drought conditions, Hurlbert examines some of the adaptive institutional principles applicable to water governance; the structure of this type of governance in the Canadian Prairie provinces; and the regulatory, management, and market instruments relevant to water. She concludes that there is an urgent need for more defined institutional boundaries, enhanced communication of the roles of water organizations, and coordination among water organizations.

Section 5 assumes a less conventional approach to drought. Our contributors propose that understanding drought and enhancing people's drought resilience is an interdisciplinary activity. However, combining the work of scholars and drought management practitioners with different areas of interest and methodological approaches with practical resilience-building activities can require new or better interactive processes to be developed. The two chapters in this section take the perspective that interdisciplinarity and transdisciplinarity are essential to understand the dynamics of global environmental change and to resolve the "wicked problems" created by this transformation (Brown 2010: 62–63).

Chapter 11, by Corkal, Morito, and Rojas, provides insight into how seemingly disparate disciplines and areas of concern can be brought together to develop vulnerability-reducing responses to drought-related stress on water resources. The chapter focuses on the idea that vulnerability is a socially constructed concept that expresses people's conceptions and ideas toward the harms that threaten them. In these terms, the issue of values is central to understanding how people conceive and react to events such as droughts and how value-analysis could be an important instrument to address conflicts that emerge in the context of water scarcities.

In Chapter 12, Pittman, Corkal, Hadarits, Harrison, Hurlbert, and Unvoas offer a transdisciplinary alternative to primarily reactive past models, incorporating not only the perspective from science but also the concerns and interests of a large number of stakeholders. These authors, coming from a variety of perspectives, describe the value of preparedness planning, anticipating the challenges presented by future droughts and working to avoid adverse impacts through vulnerability reduction. Their interest is, in part, a response to predictions that droughts in the future could be far more intense and damaging than those yet experienced by Prairie communities.

In conformity with the principle that enhancing drought preparedness and resilience in one part of the world may provide insight for those dealing with similar problems in another region, we have included examples of scholarship from two regions in Latin America. The last section of the book describes how agricultural producers in regions of Argentina and Chile deal with livelihood disruptions caused by drought and how they are confronting new prospects related to climate change. The comparison is interesting since the studies that support these last two chapters used the same conceptual and methodological approaches used in the Canadian studies. As in the Canadian case, both regions have economies that are predominantly agricultural and that depend, to a large extent, on snowpack in the mountains—the Andes in the case of Latin America and the Rockies in the case of Canada—that feeds the regional rivers, the main source of water for irrigation purposes. No less relevant to the Canadian case is the role of social and economic conditions in framing the conditions of vulnerability of local producers in these two regions. The imposition of neo-liberal policies and the institutional incapacity to secure equitable access to resources provide clear examples of how the social and political dimensions could transform a drought from being a climate hazard into a disaster for many people. In these terms, these two regions provide a glimpse of potential future conditions of vulnerability if Canada continues with the trend of neo-liberalizing its economy, restructuring its agricultural sector, and reducing its institutional support for producers (see Chapter 6 by Warren on irrigation in southwest Saskatchewan; see also Wiebe 2012; Magnan 2014; Young and Matthews 2007).

Chapter 13, by Hadarits, Santibáñez, and Pittman, discusses the implications of drought on Chilean agricultural producers in the Maule region of Chile, a region that, as in the case of Mendoza, Argentina, is mostly a wine-producing region. Similar to many other regions in Chile, the Maule region is seriously affected by droughts. The chapter describes the drought-related vulnerabilities for the regional wine industry based on a case study and a vulnerability assessment approach, as discussed in Chapter 1. It demonstrates the complexity of drought impacts, arguing that the exposure/sensitivity of the wine industry could be adverse or beneficial depending on many variables. To reduce detrimental impacts, many producers have developed a wide range of strategies, which could be important to face the near future challenges of declining precipitation and increasing demand for water due to increases in temperature. These strategies offer important lessons for Canadian agricultural producers.

The last chapter, Chapter 14 by Montaña and Boninsegna, discusses drought preparedness in the Mendoza River basin in central-western Argentina. The area is mainly dryland with large oases that depend on water provided by the Mendoza River and that support very intensive agriculture which produces world-renowned wines. The chapter discusses the climatological conditions of the region as well as the influence of its social and economic structures in shaping the vulnerabilities and adaptive capacities of agricultural producers in the basin. As in the Canadian case, the chapter illustrates another example of the weakness of short-term adaptation strategies compared with long-term solutions planning. In this vein, the chapter ends by emphasizing the need for structured policies that could improve water efficiency in this drought-prone region and provide conditions for a reduction and a more equitable distribution of the vulnerabilities.

NOTE

1 The Rural Communities Adaptation to Drought project took place between 2009 and 2013. It was carried out by a group of Canadian researchers from the universities of Regina, Saskatchewan, and Waterloo. The Institutional Adaptation to Climate Change project ran from 2004 to 2011 and involved a team of academics and drought and water management practitioners from Canada and Chile. Both projects were supported financially by the Social Sciences and Humanities Research Council of Canada. Their institutional home was the Canadian Plains Research Center at the University of Regina. The Vulnerability and Adaptation to Climate Extremes in the Americas project was launched in 2013 and it will be completed in 2016. It has been supported by International Development Research Center, the Social Sciences and Humanities Research Council of Canada, and the Natural Science and Engineering Research Council of Canada. The project team includes academics and practitioners from Canada, Chile, Columbia, Brazil, and Argentina. The Water Governance and Climate Change: The Engagement of Civil Society project was launched in 2010 and completed in 2014. For more information on these projects, visit the Prairie Adaptation Research Collaborative website at http://www.parc.ca. The project Coming Down the Mountain was supported by the Interamerican Institute for Global Change Research; it was initiated in 2008 and ended in 2011.

References

Brown, V.A. 2010. "Collective Inquiries and Its Wicked Problems." In V. Brown, J. Harris, and J. Russell (eds.), *Tackling Wicked Problems through the Transdisciplinary Action*. London: Earthscan.

Hurlbert, M., E. Andrews, Y. Tesfamariam, and J. Warren. 2015. *Governing Water: Deliberative Institutions and Adaptation*. Regina: Prairie Adaptation Research Collaborative. http://www.parc.ca/vacea/assets/PDF/reports/local%20water%20governance%20final%20report.pdf.

IACC (Institutional Adaptation to Climate Change). 2009. *Institutional Adaptation to Climate Change Project: Final Report December 2009*. Regina: Canadian Plains Research Center.

IPCC (Intergovernmental Panel on Climate Change). 2001. *Climate Change 2001: Impacts Adaptation and Vulnerability. Contribution of Working Group II to the Third Assessment Report of the Intergovernmental Panel on Climate Change*. Geneva: United Nations Environment Programme/World Meteorological Organization.

Kiparsky, M., A. Milman, and S. Vicuna. 2012. "Climate and Water: Knowledge of Impacts to Action on Adaptation." *Annual Review of Environment and Resources* 37: 163–94.
Leichenko, R., and K. O'Brien. 2008. *Environmental Change and Globalization. Double Exposures*. Toronto: Oxford University Press.
Magnan, A. 2014. "The Rise and Fall of a Prairie Giant: The Canadian Wheat Board in Food Regime History." In S. Wolf and A. Bonnano (eds.), *The Neoliberal Regime in the Agri-Food Sector. Crisis, Resilience, and Restructuring*. New York: Routledge and Earthscan.
RCAD (Rural Communities Adaptation to Drought Project). 2012. *Research Report*. Regina: Canadian Plains Research Center.
Sauchyn, D., H. Diaz, and S. Kulshreshtha. 2010. *The New Normal. The Canadian Prairies in a Changing Climate*. Regina: Canadian Plains Research Center Press.
Sheffield J., and E. Wood. 2011. *Drought: Past Problems and Future Scenarios*. London: Earthscan.
Smit, B., and J. Wandel. 2006. "Adaptation, Adaptive Capacity and Vulnerability." *Global Environment and Change* 16: 282–92.
Spry, I. 1995. *The Palliser Expedition: The Dramatic Story of Western Canadian Exploration 1857–1860*. Saskatoon: Fifth House Books.
Wheaton, E., S. Kulshreshtha, and V. Wittrock. 2010. "Assessment of the 2001 and 2002 Drought Impacts in the Prairie Provinces, Canada." In D. Sauchyn, H. Diaz, and S. Kulshreshtha (eds.), *The New Normal: The Canadian Prairies in a Changing Climate*. Regina: Canadian Plains Research Center Press.
Wiebe, N. 2012. "Crisis in the Food System: The Farm Crisis." In M. Koc, J. Summer, and A. Winson (eds.), *Critical Perspectives in Food Studies*. Toronto: Oxford University Press.
Wisner, B., P. Blaikie, T. Cannon, and I. Davis. 2003. *At Risk*. London: Routledge.
WMO (World Meteorological Organization). 2013. *The Global Climate of 2001–2010: A Decade of Climate Extremes*. WMO-No. 1103. Geneva: WMO.
Young, N., and R. Matthews. 2007. "Resource Economies and Neoliberal Experimentation: The Reform of Industry and Community in Rural British Columbia." *Area* 39, no. 2: 176–85.

PART 1

FRAMING THE BOOK

CHAPTER 1

DROUGHT AND VULNERABILITY: A CONCEPTUAL APPROACH

Johanna Wandel, Harry Diaz, Jim Warren, Monica Hadarits, Margot Hurlbert, and Jeremy Pittman

The fundamental message of this book is the need to discuss and understand drought—not just in terms of climatic parameters such as timing, duration, intensity, and geographic scope, but also relative to human exposure-sensitivity. A holistic understanding of the socio-economic conditions that define human sensitivity, vulnerability, and adaptive capacity is fundamental to grasp the implications of drought. This chapter provides the conceptual framework that contextualizes the interdisciplinary perspective informing this book and its chapters. It reviews some of the traditional approaches to drought, ranging from hydrological to socio-economic droughts, and argues for the need to understand drought in terms of contextual vulnerability and its components. By adopting this contextual approach, we are able to identify how social and economic conditions influence exposure, sensitivity, and adaptive capacity to droughts, allowing for a better understanding of how people experience and live with this hazard. Contextually based approaches are generally rooted in local cases and facilitate a comprehensive understanding of problems from a "bottom up" perspective; however, there is a need to couple this understanding with macro-scale drivers of change to devise appropriate strategies for managing drought. This perspective, with an emphasis on vulnerability, is an internationally recognized conceptual framework for assessing and understanding the social dimensions of drought and other natural hazards (see Smit and Wandel 2006 for a discussion of the conceptual framework. For examples of its application, see Turbay et al. 2014; Diaz et al. 2011; Hadarits et al. 2010).

The "Wickedness" of Droughts

Understanding droughts and their impacts has always constituted a challenge. Similar to other climate events, droughts are phenomena that take place at the centre of human-environment interactions. Droughts are natural events that have ramifications for society, affecting people, social activities, and social processes in different forms and with different consequences. Having a comprehensive understanding of droughts involves embracing all their complexities in both human and natural systems. In this way, droughts are intricate, broad, and multifaceted phenomena.

Droughts are not a simple, tame problem that can easily be explained from a single disciplinary perspective or dealt with through a simple decision-making approach. Rather, to the extent that it is difficult and complex to define and deal with their impacts, they could be considered "wicked" problems (Brown et al. 2010; Batie 2008; Conklin 2006; Rittel and Webber 1973). A wicked problem "is a complex issue that defies complete definition, for which there can be no final solution, since any resolution generates further issues, and where solutions are not true or false or good or bad, but the best that can be done at the time" (Brown et al. 2010: 4). These kinds of problems do not exist as naturally wicked events, but rather they seem to be related to our attempts to define and explain them using traditional modes of inquiry, which tend to overemphasize some aspects of these wicked problems and ignore others. The possibility of an increase in the intensity and duration of extreme climate events due to climate change or other natural drivers makes it even more urgent to expand our understanding of drought. In this perspective, there is an identified need for developing and strengthening an interdisciplinary approach to understanding these climate events (e.g., Bhaskar et al. 2010).

Droughts are climate events with characteristics that make them significantly different from other climate hazards. In comparison to other extreme weather events, such as torrential rains or tornados, droughts are known as "creeping" hazards because they tend to accumulate more slowly and over longer periods of time and may also recede at a slow pace, they have differentiated and accumulative impacts, and their spatial coverage is heterogeneous (Sheffield and Wood 2011; Kallis 2008: 3–4; Wheaton 2007).

Most definitions of drought refer to limited availability of water, relative to normal conditions, with negative consequences for humans and ecosystems. Droughts can be variable in duration, can last several weeks to several years, and can affect very small to very large areas. Water deficits have significant negative implications for human activities that are highly dependent on access to water, such as agriculture, especially when the reduction is below critical thresholds that define water requirements for plants, animals, and humans. Moreover, droughts can become self-sustaining in that the "dryness" of droughts can reduce water vapour in an area, thereby exacerbating drought conditions (Wheaton 2007: 49). Over the long term, droughts can degrade the environment and foster desertification. This notion of drought, however, is too simple. As discussed in the next section, more complex notions of drought emerge depending on the nature of the water deficit and its impacts (Sheffield and Wood 2011: 11–13). Together, they enhance our understanding of drought and improve preparedness and adaptation.

Approaching and Understanding Droughts

Defining drought is more than a semantic exercise; the lack of agreement on a common definition has hampered proactive drought management (Paulo and Pereira 2013; Wilhite et al. 2005). As indicated above, the common metric for identifying drought is a deficiency of precipitation relative to "average" conditions (Wilhite and Glantz 1985). Early discussions of the term separated definitions into two broad categories—meteorological and agricultural—with the former considering a departure from long-term mean precipitation and the latter considering the timing of precipitation relative to crop development (Glantz and Katz 1977). In recent decades, a typology based on four broad categories of drought, as first set out by

Table 1. A typology of broad conceptualizations of drought

Conceptualization	Common definitions	Metrics	Non-climatic considerations
Meteorological	Departure from the long-term mean moisture supply (Paulo and Pereira 2006)	Long-term precipitation records, precipitation indices (e.g., SPI), cumulative precipitation shortages	None
Agricultural	Timing of precipitation relative to crop needs (Glantz and Katz 1977) Declining soil moisture and precipitation failure (Mishra and Singh 2010) Availability of soil moisture to support crop growth (Wilhite and Buchanan-Smith 2005) Moisture supply below climatically appropriate moisture supply and crop production negatively affected (Quiring and Papakyriakou 2003)	Crop water stress indices (e.g., PDSI, CMI)	Crop moisture needs, soil characteristics (infiltration, moisture holding capacity)

Hydrological	Departure from average conditions in surface and subsurface supplies (Wilhite and Buchanan-Smith 2005) Inadequate surface and subsurface water resources for established water uses (Mishra and Singh 2010)	Streamflow data, surface water supply indices (e.g., SWSI)	Upstream water availability, water storage capacity, institutional allocation, legal agreements between jurisdictions (e.g., Master Agreement on Apportionment)
Socio-economic	The interplay of human activity and meteorological, agricultural, and hydrological drought (Wilhite and Buchanan-Smith 2005) Failure of water resource systems to meet demands or demand exceeds supply (Mishra and Singh 2010)	Highly contextual descriptions	Access and entitlement to water resources, perception of water availability

Note: SPI = Standardized Precipitation Index; PDSI = Palmer Drought Severity Index; CMI = Climate Moisture Index; SWSI = Surface Water Supply Index.

Wilhite and Glantz (1985), has been used to distinguish different forms of droughts. They are meteorological, agricultural, hydrological, and socio-economic droughts (Table 1).

Meteorological approaches define drought as a deficit in precipitation over a particular time period relative to the long-term mean (Mishra and Singh 2010). While metrics vary (e.g., monthly precipitation data), the meteorological approach to drought lends itself to long-term quantitative analysis of precipitation in a given region (e.g., Sauchyn et al. 2003). Frequently, drought indices are derived to evaluate duration and intensity.

For example, the Standardized Precipitation Index (SPI) uses the mean and standard deviation of precipitation over various time periods to compute probability, percentage of average, and accumulated precipitation deficits (McKee et al. 1993). Outputs of indices such as the SPI are useful for identifying statistically anomalous conditions, but they do not give insight into how much precipitation is necessary to meet the needs of stakeholders in a given area.

Hydrological approaches to drought, like meteorological ones, define the event by a departure from the long-term normal in a given area. In this case, however, the determining variables are surface and subsurface moisture availability, including lakes, reservoirs, streamflows, and soil moisture (Wilhite and Buchanan-Smith 2005), which distinguishes hydrological approaches from meteorological ones both spatially and temporally. For example, in the case of the South Saskatchewan River basin, water supplies largely depend on rivers that are affected by precipitation upstream in the Rocky Mountains (spatial variation). Both surface water and groundwater may have a lag time in response to precipitation deficits, meaning a hydrological drought can continue to have impacts after a meteorological drought has been declared over (temporal variation). Finally, hydrological approaches indirectly consider some human systems, given that upstream withdrawals from river systems or prolonged over-allocation of ground and surface water supplies can affect the severity of a drought. Common metrics for measuring hydrological drought are similar to those measuring meteorological drought in that they rely on indices. For example, the Surface Water Supply Index considers deviations from long-term conditions in reservoir storage, streamflow, snowpack, and precipitation (Mishra and Singh 2010), but it does not consider the needs of stakeholders in an area.

Agricultural approaches to drought indirectly consider stakeholder needs by analyzing deviations from long-term conditions in soil moisture to support crop and forage growth (Wilhite and Buchanan-Smith 2005). Agricultural drought is not measured as a direct function of precipitation and hydrological availability of water, because soil types vary in their water uptake and holding capacity, and crops have different moisture needs. These types of conceptualizations are thus relative not only in time and space but also to particular production systems. Agricultural drought indices range from those that use water availability and potential evapotranspiration as dominant inputs, such as the Palmer Drought Severity Index, to complex satellite-based models, such as the Integrated Surface Drought Index, which combines moisture and temperature variables with remotely sensed vegetation conditions and thus can include irrigation effects in drought definition (Wu et al. 2013).

It is also important to recognize that subsidiary categories of drought experience exist within the wider classification of agricultural drought. Recent interdisciplinary research on the adaptive capacity of Prairie farmers and ranchers demonstrates that sensitivity to drought conditions can vary considerably between production models. For example, the success of irrigated crop production can be affected by hydrological drought conditions, which may or may not coincide with localized precipitation levels (Warren and Diaz 2012; see also Chapter 6 by Warren on irrigation in southwestern Saskatchewan in this volume). Similarly, research demonstrates that the timing of precipitation events can affect field crop production differently than it does the growth of domestic forage crops and native grasses. Dry conditions early in a growing season can adversely affect forage production. However, if precipitation increases later in the season, it might still be possible to produce crops. In addition, ranchers reliant on surface water sources for cattle can be affected by hydrological drought conditions to a greater extent than farmers producing dryland crops.

The four conceptualizations of drought mentioned above are all based on variability in natural conditions (with some human modification in the case of irrigation or water withdrawals) over a given temporal and spatial extent. All of these definitions are primarily based on departures from "average" conditions and lend themselves to the identification of drought, primarily for decision makers to react and make changes to their management approaches. While objective quantification of drought is useful (and necessary) for the allocation of drought relief (e.g., for agricultural producers), it does not provide insights into how stakeholders live with and experience this hazard or how they make decisions under drought conditions, nor does it consider human perception as a factor in drought response. Furthermore, the wider social, economic, and political context is important for creating management strategies that reduce overall drought hazard. Alternative conceptualizations of drought, which include diverse considerations of human-environment systems, have been grouped in the category of *socio-economic drought*, although it should be noted that conceptualizations captured under this approach are not as homogenous as the previous ones.

The assessment of the spatial and temporal impacts of droughts on the supply and demand of water-dependent economic goods has been a significant line of work in this area (Lindesay 2003: 38–39; see also O'Meagher 2003). More recently, and in the context of climate change, efforts have focused on evaluating the costs of climate change on agricultural activities based on biophysical-agroeconomic models (Kallis 2008).

The category of socio-economic drought has also included what Wilhite and Buchanan-Smith (2005: 10) term "human-induced" drought, where "development demands exceed the supply of water available [and] may exceed supply even in years of normal precipitation." This type of drought leads to considerations of equity and differential vulnerability; for example, upstream over-allocation in the case of the southern Colorado River basin has contributed to inequities for downstream Mexican users (see Maganda 2005). Another example of how water and power come together to produce conditions of drought for those producers downstream of the river or at the bottom of the social hierarchy is provided by Montaña and Boninsegna in Chapter 14 (this volume) for the Mendoza River basin in Argentina.

In the perspective of socio-economic droughts, the issue of perception has long been recognized as a key factor in understanding and responding to drought (i.e., the way drought is perceived). Glantz and Katz (1977) noted that recent weather conditions, particularly abnormally wet conditions, influence decision making in arid and semi-arid environments more heavily than the long-term record or drought periods. This can lead to management practices being adopted that are suited only to higher-than-average moisture and result in perceived drought conditions when the wet period ends. This situation was described for the Sahel in the 1960s by Glantz and Katz (1977) and was further evaluated for northern Ethiopia by Meze-Hausken (2004). In the latter case study, farmers' perceptions of drought—that is, when they felt that a drought had occurred—were relatively poorly matched to the long-term precipitation records and were closely tied to satisfactory harvests and returns for these harvests. As livelihoods changed, so did what were considered optimal moisture conditions, and drought was determined through this lens (Meze-Hausken 2004). A related situation also applies to the Canadian Plains during the early twentieth century: an abnormally wet period relative to the long-term record led to the establishment of land claims and associated survey systems, which were maladapted to long-term conditions, including periodic drought, contributing to the failure of a wheat-based economy during the 1930s (Wandel and Marchildon 2010; see also Chapters 5 and 6 by Warren on min till and irrigation in this volume). In this case, a failure of human perception to match the long-term record captured in the indicator approaches (at scales ranging from individual to institutions) actually increased drought hazard beyond what existed in pre-settlement range-based agriculture, illustrating the importance of considering livelihoods and their exposures and sensitivities to climatic conditions. On the other hand, the perception of drought as a normal condition of the landscape contributes to a shared experience of drought among local producers that helps reduce the impacts of dry conditions. This latter argument is reinforced by Hewitt (1983), who argues against the viewpoint that a natural hazard such as a drought is an "extreme" condition, as it primarily leads to what he terms "technocratic" (i.e., engineering, science, and technological development approaches); if we accept drought as a natural part of the landscape that is considered a hazard because of human reliance on precipitation (i.e., the view that a drought is "normal"), we develop routine adaptations and consequently higher adaptive capacity to drought. For example, a recent study of the Palliser Triangle in western Canada shows that farmers in areas normally exposed to droughts tend to have higher resilience than producers residing in areas where droughts are rare (Diaz and Warren 2012). Farmers living in the core of the Palliser Triangle have greater capacity to survive long droughts relative those living outside the area, who tend to show very limited coping capacities (Diaz and Warren 2012; Warren and Diaz 2012; Wandel et al. 2009).

Under socio-economic considerations of drought, "good years" and "bad years" are not solely defined by climatic variables. For example, using the case of climatic conditions in the Okanagan grape industry, Belliveau et al. (2006) found that good years were those where both yields and market prices were high, and a year with acceptable yields may still have been considered a bad year if crop prices were low. Similarly, producers may experience a decrease in crop yields under agricultural drought but not actually see a reduction in net farm income if commodity prices are sufficiently high to compensate for lost yield. This example illustrates the importance of considering macro-economic variables and net farm returns.

Beyond perceptions and economics, we must also consider the institutional conditions that can reduce (or increase) the drought hazard. Marchildon et al. (2008) describe the development of the Special Areas of southeast Alberta as an institutional adaptation to drought. In this case, changing land-use policy has significantly reduced exposure to drought hazards. In most cases, land administered by the Special Areas Board only allows for extensive cattle grazing (Wandel et al. 2009), which has much lower moisture requirements than crop farming, meaning that the area is drought-proofed to conditions that, under a different institutional environment, would have perhaps led to a collapse in the environmental system. Hurlbert and Diaz's (2013) analysis of water governance in Chile and Canada shows a different situation, in which the adoption of a neo-liberal framework reduces the capacity of government to alleviate exposure to drought and other forms of extreme climate events.

Wilhite et al. (2005) argue for a risk management approach to drought via a ten-step planning process that incorporates stakeholder participation (and thus perception), inventories of resources, identification of needs and institutional gaps, and direct integration of science and policy with associated awareness and education programs. This sort of highly contextualized approach to drought, which is rooted in place and time, and whose primary purpose is to reduce overall drought hazard, is consistent with current approaches to vulnerability and adaptation in the climate change field.

Understanding the socio-economic impact of droughts is part and parcel with classifying droughts as natural hazards, which is a perspective assumed in this book. There is a long tradition of approaching environmental conditions that are problematic for human systems as natural hazards. Under a hazards perspective, environmental events such as flooding do not themselves represent hazards, but they become so when coupled with human occupancy and the degree to which human systems are able to manage the impacts of the event (Kates 1976). When this perspective is applied, defining drought becomes a function of both natural water availability relative to long-term normals and human activity within the region of interest (Wilhite and Buchanan-Smith 2005; see also Kallis 2008).

Adopting a hazards perspective to drought naturally leads to considerations of vulnerability and adaptation. Conceptualizations of vulnerability draw extensively on earlier environmental hazards work and maintain the view that vulnerability is a function of both natural conditions and sensitivity, as well as the ability of systems to adapt (Smit and Wandel 2006).

The hazards perspective contrasts with meteorological, hydrological, and agricultural approaches to drought, which view the event in terms of precipitation, surface and subsurface water availability, and soil moisture, respectively (Mishra and Singh 2010). These conceptualizations of drought lend themselves primarily to quantitative analyses, including indices, and foster the view of drought as an unusual circumstance as opposed to a naturally occurring hazard that is part of the long-term climate regime (Wilhite and Buchanan-Smith 2005). Treating drought as an exceptional circumstance fosters reactive and crisis-based management solutions to deal with the impacts of a particular event without necessarily decreasing the overall drought hazard, a situation termed the "hydro-illogical cycle" by Wilhite et al. (2005: 95). In impacts, vulnerability, and adaptation to climate change scholarship, similar conceptualizations occur when vulnerability is viewed as the outcome, end-point, or residual of the adaptation process—that is, the portion of the impact due to a climatic event that could not be adapted to (Smit and Wandel 2006). Similar to the hydro-illogical cycle, this lends itself to reactive management solutions rather than proactive adaptation.

Defining drought and vulnerability as naturally occurring properties, which are a function of both human and environmental systems, changes the nature of research on drought vulnerability assessment by shifting the lens to how humans interact with the environment on an ongoing basis. This in turn can help break Wilhite et al.'s hydro-illogical cycle by adopting policies of drought-preparedness that decrease overall vulnerability to drought.

Living with Drought: Vulnerability and Adaptive Capacity

Early conceptualizations of vulnerability to climate change have been categorized variously as "vulnerability as an end point," "outcome vulnerability," or "residual impact" (Fussel and Klein 2006; Smit et al. 2000; Kelly and Adger 2000). These conceptualizations grew out of first-order climate impact assessments and take the methodological approach of first projecting future climate, then modelling impacts of future conditions, and then identifying adaptations to moderate the harm (or exploit beneficial opportunities). In this case, "vulnerability" becomes those impacts that cannot be compensated for by adaptation. This early conceptualization, although still in use in narrowly defined crop yield models (e.g., Osborne et al. 2013), has been criticized for its lack of consideration of a full suite of flexible adaptation strategies beyond those that respond to a projected impact (Ortiz-Bobea and Just 2012; Schneider et al. 2000).

The Intergovernmental Panel on Climate Change's Third Assessment Report argued for the consideration of vulnerability as a system property and drew on environmental hazards and international development work to define the concept as the product of both physical exposure to climate stresses and ability to cope with the impacts of that exposure (Smit and Pilifosova 2001). Associated terms such as sensitivity, susceptibility, coping ability, adaptability, and adaptive capacity, among others, were proposed to capture what others have termed "social vulnerability." Since then, conceptualizations that have variously been framed as "vulnerability as a starting point" or "contextual vulnerability" (O'Brien et al. 2007) have gained traction. In this framing of the concept, an understanding of vulnerability goes beyond its treatment relative to a narrow suite of climatic stimuli to "the context of political, institutional, economic and social structures and changes, which interact dynamically with contextual conditions associated with a particular 'exposure unit'" (O'Brien et al. 2007: 76). This alternate framing guides how questions are asked about vulnerability, and in turn the methods used for vulnerability assessment, providing us with an understanding of the "lived experience" of drought. Frequently, empirical analyses are conducted at the scale at which multiple stresses in the context of climate change are experienced, and a growing body of scholarship on community-based case studies has emerged (e.g.,

Westerhoff and Smit 2009; Brouwer et al. 2007; Ford et al. 2006; Stehlik 2003). However, as recognized by Adger et al. (2005), adaptation occurs across scales, and thus contextual vulnerability can be seen as a nested hierarchy where local adaptation actions are made within a broader set of determinants of exposure, sensitivity, and adaptive capacity. Contextual vulnerability has been applied by the Intergovernmental Panel on Climate Change (IPCC), in conjunction with its efforts to enhance community sustainability in response to the challenges presented by climate change, including the prospect of more severe and prolonged droughts on the Canadian Prairies and other world regions.

Most chapters in this book (see Chapters 4–14) are based on empirical studies framed within the contextual vulnerability approach. In many of these analyses, the "exposure unit" for empirical analysis is the rural community or the agricultural production unit, with an implicit recognition that adaptation decisions are made within a broader institutional, governance, and political environment (see Chapters 5, 6, 8, 9, 10, and 12 on institutional context).

Following the IPCC, vulnerability is defined in this volume as "the degree to which a system is susceptible to, and unable to cope with, adverse effects of climate change, including climate variability and extremes. Vulnerability is a function of the character, magnitude and rate of climate change and variation to which a system is exposed, its sensitivity, and its adaptive capacity" (McCarthy et al. 2001: 6). In this volume, we use a socio-economic conceptualization of drought as a lens for analyzing climate variability, extremes, and change, and examine the various perceptions, values, and enabling and constraining factors by scaling out from community-based vulnerability assessments.

An important point of departure for most chapters in this volume is the recognition that all agricultural producers are exposed to the extremes of climate variability, but not all of them are vulnerable to the same degree. Differences in vulnerabilities are closely related to a variety of social, economic, and political conditions and capacities, which either facilitate or constrain, for example, the ability of farmers and ranchers to cope with harsh climate conditions. However, it is important to remember that even in those situations in which producer communities have adopted practices that increase their capacity to cope with drought, their resilience is based on experience with past droughts. Should future droughts exceed the thresholds for severity and duration of those experienced in the past, as climate change science suggests is likely, current levels of adaptive capacity may no longer be sufficient to sustain current practices (see Chapter 3 by Wheaton et al. in this volume). Understanding the processes associated with successful adaptations in the past can provide insights into how communities might adapt to future conditions (see Chapter 5 by Warren on "min till" (minimal tillage) and see Chapter 8 by Marchildon in this volume). Similarly, observing and assessing how communities in other regions of the world have adapted to drought conditions can provide useful lessons for other localities. Chapters 13 and 14, which discuss adaptation to drought in Latin America, reflect this principle.

As indicated in IPCC's definition, vulnerability combines two dimensions: first, exposure to climate hazards and its impacts on social systems; and second, social conditions that determine the sensitivity of a ranch or a farm—the degree to which they are affected by climate-related stimuli—as well as the system's adaptive capacity (i.e. the ability of the system, such as a production unit, to adjust to climate risks and opportunities by increasing its adaptive range). Figure 1 represents these two dimensions of vulnerability. Exposure is a characteristic of a climate system, and it refers to climate hazards—that is, droughts, storms, and others—and their attributes—such as intensity, duration, and coverage—that define the magnitude of their impact on social systems. Sensitivity and adaptive capacity, on the other hand, are characteristics of the social system defined by access to and control of a variety of resources. In this perspective, vulnerability is a characteristic of a social system that emerges when a natural hazard impacts human systems. In very simple terms, a social system that is characterized by limited resources is more vulnerable and consequently more susceptible to being impacted by climate hazards. Figure 1 lists these resources, defined by the IPCC as "the determinants of adaptive capacity" (McCarthy et al. 2001: 893). Access to and control of these resources are important to reduce vulnerabilities, but it is the capabilities of actors to organize them into adaptive activities that define the balance between sensitivity (determined by lack of or limited resources) and adaptation (defined by the existence of resources that could be mobilized to reduce sensitivity).

These determinants of adaptive capacity—also called assets or "capitals" (Department of International Development 2000)—are resources that could be used to ensure the sustainability of farms and ranches in

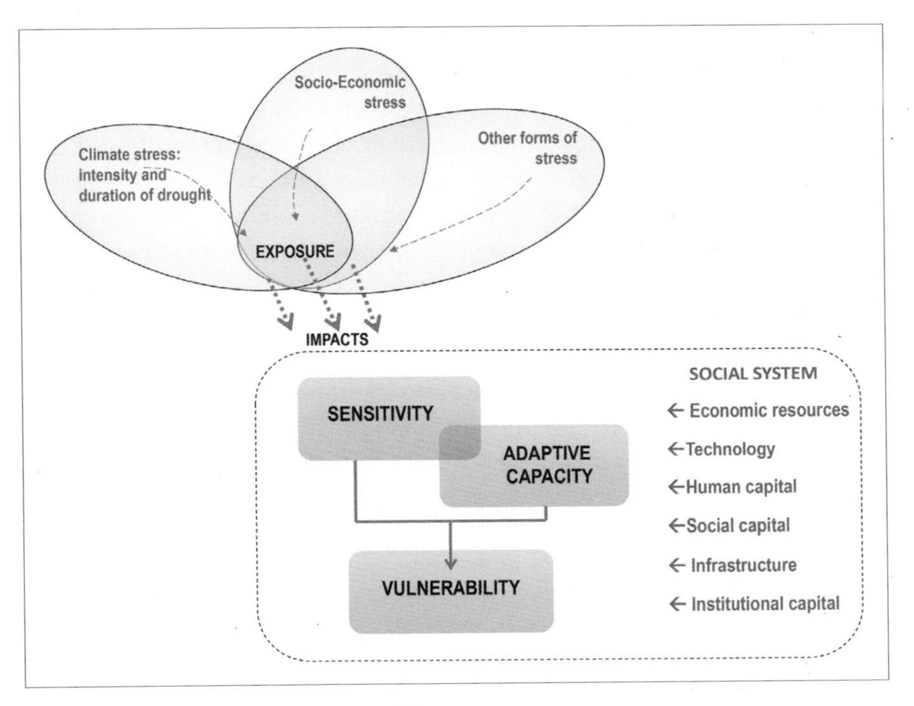

Figure 1. The dimensions of vulnerability.

contexts other than climate change. Economic assets refer to financial resources, such as cash, credit, productive resources (machinery, buildings), and other forms of economic capital, that could be mobilized to sustain a livelihood. These resources are undoubtedly central to secure the conditions that enhance the sustainability of a farm or ranch, but there are other resources no less significant. Access to good infrastructure (proper housing conditions, drainage systems, weather-resistant roads, coastal defence, and others forms of infrastructure) and to technology (irrigation systems, flood control measures, warning systems, and others) is fundamental to sustain productivity in the face of increasing climate-related risks. No less relevant is access to natural capital—those basic ecosystem services, such as water and soil, which are fundamental to the viability of rural livelihoods. The quantity and quality of these natural resources are, obviously, two important aspects that secure the success of agricultural activities.

Also relevant are other elements such as human capital—the educational experiences, knowledge, skills, and expertise of a person. This capital includes not only knowledge obtained in the formal educational system but also local knowledge and experiences that could be used to employ, modify, and develop other types of resources. In this context of human capital, the capacities to wisely manage materials and human resources, the ability to learn from experience, and the ability to gain access to and process information are important.

In the same perspective, institutional capital, defined as those resources that exist at the level of local, regional, and national institutions, is important. The process of generating and maintaining an adaptive capacity at the level of the farm or the ranch is always related to the existence of collective resources and capacities that support and multiply individual efforts. Established institutions, such as government agencies, facilitate the management of a variety of risks—such as the existence and availability of insurance services, water conservation programs, and others—which reinforce the adaptive capacity of the population. Previous studies in the area of climate vulnerability have shown that adaptation of communities is nested in larger institutional contexts, from where a myriad of resources, programs, and policies are provided to individuals and local communities (Hurlbert and Diaz 2013; Diaz et al. 2011; Diaz et al. 2009; Hurlbert et al. 2009; see also Chapter 8 by Marchildon and Chapters 9 and 10 by Hurlbert in this volume). How this institutional capital interacts, or how governments, organizations, producers, and other entities make decisions and share power, exercise responsibility, and ensure accountability, is the essence of governance (Cundill and Fabricius 2010) and an important component of adaptive capacity (Gupta et al. 2010; Folke et al. 2005). In the same vein, institutional capacities are not limited to formal agencies and organizations that exist beyond the local community. Local institutional capital—whether in the form of local government or local organizations—is also relevant as a form of capital that could be mobilized to reduce sensitivities to a variety of stressors (Wandel et al. 2009). There is also increasing evidence that informal local institutions—such as social capital based on friendship or kinship—strengthen the capacity to reduce the stress of natural and economic hazards (learning from experience, capacity for innovation, flexibility) and are important for organizing these assets into adaptation actions (Warren and Diaz 2012).

Figure 1 also shows that climate is not the single determinant of a system's vulnerability. Rather, climate and water stresses are part of a suite of stresses that individual producers and rural communities must manage in their everyday lives. Rural people are exposed to several non-climatic stressors—such as market conditions, political processes, domestic catastrophes, and others—which are frequently more relevant to them than extreme climate events. Particularly problematic for them is the combination of climatic and non-climatic vulnerabilities at a single moment in time, such as the case of a drought at a moment in which market crop prices are low. It is this combination of stressors that multiplies the negative impacts of risks leading to double exposures (Leichenko and O'Brien 2008; see also Chapter 4 by Kulshreshtha et al. and Chapter 13 by Hadarits et al. in this volume). In addition, the nature of production systems creates specific conditions of vulnerability for different types of agricultural producers. For example, water demands vary between farmers and ranchers, as well as among different production units. No less relevant is the localization of the production units within a region. Non-existent or limited access to irrigation is a fundamental issue for agricultural producers in the context of increasing water scarcities (see Chapter 6 by Warren on irrigation in this volume). Similarly, having a farm in certain areas of a region or water basin may limit access to water (the case of the Mendoza River basin is a good example of this situation; see Chapter 14 by Montaña and Boninsegna in this volume).

As expected, vulnerabilities—and associated adaptive capacity—tend to be unequally distributed. These unequal conditions are associated with processes of economic differentiation, which allow some producers to have access to more and better resources than others. This differentiation results not only from the economic conditions generated by competition and a process of globalization but also from institutional failures, which result in an unequal distribution of resources vital to adaptive capacity (Hurlbert and Diaz 2013). In other words, some rural people have greater adaptive capacity than others because of greater access and control of the different forms of capital discussed above.

Vulnerability is not an unalterable condition but rather is subject to change depending on the intensity of the stressor and the quality and quantity of resources that are available to rural people. In other words, vulnerability must be considered as a fluid process. In the case of resources, they are obviously subject to change depending on the intensity of the stressor and the quality and quantity of the different forms of capital available to the local community. In other words, vulnerabilities, and adaptive capacities, are not a given condition but rather are subject to a myriad of processes that could increase or reduce the quantity and quality of resources. Thus, when resources are limited and they are used unwisely, the capacity of a farm or ranch to face future risks declines (Pelling 2011). The wise management of resources is therefore essential to the sustainability of livelihoods.

Conclusion

This chapter has presented a case for considering both human and natural systems—and their interactions—when assessing vulnerability to drought. Both socio-economic definitions of drought and conceptual approaches to contextual vulnerability, by definition, incorporate this dynamic. Moreover, they allow researchers to approach and understand droughts and other climate events from a people-centred focus, providing an opportunity to grasp how rural people live and experience droughts, and how differential access to resources promotes or reduces the resilience of producers.

This knowledge is fundamental to developing appropriate climate governance approaches that could facilitate a move beyond the assumptions of homogeneity which characterize the policy landscape. In facing the increasing threat of climate change, drought policies and programs need to incorporate a deeper understanding of local vulnerabilities to develop and implement more focused, targeted, and relevant drought management strategies. This so-called "bottom-up" knowledge also helps expand our scientific understanding of the complexities of droughts and adds to our existing knowledge of the biophysical elements that contribute to and characterize droughts. By adding new dimensions to our knowledge, we move another step forward in taming the wickedness of droughts.

References

Adger, N., N. Arnell, and E. Tompkins. 2005. "Adapting to Climate Change: Perspectives across Scales." *Global Environmental Change* 15, no. 2: 75–76.

Batie, S. 2008. "Wicked Problems and Applied Economics." *American Journal of Agricultural Economics* 5: 1176–91.

Belliveau, S., B. Smit, and B. Bradshaw. 2006. "Multiple Exposures and Dynamic Vulnerability: Evidence from the Grape and Wine Industry in the Okanagan Valley, British Columbia, Canada." *Global Environmental Change* 16, no. 4: 364–78.

Bhaskar, R., C. Frank, K.G. Høyer, P. Naess, and J. Parker. 2010. *Interdisciplinarity and Climate Change. Transforming Knowledge and Practice for our Global Future.* London: Routledge.

Brouwer, R., S. Akter, L. Brander, and E. Haque. 2007. "Socio-economic Vulnerability and Adaptation to Environmental Risk: A Case Study of Climate Change and Flooding in Bangladesh." *Risk Analysis* 27, no. 2: 313–26.

Brown, V.A., P.M. Deane, J.A. Harris, and J.Y. Russell. 2010. "Towards a Just and Sustainable Future." In V. Brown, J. Harris, and J. Russell (eds.), *Tackling Wicked Problems through the Transdisciplinary Action.* London: Earthscan.

Conklin, J. 2006. "Wicked Problems and Social Complexity." Chapter 1 in J. Conklin, *Dialogue Mapping: Building Shared Understanding of Wicked Problems.* West Sussex: John Wiley & Sons. http://cognexus.org/wpf/wickedproblems.pdf.

Cundill, G., and C. Fabricius. 2010. "Monitoring the Governance Dimension of Natural Resource Co-management." *Ecology and Society* 15, no. 1. http://www.ecologyandsociety.org/vol15/iss1/art15/main.html.

Department for International Development (UK). 2000. "Sustainable Livelihoods Framework." In *Livelihoods Connect: Sustainable Livelihoods Guidance Sheets from DFID.* London: Department for International Development.

Diaz, H., R. Garay-Fluhmann, J. McDowell, E. Montaña, B. Reyes, and S. Salas. 2011. "Vulnerability of Andean Communities to Climate Variability and Climate Change." Pp. 209–24 in W. Leal Filho (ed.), *Climate Change and the Sustainable Management of Water Resources.* Berlin: Springer-Verlag.

Diaz, H., M. Hadarits, and P. Barrett-Deibert (eds.). 2009. *IACC Final Report.* Regina: Canadian Plains Research Center. http://www.parc.ca/mcri/int01.php. Accessed 5 April 2010.

Diaz, H., and J. Warren. 2012. *Final Research Report of the Rural Communities Adaptation to Drought.* Regina: Canadian Plains Research Center. http://www.parc.ca/vacea/assets/PDF/rcadreport2012.pdf.

Folke, C., T. Hahn, P. Olsson, and J. Norberg. 2005. "Adaptive Governance of Social-Ecological Systems." *Annual Review of Environmental Resources* 30: 411–73.

Ford, J., B. Smit, and J. Wandel. 2006. "Vulnerability to Climate Change in the Arctic: A Case Study from Arctic Bay, Nunavut." *Global Environmental Change* 16, no. 2: 145–60.

Fussel, H., and R. Klein. 2006. "Climate Change Vulnerability Assessments: An Evolution of Conceptual Thinking." *Climatic Change* 75: 301–29.
Glantz, M.H., and R.W. Katz. 1977. "When is a Drought a Drought?" *Nature* 267: 192–93.
Gupta, J., C. Termeer, J. Klostermann, S. Meijerink, M. van den Brink, P. Jong, S. Nootboome, and E. Bergsma. 2010. "The Adaptive Capacity Wheel: A Method to Assess the Inherent Characteristics of Institutions to Enable the Adaptive Capacity of Society." *Environment Science and Policy* 13: 459–71.
Hadarits, M., B. Smit, and P. Diaz. 2010. "Adaptation in Viticulture: A Case Study of Producers in the Maule Region of Chile." *Journal of Wine Research* 21, nos. 2–3: 167–78.
Hewitt, K. 1983. "The Idea of Calamity in a Technocratic Age." Pp. 4–32 in K. Hewitt (ed.), *Interpretations of Calamity from the Viewpoint of Human Ecology.* Winchester: Allen and Unwin.
Hurlbert, M., and H. Diaz. 2013. "Water Governance in Chile and Canada: A Comparison of Adaptive Characteristics." *Ecology and Society* 18, no. 4: 61.
Hurlbert, M., H. Diaz, D.R. Corkal, and J. Warren. 2009. "Climate Change and Water Governance in Saskatchewan, Canada." *International Journal of Climate Change Strategies and Management* 1, no. 2: 118–32.
Kallis, G. 2008. "Droughts." *Annual Review of Environment and Resources* 33: 3.1–3.34.
Kates, R.W. 1976. "Experiencing the Environment as Hazard." Pp. 133–56 in S. Wapner, S. Cohen, and B. Kaplan (eds.), *Experiencing the Environment.* New York: Plenum Press.
Kelly, M., and N. Adger. 2000. "Theory and Practice in Assessing Vulnerability to Climate Change and Facilitating Adaptation." *Climatic Change* 47: 325–52.
Leichenko, R., and K. O'Brien. 2008. *Environmental Change and Globalization. Double Exposures.* New York: Oxford University Press.
Lindesay, J. 2003. "Climate and Drought in Australia." Pp. 21–48 in L. Botterill and M. Fisher (eds.), *Beyond Drought: People, Policy, and Perspectives.* Collingwood, AU: CSIRO Publishing.
Maganda, C. 2005. "Collateral Damage: How the San Diego-Imperial Water Agreement Affects the Mexican Side of the Border." *Journal of Environment and Development* 14: 486–506.
Marchildon, G.P., S. Kulshreshtha, E. Wheaton, and D. Sauchyn. 2008. "Drought and Institutional Adaptation in the Great Plains of Alberta and Saskatchewan, 1914–1939." *Natural Hazards* 45, no. 3: 391–411.
McCarthy, J.J., O.F. Canziani, N.A. Leary, D.J. Dokken, and K.S. White (eds.). 2001. *Climate Change 2001: Impacts, Adaptation, and Vulnerability. Contribution of Working Group II to the Third Assessment Report of the Intergovernmental Panel on Climate Change.* Cambridge, UK: Cambridge University Press.
McKee, T.B., N.J. Doesken, and J. Kleist. 1993. "The Relationship of Drought Frequency and Duration to Time Scales." Pp. 179–184 in *Proceedings of the Eighth Conference on Applied Climatology.* Anaheim, CA, 17–22 January 1993.

Meze-Hausken, E. 2004. "*Contrasting Climate Variability* and Meteorological Drought with Perceived Drought and *Climate Change* in Northern Ethiopia." *Climate Research* 27, no. 1: 19–31.
Mishra, A.K., and V.P. Singh. 2010. "A Review of Drought Concepts." *Journal of Hydrology* 391, nos. 1–2: 202–16.
O'Brien, K., S. Eriksen, L.P. Nygaard, and A. Schjolden. 2007. "Why Different Interpretations of Vulnerability Matter in Climate Change Discourses." *Climate Policy* 7, no. 1: 73–88.
O'Meagher, B. 2003. "Economic Aspects of Drought and Drought Policy." In J. Courtenay and M. Fisher (eds.), *Beyond Drought: People, Policy, and Perspectives*. Collingwood, AU: CSIRO Publishing.
Ortiz-Bobea, A., and R.E. Just. 2012. "Modeling the Structure of Adaptation in Climate Change Impact Assessment." *American Journal of Agricultural Economics, Agricultural and Applied Economics Association* 95, no. 2: 244–51.
Osborne, T., G. Rose, and T. Wheeler. 2013. "*Variation in the Global-scale Impacts of Climate Change on Crop Productivity Due to Climate Model Uncertainty and Adaptation.*" *Agricultural and Forest Meteorology* 170: 183–94.
Paulo, A., and L.S. Pereira. 2006. "Drought Concepts and Characterization." *Water International* 31, no. 1: 37–49.
Pelling, M. 2011. *Adaptation to Climate Change: From Resilience to Transformation*. London: Routledge.
Quiring, S.M., and T.N. Papakyriakou. 2005. "Characterizing the Spatial and Temporal Patterns of June–July Moisture Conditions in the Canadian Prairies." *International Journal of Climatology* 25: 117–38.
Rittel, H., and M. Webber. 1973. "Dilemmas in a General Theory of Planning." *Policy Sciences* 4: 155–69.
Sauchyn, D.J., J. Stroich, and A. Beriault. 2003. "A Paleoclimatic Context for the Drought of 1999–2001 in the Northern Great Plains." *The Geographical Journal* 169, no. 2: 158–67.
Schneider, S.H., W.E. Easterling, and L.O. Mearns. 2000. "Adaptation: Sensitivity to Natural Variability, Agent Assumptions and Dynamic Climate Changes." *Climatic Change* 45: 203–21.
Sheffield, J., and E. Wood. 2011. *Drought: Past Problems and Future Scenarios*. London: Earthscan.
Smit, B., B. Burton, R. Klein, and J. Wandel. 2000. "An Anatomy of Adaptation to Climatic Change and Variability." *Climatic Change* 45: 223–51.
Smit, B., and O. Pilifosova. 2001. "Adaptation to Climate Change in the Context of Sustainable Development and Equity." In J.J. McCarthy, O.F. Canziani, N.A. Leary, D.J. Dokken, and K.S. White (eds.), *Climate Change 2001: Impacts, Adaptation, and Vulnerability. Contribution of Working Group II to the Third Assessment Report of the Intergovernmental Panel on Climate Change*. Cambridge, UK: Cambridge University Press.
Smit, B., and J. Wandel. 2006. "Adaptation, Adaptive Capacity and Vulnerability." *Global Environmental Change* 16, no. 3: 282–92.

Stehlik, D. 2003. "Australian Drought as a Lived Experience: Social and Community Impacts." Pp. 87–108 in L. Botterill and M. Fisher (eds.), *Beyond Drought: People, Policy, and Perspectives*. Collingwood, AU: CSIRO Publishing.

Turbay, S., B. Nates, F. Jaramillo, J.J. Vélez, and O.L. Ocampo. 2014. "Adaptación a la variabilidad climática entre los caficultores de las cuencas de los ríos Porce y Chinchina, Colombiá." *Investigaciones Geograficas*, Boletín del *Instituto de Geografía, UNAM* 85: 95–112.

Wandel, J., and G. Marchildon. 2010. "Institutional Fit and Interplay in a Dryland Agricultural Social-Ecological System in Alberta, Canada." Pp. 179–98 in D. Armitage and R. Plummer (eds.), *Adaptive Capacity and Environmental Governance*. Heidelberg: Springer-Verlag.

Wandel, J., G. Young, and B. Smit. 2009. "Vulnerability and Adaptation to Climate Change: The Case of the 2001–2002 Drought in Alberta's Special Areas." *Prairie Forum* 34, no. 1: 211–34.

Warren, J., and H. Diaz. 2012. *Defying Palliser. Stories of Resilience from the Driest Region of the Canadian Prairies*. Regina: Canadian Plains Research Center Press.

Westerhoff, L., and B. Smit. 2009. "The Rains Are Disappointing Us: Dynamic Vulnerability and Adaptation to Multiple Stressors in the Afram Plains, Ghana." *Mitigation and Adaptation Strategies for Global Change* 14, no. 4: 317–37.

Wheaton, E. 2007. "Drought." Pp. 40-52 in B.D. Thraves, M.L. Lewry, J.E. Dale, and H. Schlichtmann (eds.), *Saskatchewan: Geographic Perspectives*. Regina: Canadian Plains Research Center Press.

Wilhite, D.A., and M. Buchanan-Smith. 2005. "Drought as a Natural Hazard: Understanding the Natural and Social Context." Pp. 3–29 in D.A. Wilhite (ed.), *Drought and Water Crises: Science, Technology, and Management Issues*. New York: Taylor and Francis.

Wilhite, D.A., and M.H. Glantz. 1985. "Understanding the Drought Phenomenon: The Role of Definitions." *Water International* 10, no. 3: 111–20.

Wilhite, D.A., M.J. Hayes, and C.L. Knutson. 2005. "Drought Preparedness Planning: Building Institutional Capacity." Pp. 93–135 in D. Wilhite (ed.), *Drought and Water Crises: Science, Technology and Management Issues*. New York: Taylor and Francis.

Wu, D., D.-H. Yan, G.-Y. Yang, S.-G. Wang, W.-H. Xiao, and H.-T. Zhang. 2013. "Assessment on Agricultural Drought Vulnerability in the Yellow River Basin Based on a Fuzzy Clustering Iterative Model." *Natural Hazards* 67, no. 2: 919–36.

PART 2

PAST AND FUTURE DROUGHT: LESSONS FROM CLIMATE SCIENCE

CHAPTER 2

CANADIAN PRAIRIES DROUGHT FROM A PALEOCLIMATE PERSPECTIVE

David Sauchyn and Samantha Kerr

Introduction

Recurring drought is characteristic of the climate of the Canadian Prairies (Bonsal et al. 2011). It has serious consequences given the sub-humid climate (potential evapotranspiration exceeds precipitation in an average year) and predominance of agricultural land use in this region, which accounts for more than 80% of Canada's agricultural land. The impacts of snowpack deficits, soil moisture depletion, and decreased streamflow and lake levels on the agricultural sector, and on water supplies in general, are well documented (e.g., Bonsal et al. 2011; Wheaton et al. 2008). Prolonged drought is especially damaging because its impacts are cumulative and can lower the resistance of ecosystems and soil landscapes to disturbance from hydroclimatic events to a point that thresholds of landscape change are exceeded and recovery of natural systems can take decades or centuries (Wolfe et al. 2001).

Drought is understood as a deficit of water: "Drought originates from a deficiency of precipitation over an extended period of time—usually a season or more—resulting in a water shortage for some activity, group, or environmental sector" (National Drought Mitigation Center 2006). According to this typical definition, a drought exists when the water deficit crosses a threshold in terms of duration and degree. These thresholds are a function of regional social and historical circumstances: sensitivity to water shortages and the adaptive capacity to deal with their adverse impacts. While precipitation and water level data are used to measure *meteorological* and *hydrological* drought, respectively, whether a drought is occurring depends on whether it is having an impact. The impacts of *socio-economic drought*, and specifically *agricultural drought*, range from a lack of soil moisture for dryland farming to the eventual depletion of water stored for irrigation.

In the Canadian Prairies, aridity and drought define the landscape and have punctuated the human history with periodic impacts and adaptation. Since ecosystems and rural communities in the driest areas are adapted to drought, and weeks without rain are characteristic of the summer climate, a season is probably an appropriate minimum duration for defining drought in this region. Summer water deficits are the norm for a semi-arid climate, and thus a season without much rain is tolerable, provided there is either access to irrigation or adequate soil moisture early in the season to enable germination and emergence of the crop. The impacts of a water deficit will therefore depend almost entirely on how much it lasts beyond one season—the "more" in "a season or more." In recent years, droughts have only rarely persisted for more than several seasons or at most two to three years. These droughts were recorded by water and weather gauges and thus both perceived and defined meteorologically as the "normal" maximum duration. As a result, droughts of longer duration would conceivably exceed the adaptive capacity of Prairie agricultural communities. This chapter explores the question of whether droughts recorded and experienced by agrarian communities in western Canadian are as bad as they can get or whether we can expect droughts of greater intensity and longer duration based on our knowledge of past droughts. Thus, this chapter puts our recent experience with drought in a paleoclimatic context. Prairie drought is a recurring theme in paleo-environmental research (Sauchyn and Bonsal 2013; Bonsal et al. 2012; Lapp et al. 2012; St. George et al. 2009; Sauchyn et al. 2002, 2003). This chapter provides an overview of this research and presents a case study of paleodrought based on the reconstruction of the annual flow of Swift Current Creek (Saskatchewan) over the past four centuries.

Future climate will be a combination of natural climate cycles and the effects of anthropogenic climate change. Because the period of weather observations, since the 1880s, is short relative to some natural climate cycles and the return period of rare severe events, knowledge of pre-instrumental climate is required to determine the full range of variability and extremes in the regional climate and hydrology. Longer proxy hydroclimate records provide water resource managers and engineers with a historical context to evaluate 1) baseline conditions and water allocations, 2) worst-case scenarios in terms of severity and duration of drought, 3) long-term probability of hydroclimate conditions exceeding specific thresholds, 4) scenarios of water supply under climate change, 5) variability of water levels to assess reliability of water supply systems under a wider range of flows than recorded by a gauge, and 6) geographic extent of multi-year periods of low-and-high flows, including the synchronicity of droughts in adjacent watersheds (St. George and Sauchyn 2006).

Drought Proxies

The climate of the past, or paleoclimate, is preserved in biological and geological archives. Ecosystems and soil landscapes evolve under a certain range and seasonality of heat and moisture, and thus they correspond to regional climate regimes. But climate is not static, and as it varies, ecosystems and sediments preserve the climate changes and variability; they act as recorders of environmental change enabling the reconstruction of past climate variability on seasonal, yearly, and century-long time scales. This reconstruction of environmental history, irrespective of the proxy, is based on an understanding of the natural systems and their relationship to the current climate: "the present is the key to the past." Thus, the interpretation of proxy data is only as good as the contemporary ecological, hydrological, meteorological, and geological data used to calibrate and interpret the proxy. There must be an appropriate measure of drought if environmental history is to be reconstructed.

Any systematic analysis of the intensity, duration, timing, frequency, and spatial extent of drought, including the inference of these characteristics from biological and geological archives, requires an operational definition based on one or more drought-related variables (Zargar et al. 2011). Among the quantitative expressions of drought, the most popular has been the Palmer Drought Severity Index (PDSI), which is based on precipitation, temperature (evapotranspiration), and soil water recharge rates. One complaint about the PDSI is that scaling of the index is sensitive to the soil moisture balance component. As McKee et al. (1993) pointed out, all measures of drought frequency, duration, and intensity are a function of implicitly or explicitly established time scales. They introduced the Standardized Precipitation Index (SPI) and demonstrated its applicability over intervals of 3 to 48 months. McKee et al. (1993: section 3.0) found a maximum correlation between the PDSI and SPI at 12 months, "suggesting that the PDSI does indeed have an inherent time scale even though it is not explicitly defined." Vicente-Serrano et al. (2010) added a temperature term to the SPI to create the Standardized Precipitation-Evapotranspiration Index (SPEI). They concluded that "the PDSI is not a reliable index for identifying either the shortest or the longest time-scale droughts, which can have greater effects on ecological and hydrological systems than droughts at the intermediate time scales . . . only hydrological and economic systems that respond to water deficits at time scales of 9–18 months can be monitored using the PDSI" (2010: 9).

No common index or definition is available for paleodrought. The PDSI is the basis for most previous studies of recent and past prairie drought (Bonsal et al. 2012; Lapp et al. 2012; Sauchyn and Skinner 2001) and for the *North American Drought Atlas*, a continent-wide reconstruction of past drought from thousands of tree-ring chronologies (Cook et al. 2007). Use of the SPI or SPEI is likely to yield similar results as the PDSI, but at least these indices have the advantage of explicit time scales over which the applicability of the index is consistent (Vicente-Serrano et al. 2010; McKee et al. 1993). When the SPI is averaged over periods of up to 48 months, drought occurs with decreasing frequency, although these infrequent droughts of long duration represent the integration of series of water deficit events, with intervening periods of precipitation that are insufficient to overcome the accumulating water deficit. This statistical averaging is analogous to the integration of weather events, and the smoothing of short-term hydroclimatic variability, by drought proxies, whether annual tree growth or the gradual accumulation of plant and animal remains at the bottom of a lake. The higher the resolution of a proxy, the closer it comes to capturing a discrete drought event.

Drought indices were developed for analyzing instrumental meteorological and hydrometric time series and expressing the frequency and severity (intensity and duration) of water deficits. Their use for calibrating drought proxies is a unique application. Normally the use of a drought index, and particularly the choice of an averaging period, depends on the sensitivity of a system to water deficits of varying duration and intensity (Maliva and Missimer 2012). With drought proxies, however, that "averaging period" is a function of the sampling of the ecological or geological archive. The use of numerical drought indices is best suited for proxies, such as laminated sediments and tree rings, where the temporal resolution is high (years) and consistent, and the proxy is a measured physical or chemical property of the natural archive. Where the temporal resolution is low (decadal), and the indicator is simply the relative abundance of a climate proxy (e.g., plant pollen), inferred drought is typically described as an interval of dry climate or low water levels. Long paleo-environmental records encompass changes in climate, including shifts in aridity, which is a permanent water deficit as opposed to the temporary weather condition of drought (Maliva and Missimer 2012).

Each climate proxy represents a unique response of natural systems and processes to environmental change. Therefore, there is no universal definition of paleodrought in terms of duration and intensity, and human impacts are not considered unless there is an archaeological component to a paleoclimate study. Each proxy is a signal of a particular scale and aspect of climate, from the response of terrestrial (upland) vegetation to regional temperature and precipitation over multiple years, to the sensitivity of aquatic organisms to lake salinity, and carbonate mineralogy to lake water chemistry and temperature. The use and interpretation of climate proxies are subject to the following universal limitations and factors.

Location and timing: The sensitivity of natural systems to climate change and variability fluctuates over time and space. On the margins of regional ecosystems, species are at the limits of their ranges and thus are climatically sensitive. Island forests and permanent wetlands in the Prairie Ecozone provide a valuable source of information on environmental change, because they exist in an otherwise semiarid region and thus the terrestrial and aquatic species in these forest and wetlands are living on the margins climatically. The availability of indications of environmental change also varies over time. For example, lake sediments can yield detailed information about climate during dry periods (i.e., when lakes are low and sensitive), but during wet periods, lake sediments tend to yield less information about climate, because high water levels buffer the effects of fluctuations in temperature and precipitation. For 21 lakes in central Saskatchewan, Pham et al. (2008) found a coherent response to climatic variability in dry years but a lack of synchrony in wet years. Similarly, where heat, light, moisture, and nutrients are sufficient, tree growth is complacent—the rings have consistent width and no signal of inter-annual climate variability.

Resolution: Temporal resolution varies among proxy records according to the time span represented by individual samples and measurements. For example, in shallow and dry prairie lake basins, a single sample of lake sediment can represent material accumulated over decades, because sediments are re-suspended in the water on windy days and also lakes can periodically disappear. Some unusually deep freshwater lakes, on the other hand, contain continuous, undisturbed, and in some cases, annually laminated sediments (St. Jacques et al. 2015).

Non-climatic controls: Significant variations in proxy data can reflect the response of natural systems to internal thresholds or to events that are indirectly or not related to climate. Forests expand and become denser, and lakes fill with sediment and evolve chemically; proxy data from these systems can contain a signature of these processes. Land-use change also has a strong influence on the pollen and chemical record from prairie lakes (Pham et al. 2008).

Chronological control: Establishing the timing of climatic changes and resolving climatic variability depend entirely on chronological control, typically based on radiometric dating of organic and mineral carbon. Only tree rings and varves (annually laminated sediments) can be assigned to individual calendar years. Even these often represent floating chronologies, which must be dated by other means or correlated (cross-dated) with modern samples or strata to obtain absolute dates.

In the northern Great Plains of North America, changes in climate are recorded in the shifting of vegetation, fluctuations in the level and salinity of lakes, patterns of tree rings, and the age and mobility of sand dunes. In this dry environment, where lake water levels and chemistry, prairie vegetation, and rates of runoff and erosion reflect the soil and surface water balance, most proxies record fluctuations in hydroclimate, including periods of water deficits. Prairie paleodrought is identified mainly from studies of sediments, archival documents, and tree rings.

Lake and Terrestrial Sediments

Soils and sediments are ubiquitous climate proxies. They provide paleoclimate records that span the geologic history of a surface or sediment sink, although the age and origin of sediments usually can be resolved only to within decades (with the rare exception of annually laminated sediments). The sediments in permanent lakes represent a continuous accumulation of mineral and biological proxies. Although lakes are less common in semi-arid environments, they are important climate archives where drought is frequent and has ecological consequences. Pham et al. (2009) determined that the long-term chemical characteristics of prairie lakes were regulated mainly by changes in winter precipitation or groundwater flux. This finding has important implications for the hydro-climatic interpretation of the abundance of type of organisms found in lake sediments.

The postglacial climate history of the Canadian Prairies is known mostly from the analysis and interpretation of the type and abundance of fossil plants and aquatic organisms found in lake sediments. The analysis of bulk samples representing multiple decades limits the inference of hydroclimate to indications of relative aridity rather than drought. More recently, the continuous sampling and precise dating of lake sediments at fine intervals has yielded time series of higher resolution. The fine sampling of diatom assemblages from prairie lakes has revealed droughts embedded in multi-centennial shifts in moisture regimes (Michels et al. 2007; Laird et al. 2003). Using paleo-environmental information from the Peace–Athabasca Delta (PAD), Wolfe et al. (2008) determined that the levels of Lake Athabasca have fluctuated systematically over the past millennium. The lowest levels were during the eleventh century, whereas the highest lake levels coincided with maximum glacier extent during the Little Ice Age (sixteenth to nineteenth centuries). This important work has demonstrated that recent water level fluctuations on the PAD are within the range of long-term natural variability and therefore are very unlikely to be caused by the impoundment of water upstream (Wolfe, Hall, et al. 2012).

The frequency and duration of droughts also has been inferred from the age and origin of sand dune deposits (Wolfe, Hugenholtz, et al. 2012). Dry periods lasting years to decades will trigger the reactivation of a dune field; but the most severe droughts may not be detectable if continuous and extensive sand dune activity prevents the preservation of biological or geological evidence. From the precise optical dating of quartz grains, Wolfe et al. (2001) identified widespread reactivation of sand dunes in southwestern Saskatchewan about 200 years ago and correlated this geomorphic activity with tree-ring records of prolonged drought during the mid-to-late eighteenth century. A lag occurred between peak dryness around 1800 and the onset of dune activity at about 1810. Dune stabilization has occurred since 1890. The droughts of the 1930s and 1980s were insufficient to reactivate dunes.

Historical (Archival) Records

The Euro-Canadian (non-Aboriginal) history of the northern plains is several centuries longer than the instrumental observation of weather that began with agrarian settlement. Explorers and fur traders reported extreme weather and related events (e.g., fires, floods, ice cover). These documents are archived in libraries, museums, government repositories, and notably in the Hudson's Bay Company Archives in Winnipeg. This archival information is valuable for verifying paleoclimate data from other sources (Rannie 2006; Blair and Rannie 1994). Severe, and at times prolonged, drought in the late eighteenth and mid-nineteenth centuries, evident in tree-ring and sand-dune chronologies, are described by explorers and fur traders. The archives of the Hudson's Bay Company contain this report:

> At Edmonton House, a large fire burned "all around us" on April 27th (1796) and burned on both sides of the river. On May 7th, light canoes arrived at from Buckingham House damaged from the shallow water. Timber intended to be used at Edmonton House could not be sent to the post "for want of water" in the North Saskatchewan River. On May 2nd, William Tomison wrote to James Swain that furs could not be moved as *"there being no water in the river."* (Johnson 1967: 33–39, 58)

At the end of this dry decade, reports from Fort Edmonton House describe poor trade with both the Slave and Southern Indians due to "*the amazing warmness of the winter*" (Johnson 1967: 33-39, 58) diminishing both the bison hunt and creating a "*want of beaver.*" There were reports of smoke that almost obscured the sun and remarks like "*the country all round is on fire.*" The "*amazing shallowness of the water*" prevented the shipment of considerable goods from York Factory (the headquarters of the Hudson's Bay Company on Hudson Bay).

In the 1850s, Captain John Palliser was dispatched from London by the Royal Geographical Society to evaluate the potential for British settlement of western Canada. He concluded that

> this large belt of country embraces districts, some of which are valuable for the purposes of the agriculturalist, while others *will forever be comparatively useless*. . . . The least valuable portion of the prairie country has an extent of about 80,000 square miles, and is that lying along the southern branch of the Saskatchewan, and southward from thence to the boundary line [the US border]. (Palliser 1862: n.p.)

Palliser filed these remarks in 1860 in the midst of a 25-year drought. Despite his warning, settlers were drawn from Europe, eastern Canada, and the United States. The railroad and communities like Medicine Hat, Alberta, were built. In the very first edition of the *Medicine Hat Times*, dated 5 February 1891, an editorial entitled "Our True Immigration Policy" stated, "It would be almost criminal to bring settlers here to try to make a living out of straight farming" (Jones 2002: 18). As it turned out, the next several decades were relatively wet, settlers flooded in, and the populations of Saskatchewan and Alberta increased by nearly 500% in one decade. Certainly they were not aware of decadal-scale climatic variability and the fact the climate would later flip again and bring the devastating droughts of the 1920s and "Dirty 30s."

Tree Rings

Tree rings provide a source of hydroclimatic data, such as data on available water and heat, and a chronology with absolute annual resolution spanning centuries to millennia. During the summer growing season in

Canada's western interior, there is usually more than enough light and heat. In this dry continental climate, soil moisture is the most limited determinant of tree growth. Therefore, the increment of annual growth is a proxy of hydrological or agricultural drought; dry years consistently produce narrow rings. Tree rings from living and dead trees that were growing at the same time for at least a few decades can be cross-dated, and calendar years can be transferred from the living to the dead trees. This process has produced tree-ring chronologies spanning the past millennium in western Canada. The mathematical relationship between standardized tree-ring widths and hydroclimatic data from nearby gauges is applied to the tree-ring data to reconstruct the relative moisture levels each year for the entire tree-ring record. Because the soil moisture that supports tree growth is derived mainly from melting snow and early-season rain, and because winter precipitation is strongly linked to large-scale climate oscillations, tree rings capture these teleconnections and the associated inter-annual to multi-decadal climatic variability, including periodic severe and prolonged drought.

Over the past 25 years, researchers in the Tree-Ring Lab at the University of Regina have collected more than 8,000 samples from old trees at more than 170 sites in the boreal, montane, and island forests of the Rocky Mountains and northern Great Plains. This network of tree-ring sites encompasses the semi-arid Prairie Ecozone. Because the tree rings were collected at dry sites (south- and west-facing slopes, sandy soils, ridge crests), where tree growth is moisture-sensitive, a strong correlation exists between the ring-width chronologies and drought and moisture indices. These tree-ring data have been the basis for a series of studies of Prairie drought (e.g., Bonsal et al. 2012; Lapp et al. 2012; St. George et al. 2009; Sauchyn et al. 2002, 2003; Sauchyn and Skinner 2001), including recent research by Kerr (2013), which is the source of the following case study.

A Tree-Ring Reconstruction of Hydrological Drought

Throughout the world, agriculture is the dominant use of water, and the major impacts of drought are directly or indirectly related to food production. Therefore, most indices of agricultural drought are expressions of the soil moisture balance, which unlike precipitation is not routinely measured. The best index of hydrological drought is streamflow; it is extensively monitored and integrates the net precipitation (in excess of evapotranspiration) over time (days to seasons) and watersheds. There is a relatively dense network of water-level gauges in the southern Prairies, since there is a strong demand for a limited surface water supply. This hydrometric network was originally established in the early twentieth century—not for the study of hydrology or climate, but rather to identify supplies of water initially for steam locomotives and irrigation (Greg McCullough, Water Survey of Canada, personal communication, June 2011). Therefore, just a few gauges have operated continuously for more than 50 years, recording only a few periods of sustained low water levels.

Tree rings are an effective streamflow proxy; they record the timing and duration of high and low water levels, and they have a similar muted response to episodic inputs of precipitation. When watersheds are wet (dry), streams rise (fall) and tree growth is enhanced (suppressed). Tree rings usually underestimate hydrological peaks, because there is a maximum positive biological response to available moisture; other factors constrain growth when soil moisture is not lacking. Thus, tree-ring data from moisture-sensitive ring-width chronologies are a better proxy of drought than of excess moisture.

Recently, Kerr (2013) completed a study of paleohydrology in the dry core of the northern Great Plains. Much of this region, at the junction of Alberta, Saskatchewan, and Montana (Figure 1), receives less than 330 mm of annual precipitation. Wetter conditions prevail in the uplands, so they contain island forests and the headwaters of all local rivers and streams. Kerr (2013) augmented and updated a network of tree-ring chronologies derived from lodgepole pine (*Pinus contorta*) and white spruce (*Picea glauca*) in the Cypress Hills (Alberta and Saskatchewan) and from Douglas fir (*Pseudotsuga menziesii*) and ponderosa pine (*Pinus ponderosa*) in the Sweet Grass Hills and Bears Paw Mountains of north-central Montana. Statistical tree-ring models explained 40%–55% of the recorded summer and annual flow of the Frenchman River, Battle Creek, and Swift Current Creek in southwestern Saskatchewan. The water-year (October–September) data from a gauge on Swift Current Creek are plotted in Figure 2 along with the flow predicted by a statistical tree-ring model for the same period (1979–2009). The two curves match in terms of the timing of high and low flows, although the tree rings underestimate the magnitude of the highest flows. Thus, they are a better proxy of drought than excess water.

Figure 1. Tree-ring sites (triangles) and streamflow gauges (squares) for a study of paleohydrology in the dry core of the northern Great Plains
(Source: Kerr 2013)

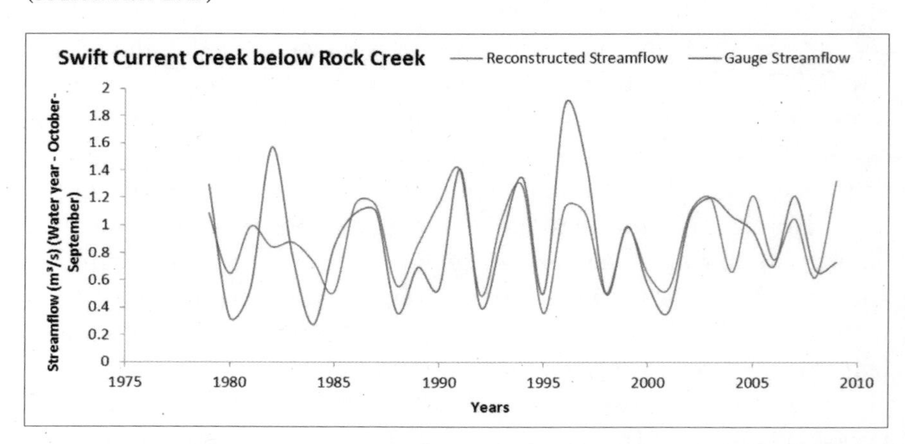

Figure 2. A plot of water-year (October–September) streamflow (m3/sec), from 1979 to 2009, as recorded at the gauge on Swift Current Creek below Rock Creek and reconstructed using tree rings.

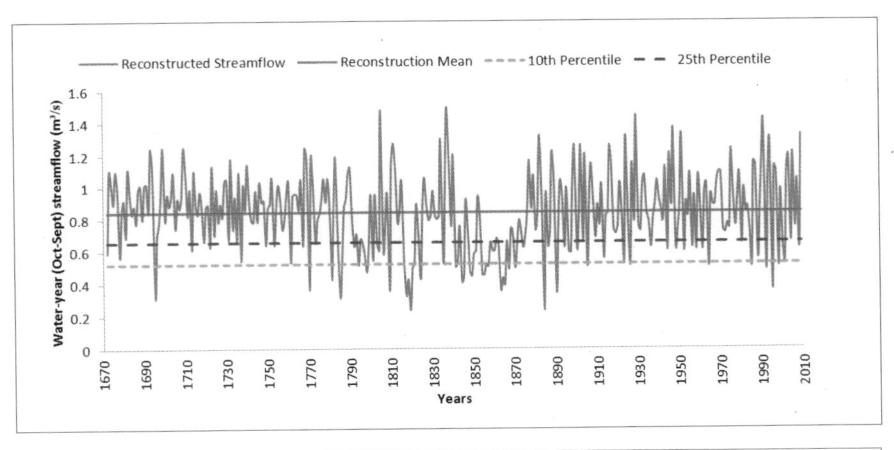

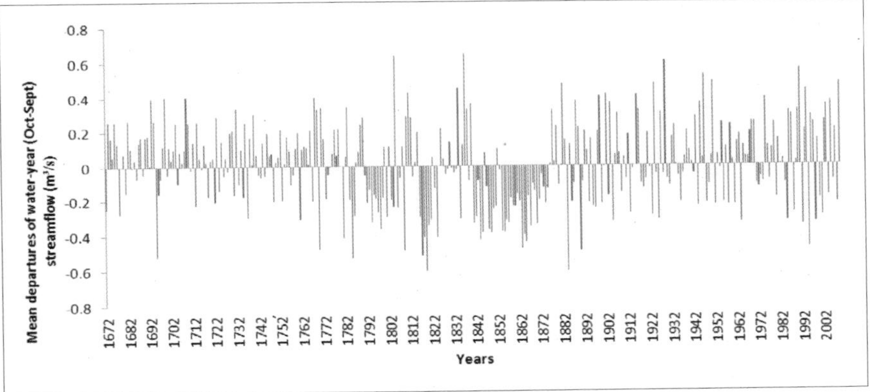

Figure 3. A tree-ring reconstruction of the flow of Swift Current Creek since 1672. Top plot: Water-year (October–September) streamflow (m^3/sec) showing the mean flow and 10th and 25th percentiles. Bottom: Water-year (October–September) streamflow (m^3/sec) plotted as departures from the mean reconstructed value. (Source: Kerr 2013)

By applying the statistical tree-ring model of streamflow to the entire length of the tree-ring chronologies, water-year flow from 1672 to 2009 was reconstructed. Two versions of this paleo-flow time series are plotted in Figure 3. The top plot shows the inferred annual flow, mean flow, and two thresholds of low flow—the 10th and 25th percentiles. In the bottom plot, departures from the reconstructed mean flow highlight the inter-annual variability and inter-decadal pattern, with extended periods of low flow evident in the 1790s to early 1800s and the late 1840s through 1870s.

Table 1. Hydrological droughts (<25th percentile) with severe droughts (<10th percentile) indicated in bold for average water-year flows at Swift Current Creek, below Rock Creek (1670–2009). Red text indicates five or more consecutive years of drought.

Single-year event	Two or more consecutive-year events
1672, 1678, **1695**	
1713, 1722, 1737, 1749, 1753, 1761, 1767, **1770**, 1773, **1781**, 1792, **1794**	1784, **1785**, 1786, 1796, 1797, **1798**
1801, 1806, **1809**, **1824**, 1835, **1867**, 1874, **1884**, 1886, **1890**, 1894	1803, 1804, 1816, **1817**, **1818**, **1819**, **1820**, 1821, **1841**, 1842, **1844**, **1845**, **1848**, **1849**, 1850, 1851, **1854**, **1855**, 1856, **1857**, 1858, 1859, 1860, 1862, **1863**, **1864**, **1865**, **1870**, 1871, 1896, 1897
1900, **1905**, 1913, 1923, 1926, 1936, 1944, 1948, 1952, 1956, 1958, 1961, **1964**, 1980, **1985**, 1988, **1992**, **1995**, 1998	
2008	2000, 2001

Because much of the unexplained variance in the calibration period (1979–2009; Figure 2) can be attributed to the underestimation of high flows, more confidence can be applied to the interpretation of low flows, which consistently correspond to narrow tree-rings, capturing the timing and duration of drought. The late eighteenth through mid-nineteenth centuries have the most sustained low flows. Severe hydrological droughts occurred from 1794 to 1798, 1816 to 1821, and 1854 to 1860 (Table 1). The repetitive nature of moisture surpluses and deficits in the streamflow reconstruction suggests some quasi-cyclical behaviour in the hydroclimatic regime. Spectral analyses provided evidence of this periodic hydroclimatic variability at the inter-annual (~2–6 years) and multi-decadal (~20–30 years) scales corresponding to the dominant frequencies of the El Niño–Southern Oscillation (ENSO) and the Pacific Decadal Oscillation (PDO).

Hydrologic droughts in the Swift Current Creek paleohydrology coincide with low flows and below-average precipitation in other paleoclimate records from western North America. The early to mid-1700s was a period of prolonged drought documented by paleoclimatic investigations across western North America (Cook et al. 1999; Woodhouse and Overpeck 1998; Laird et al. 1996; Stockton and Meko 1983). Various paleoclimatic studies emphasize the sustained nature of drought during the mid-nineteenth century, with very little relief in a few scattered wet years. This intense, long-lasting drought is well documented as occurring from the 1840s through mid-1860s throughout the western United States, Canada, and Mexico (Stahle and Dean 2011; Stahle and Cleaveland 1988; Fritts 1983; Stockton and Meko 1983; Hardman and Reil 1936).

This case study of the paleohydrology of southwestern Saskatchewan demonstrates tree rings are an effective proxy of annual streamflow. The proxy record for Swift Current Creek reveals periods of sustained low flow, including pre-settlement droughts that exceed in intensity and duration the worst conditions that have affected modern agriculture on the northern plains. Water deficits of this severity will reoccur in the future in a climate of rising temperatures.

Conclusion

Seasonal, and sometimes prolonged, moisture deficits are so characteristic of the climate of the Canadian Prairies that they define the region ecologically and ultimately limit forest and farmland productivity. Severe drought of high intensity and/or long duration, with serious consequences, is relatively infrequent. Thus over the past 13 decades, since Euro-Canadians first came to ranch and farm, people have experienced and recorded relatively few severe droughts. Although the meteorology and socio-economic impacts of these relatively few severe droughts have been studied extensively (e.g., Bonsal et al. 2011; Wheaton et al. 2008), the sample size is too small to analyze the frequency of these events in relation to regional climate variability and change. Thus, in this chapter, we examined the paleoclimate record of drought extending back over the past millennium.

A robust conclusion of the paleoclimate research on the Prairies is that the climate of the instrumental period is representative of the longer-term frequency of one- to two-year droughts but does not capture the full range of intensity and duration. The dry periods of greatest severity and duration occurred before the Prairies were settled. These include the intense drought years of the late eighteenth century (and the sand dune activity described above) and the sustained drought of the 1840–60s. Thus, the proxies suggest that the climate of the twentieth century (especially since the 1930s) was relatively favourable for the settlement of the Prairies, because the region has lacked the sustained droughts of preceding centuries. While the twentieth-century droughts may have been characterized by relatively modest precipitation deficits compared to earlier events, they have been hotter droughts than the cooler moisture-deficient periods of preceding centuries. This finding has important implications for studying and projecting future drought in a period of rapid global warming. The most serious impact of a warming climate in this region would be realized if the droughts of the 1790s or 1850s, and associated natural forcing, were to reoccur in the much warmer greenhouse gas climate of the twenty-first century.

The paleoclimatic records capture the tempo of natural climate variability, including the near-regularity of wet and dry cycles at certain frequencies. They show that the hydroclimatic regime periodically shifts from predominantly interannual variation to intervals with extended wet and dry spells and that there is a significant difference in the likelihood of drought according to phase of ocean-atmosphere oscillations (ENSO and PDO). This knowledge of long-term climate variability contributes to our understanding of the climate system at scales that exceed the length of instrumental records. The longest and most intense droughts, and the factors that cause them, reoccur so infrequently that a pre-instrumental paleoclimate perspective is required to validate the modelling and prediction of these events.

Our capacity to withstand and prepare for water scarcity has developed in response to the droughts that have occurred since the Prairies were first settled for agriculture, which have been shown to be much less intense than those that occurred before the Prairies were settled (i.e., those in the paleoclimate record presented here). Greater adaptive capacity will be required if future drought conditions are more intense or prolonged than those previously experienced. Significant adaptations may be required, particularly to water management practices and policies, starting with a scientific knowledge base that extends beyond instrumental records and the scale at which water supplies seem relatively secure and stationary, and then encompassing the longer view provided by paleoclimate records and model projections of future climate. Communities and governments are investing effort and resources in adaptation planning, in large part to mitigate the potential impacts of a warmer and more extreme climate. To inform this process, and be perceived as a credible source of information on exposure to drought, reconstructions of long-term climatic variability must be based on definitions of drought that are applicable to agriculture and water resource management.

The characteristics of drought detected using natural and historical archives must be related to droughts of recent experience and to the nearest modern analogues. To communicate the severity of paleodroughts, an example might be given of a situation in which one intense recent drought is followed immediately by another similarly intense drought. In this way, the characteristics of the megadroughts in the paleoclimate record can be translated into terms that are used and understood by other natural and social scientists, and by engineers and resource managers responsible for monitoring and managing drought.

Acknowledgments

Funding for the case study was provided by the Rural Community Adaptation to Drought project through the Social Sciences and Humanities Research Council and by the Prairie Adaptation Research Collaborative. Jessica Vanstone, Cesar Perez-Valdivia, Ben Brodie, and Tiffany Vass assisted Samantha Kerr with field and laboratory work.

References

Blair, D., and W.F. Rannie. 1994. "Wading to Pembina: 1849 Spring and Summer Weather in the Valley of the Red River of the North and Some Climatic Implications." *Great Plains Research* 4, no. 1: 3–26.

Bonsal, B.R., R. Aider, P. Gachon, and S. Lapp. 2012. "An Assessment of Canadian Prairie Drought: Past, Present, and Future." *Climate Dynamics*. doi: 10.1007/s00382-012-1422-0.

Bonsal, B., E. Wheaton, A. Chipanshi, C. Lin, D. Sauchyn, and L. Wen. 2011. "Drought Research in Canada: A Review." *Atmosphere-Ocean* 4: 303–19. http://dx.doi.org/10.1080/07055900.2011.555103.

Cook, E.R., D.M. Meko, and D.W. Stahle. 1999. "Drought Reconstruction for the Continental United States." *Journal of Climate* 12: 1145–62.

Cook, E., R. Seager, M. Cane, and D. Stahl. 2007. "North American Drought: Reconstructions, Causes, and Consequences." *Earth-Science Reviews* 81: 93–134.

Fritts, H. 1983. "Tree-ring Dating and Reconstructed Variations in Central Plains Climate." *Transactions of the Nebraska Academy of Science and Affiliated Societies* 11: 37–41.

Hardman, G., and O.E. Reil. 1936. "The Relationship between Tree-growth and Stream Runoff in the Truckee River Basin, California–Nevada." *Agricultural Experiment Station* 41: 1–38.

Johnson, A.M. (ed.). 1967. *Saskatchewan Journals and Correspondence: Edmonton House and Chesterfield House, 1795–1800.* London: Hudson's Bay Record Society.

Jones, D.C. 2002. *Empire of Dust: Settling and Abandoning the Prairie Dry Belt.* Calgary: University of Calgary Press.

Kerr, S.A. 2013. "A Dendroclimatic Investigation of Southwestern Saskatchewan." M.Sc. thesis, University of Regina.

Laird, K.R., B.F. Cumming, S. Wunsam, O. Olson, J.A. Rusak, R.J. Oglesby, S.C. Fritz, and P.R. Leavitt. 2003. "Lake Sediments Record Large-Scale Shifts in Moisture Regimes across the Northern Prairies of North America during the Past Two Millennia." *Proceedings of the National Academy of Sciences* 100: 2483–88.

Laird, K.R., S.C. Fritz, K.A. Maash, and B.F. Cumming. 1996. "Greater Drought Intensity and Frequency before A.D. 1200 in the Northern Great Plains." *Nature* 384: 551–54.

Lapp, S., J.-M. St. Jacques, D. Sauchyn, and J. Vanstone. 2012. "Forcing of Hydroclimatic Variability in the Northwestern Great Plains since AD 1406." *Quaternary International* 310: 47–61.

Maliva, R., and T. Missimer. 2012. *Arid Lands Water Evaluation and Management, Environmental Science and Engineering.* Berlin Heidelberg: Springer-Verlag.

Michels, A., K.R. Laird, S.E. Wilson, D. Thomson, P.R. Leavitt, R.J. Oglesby, and B.F. Cumming. 2007. "Multi-Decadal to Millennial-Scale Shifts in Drought Conditions on the Canadian Prairies over the Past Six Millennia: Implications for Future Drought Assessment." *Global Change Biology* 13: 1295–1307.

McKee, T.B., N.J. Doesken, and J. Klesit. 1993. "The Relationship of Drought Frequency and Duration to Time Scales." Pp. 179–84 in *Proceedings of the Eighth Conference on Applied Climatology.* Anaheim, CA, 17–22 January 1993.

National Drought Mitigation Center. 2006. "What is Drought? Understanding and Defining Drought." National Drought Mitigation Center website. http://drought.unl.edu/DroughtBasics/WhatisDrought.aspx.

Palliser, J. 1862. "Journals, Detailed Reports and Observations Relative to Captain Palliser's Exploration of a Portion of British North America." In I.M. Spry (ed.) (1968), *The Papers of the Palliser Expedition.* Toronto: The Champlain Society.

Pham, S.V., P.R. Leavitt, S. McGowan, and P. Peres-Neto. 2008. "Spatial Variability of Climate and Land-use Effects on Lakes of the Northern Great Plains." *Limnology and Oceanography* 53, no. 2: 728–42.

Pham, S.V., P.R. Leavitt, S. McGowan, B. Wissel, and L.I. Wassenaar. 2009. "Spatial and Temporal Variability of Prairie Lake Hydrology as Revealed Using Stable

Isotopes of Hydrogen and Oxygen." *Limnology and Oceanography* 54, no. 1: 101–18.
Rannie, W.F. 2006. "A Comparison of 1858–59 and 2000–01 Drought Patterns on the Canadian Prairies." *Canadian Water Resources Journal* 31, no. 4: 263–74.
Sauchyn, D.J., E.M. Barrow, R.F. Hopkinson, and P. Leavitt. 2002. "Aridity on the Canadian Plains." *Géographie physique et Quaternaire* 66: 247–69.
Sauchyn, D.J., and B. Bonsal. 2013. "Climate Change and North American Great Plains' Drought." In A.-H. El-Shaarawi and W. Piegorsch (eds.), *Encyclopedia of Environmetrics*. Chichester, UK: John Wiley & Sons. doi: 10.1002/9780470057339.vnn123.
Sauchyn, D.J., and W.R. Skinner. 2001. "A Proxy PDSI Record for the Southwestern Canadian Plains." *Canadian Water Resource Journal* 26, no. 2: 253–72.
Sauchyn, D.J., J. Stroich, and A. Beriault. 2003. "A Paleoclimatic Context for the Drought of 1999–2001 in the Northern Great Plains." *The Geographical Journal* 169, no. 2: 158–67.
St. George, S., D.M. Meko, M.P. Girardin, G.M. MacDonald, E. Nielsen, G.T. Pederson, D.J. Sauchyn, J.C. Tardif, and E. Watson. 2009. "The Tree-Ring Record of Drought on the Canadian Prairies." *Journal of Climate* 22: 689–710.
St. George, S., and D. Sauchyn. 2006. "Paleoenvironmental Perspectives on Drought in Western Canada." *Canadian Water Resources Journal* 31, no. 4: 197–204.
St. Jacques, J.-M., B.F. Cumming, D.J. Sauchyn, and J.P. Smol. 2015. "The Bias and Signal Distortion Present in Conventional Pollen-based Climate Reconstructions as Assessed by Early Climate Data from Minnesota, USA." *PLoS One* 10, no. 1: e0113806.
Stahle, D.W., and M.K. Cleaveland. 1988. "Texas Drought History Reconstructed and Analyzed from 1698 to 1980." *Journal of Climate* 1: 59–74.
Stahle, D., and J. Dean. 2011. "North American Tree Rings, Climatic Extremes, and Social Disasters: Social Impacts of Climate Extremes during the Historic Era." In M. Hughes, T. Swetnam, and H. Diaz (eds.), *Dendroclimatology Progress and Prospects. Developments in Paleoenvironmental Research*. Volume 11. London/New York: Springer Dordrecht Heidelberg.
Stockton, C., and D. Meko. 1983. "Drought Recurrence in the Great Plains as Reconstructed from Long-term Tree-ring Records." *Journal of Climate and Applied Meteorology* 22: 17–29.
Vicente-Serrano, S., S. Beguería, and J. López-Moreno. 2010. "A Multiscalar Drought Index Sensitive to Global Warming: The Standardized Precipitation Evapotranspiration Index." *Journal of Climate* 23: 1696–1718.
Wheaton, E., S. Kulshreshtha, V. Wittrock, and G. Koshida. 2008. "Dry Times: Hard Lessons from the Canadian Drought of 2001 and 2002." *Canadian Geographer* 52: 241–62.
Wolfe, B.B., R.I. Hall, T.W.D. Edwards, S.R. Jarvis, R.N. Sinnatamby, Y. Yi, and J.W. Johnston. 2008. "Climate-Driven Shifts in Quantity and Seasonality of River Discharge over the Past 1000 Years from the Hydrographic Apex of North America." *Geophysical Research Letters* 35: 1–5.

Wolfe, B.B., R.I. Hall, T.W.D. Edwards, and J.W. Johnston. 2012. "Developing Temporal Hydroecological Perspectives to Inform Stewardship of a Northern Floodplain Landscape Subject to Multiple Stressors: Paleolimnological Investigations of the Peace–Athabasca Delta." *Environmental Reviews* 20: 1–20.

Wolfe, S.A., C.H. Hugenholtz, and O.B. Lian. 2012. "Palliser's Triangle: Reconstructing the 'Central Desert' of the Southwestern Canadian Prairies during the late 1850s." *The Holocene* 23: 669–707.

Wolfe, S.A., D.J. Huntley, P.P. David, J. Ollerhead, D.J. Sauchyn, and G.M. Macdonald. 2001. "Late 18th Century Drought-induced Sand Dune Activity, Great Sand Hills, Southwestern Saskatchewan." *Canadian Journal of Earth Sciences* 38: 105–17.

Woodhouse, C., and J. Overpeck. 1998. "2000 Years of Drought Variability in the Central United States." *Bulletin of the American Meteorological Society* 79, no. 12: 2693–2714.

Zargar, A., R. Sadiq, B. Naser, and F. Khan. 2011. "A Review of Drought Indices." *Environmental Reviews* 19: 333–49.

CHAPTER 3

FUTURE POSSIBLE DROUGHTS

Elaine Wheaton, David Sauchyn, and Barrie Bonsal

Background and Rationale

Nothing is definite in the future, but drought is certain to play a role, as it is a part of the climate of the Canadian Prairies. Droughts can be casually defined as a worrisome lack of water or more formally defined as a prolonged period of abnormally dry weather that depletes water resources for human and environmental needs (Meteorological Service of Canada 1986). Droughts occur in many regions of North America and the world, but the agricultural region of the Canadian Prairie provinces of Alberta, Saskatchewan, and Manitoba is among the most susceptible to droughts and is the focus of this chapter.

Prairie people have considerable experience with climate extremes, such as drought and heat, but these extremes can still cause concern and damage. Drought is more costly than any other form of natural disaster (Wilhite 2000). This is especially true of the Canadian Prairies, where drought is very damaging to the economy, society, and the environment, even in recent years (e.g., Wheaton et al. 2008). Drought occurs in most years in some part of the Canadian Prairies, but it is the longer-duration and larger-area droughts that have the most severe impacts and provide the greatest challenges for adaptation. At least five major droughts have occurred in the Canadian Prairies during the past 120 years. These include multi-year droughts in the 1890s, 1910s, 1930s, and 1980s, and in 2000–2004 (Bonsal, Wheaton, Chipanshi, et al. 2011; Bonsal, Wheaton, Meinert, et al. 2011). During this last major drought, parts of the Prairies had some of the driest conditions in the historical record, and it was one of the first documented coast-to-coast droughts in Canada.

Each major drought on the Prairies appears to have several unique characteristics, including duration, area of coverage, intensity, and cause, but Bonsal, Wheaton, Meinert, et al. (2011) documented several similarities among the droughts, including their origin in the US northern plains and subsequent migration into the Canadian Prairies. The authors devised and used a six-stage drought classification system to compare the major droughts. A key difference of the 2000–2004 drought is that its peak in terms of area of severe drought was during winter, whereas the others peaked during the May-to-August growing season. Most of these major droughts lasted almost two years, but the 1928–32 drought lasted over 40 months.

More recently, a less severe and shorter drought occurred from 2008 to 2010 in the Canadian Prairies (Wittrock et al. 2010). A core of well-below-average rainfall appeared around Edmonton, Alberta, and northward in the summer of 2008. By that autumn, this core area had expanded westward to the British Columbia border and eastward into Saskatchewan. The drought intensified in the winter of 2008–9, and the most severe and largest dry area appeared in spring 2009. Rainfall eased the dryness by autumn 2009, but dryness continued in areas of Alberta. Spring rains in 2010 ended the meteorological drought for most areas, but effects lingered.

Reducing the negative impacts of drought requires considerable planning and preparation so that society can effectively adapt to future droughts. These activities require understanding the nature of future possible droughts, which is the rationale for work that projects important climate extremes, especially future drought. It is prudent and critical to advance knowledge of future possible droughts for effective adaptation, that is, to decrease these massive costs and to take advantage of any opportunities. Such opportunities can result from numerous benefits of drought, including increased quality of grain and hay, reduced levels of some insects and diseases, and fewer delays for construction of roads and buildings (Wheaton et al. 2011).

If the past were the only guide to the future, past information would suffice for estimating future droughts. However, the current risk of drought is changing, perhaps fairly rapidly, and research indicates that dry areas (such as the Prairies) are expected to become drier. The potential for future drought risk is increasing largely because of human-induced climate change (IPCC 2012).

Objectives and Methods

Information is needed regarding the nature of future droughts to facilitate adaptation and reduce vulnerability. Critical questions to address include: How will droughts change in terms of characteristics such as severity, duration, frequency, timing, cause, and area? The objective of this chapter is to address these questions by reviewing drought literature focused on the Canadian Prairies. Much work has emphasized the global scale, but considering that drought is an important hazard for the region, several papers have focused at the scale of the Prairie provinces. We attempt to emphasize the near future to about mid-century, but the relevant literature tends to use the following standard periods: the 2020s (2011–40), 2050s (2041–70), and the 2080s (2071–2100); some literature uses other scales, such as time-series results, for the period up to 2100.

For each study reviewed in this chapter, its methods are briefly described to help assess its results. More recent literature is used, where possible, with some reference to earlier literature for perspective. Generally, drought characteristics are measured using several indicators. The most common of these indicators are the moisture deficit (e.g., precipitation minus potential evapotranspiration), Palmer Drought Severity Index (PDSI), and Standardized Precipitation Index (SPI). Newer indices, such as the Standardized Precipitation Evapotranspiration Index (SPEI), are the subject of current research.

Future Possible Droughts of the Canadian Prairie Agricultural Region

A few studies have focused on the nature of future droughts in the Canadian Prairie provinces. Work at the global scale hints at future drought conditions on the Prairies, but focused assessments done at a finer resolution provide more detailed information. The reason for estimating future possible droughts is to improve adaptation as droughts have serious impacts and require considerable adaptation to reduce vulnerability (Kulshreshtha et al., this volume). In this section, we present methods and findings of recent work on future droughts in the Canadian Prairies and discuss this research within the context of earlier work.

One of the studies with findings for the Canadian Prairies is by Barrow (2010). She analyzed output from a set of regional climate models (RCMs) to determine characteristics of potential evapotranspiration (PET) and moisture deficit. PET was calculated using two methods for comparison: Thornthwaite and Penman-Monteith. RCMs were shown to simulate observed precipitation values better than global climate models (GCMs), as the RCMs have finer spatial resolution. All but one of the models used by Barrow (2010) showed increases in both intensity and area of moisture deficit (water-year October to September) for the 2041–70 period. Time-series analyses project increases in evaporative demand over time for all simulations driven by the projected temperature increases. For the Canadian Regional Climate Model (CRCM) results, annual moisture deficits for this future period range from about -400 mm in southwest Saskatchewan to -200 mm just north of about 50° North (Figure 1). These values are about double the annual moisture deficit for the 1971–2000 period. Also, the area with an annual moisture deficit of -200 to -400 mm expands considerably into the future, migrating from a narrow ribbon along the US border in southwestern Saskatchewan and southeastern to central Alberta to cover all of southern Saskatchewan past Regina.

Barrow's (2010) findings of future expansion of arid areas confirm earlier work. For example, Sauchyn et al. (2005) used the aridity index (ratio of annual precipitation to PET) for the Canadian Prairies, and found that the area of aridity (ratio less than 0.65) increased by 50% and expanded northward.

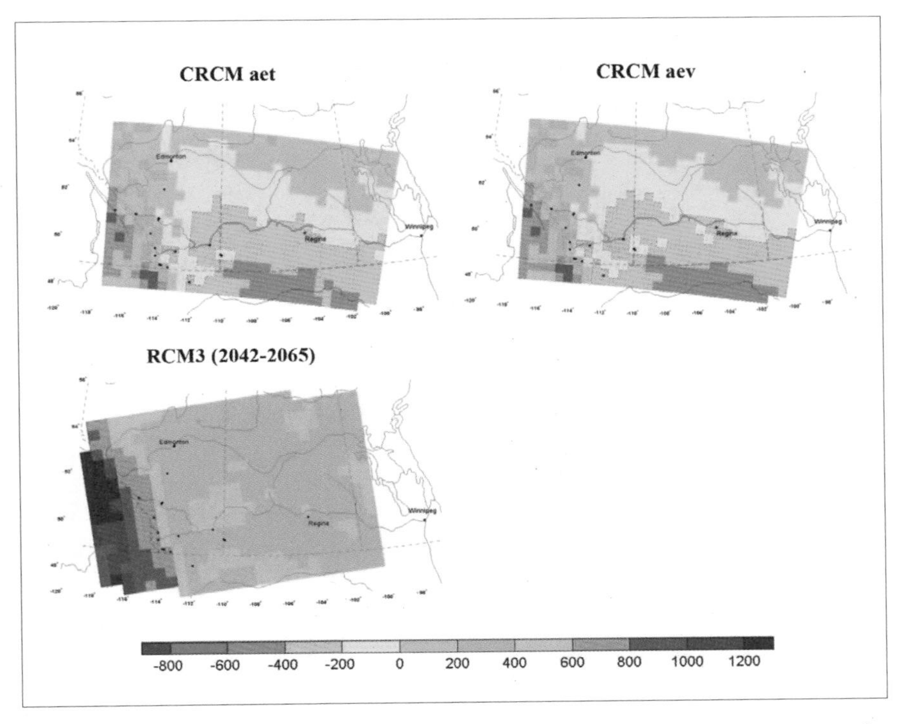

Figure 1. Spatial pattern of the future annual moisture deficit for the water-year of October to September (2041–70, precipitation minus potential evapotranspiration, mm). The two different experiment identifiers of the Canadian Regional Climate Model (CRCM) are labeled as aev and aet
(Source: Barrow 2010: 21)

Another study focusing on the Canadian Prairies by Thorpe (2011) used a range of climate change scenarios from several GCMs to estimate future PET for the Prairie Ecozone. He also found that PET increases in the future. The average Prairie Ecozone PET for Saskatchewan and Alberta is about 550 mm for the baseline climate of 1961–90, increases to about 600 mm at about 2020, reaches almost 700 mm by 2040, and increases even more rapidly thereafter for the warm scenario (Figure 2). Changes in annual precipitation are projected to vary from only small increases in the warm scenario to small decreases in the cooler scenario. The changes in precipitation are projected with much lower confidence than for temperature. These potential increases in precipitation are insufficient to compensate for the increased atmospheric water demand, producing the

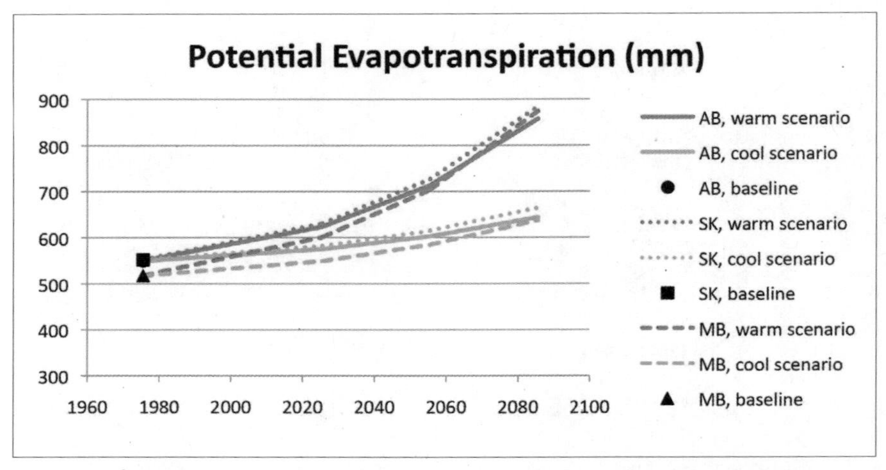

Figure 2. Average potential evapotranspiration for the Prairie Ecozone of Alberta (AB), Saskatchewan (SK), and Manitoba (MB) for the baseline climate (1961–90) and for two future scenarios
(Source: Thorpe 2011: 5)

greater moisture deficits as estimated by Barrow (2010), for example. Williams and Wheaton (1998) calculated that increased annual precipitation of about 7%–10% is needed to compensate for an increase in mean annual temperature of 3°C.

Sushama et al. (2010) used the CRCM and the number of dry spells or dry days with precipitation less than 2 mm (and other thresholds) to explore future drought characteristics. Results indicate that the number of dry days will increase by up to about five days in the 2050s for southern Saskatchewan. The 10- and 30-year return levels of maximum dry spell length are projected to increase during the 2050s and 2080s in the Canadian Prairies, especially in the south.

Price et al. (2011) developed high-resolution climate scenarios for Canada from several GCMs. Besides increases in temperature and only modest increases in precipitation, they project solar radiation levels will increase slightly during summer in the Canadian Prairies' semi-arid ecozone. These changes would contribute to increasing dryness. Vapour pressure levels are also projected to increase and would offset some of the effects of warming on evaporative demand, but overall evaporation rates are expected to increase. Generally, Price et al. (2011) estimate that temperature, precipitation, and solar radiation will increase, along with

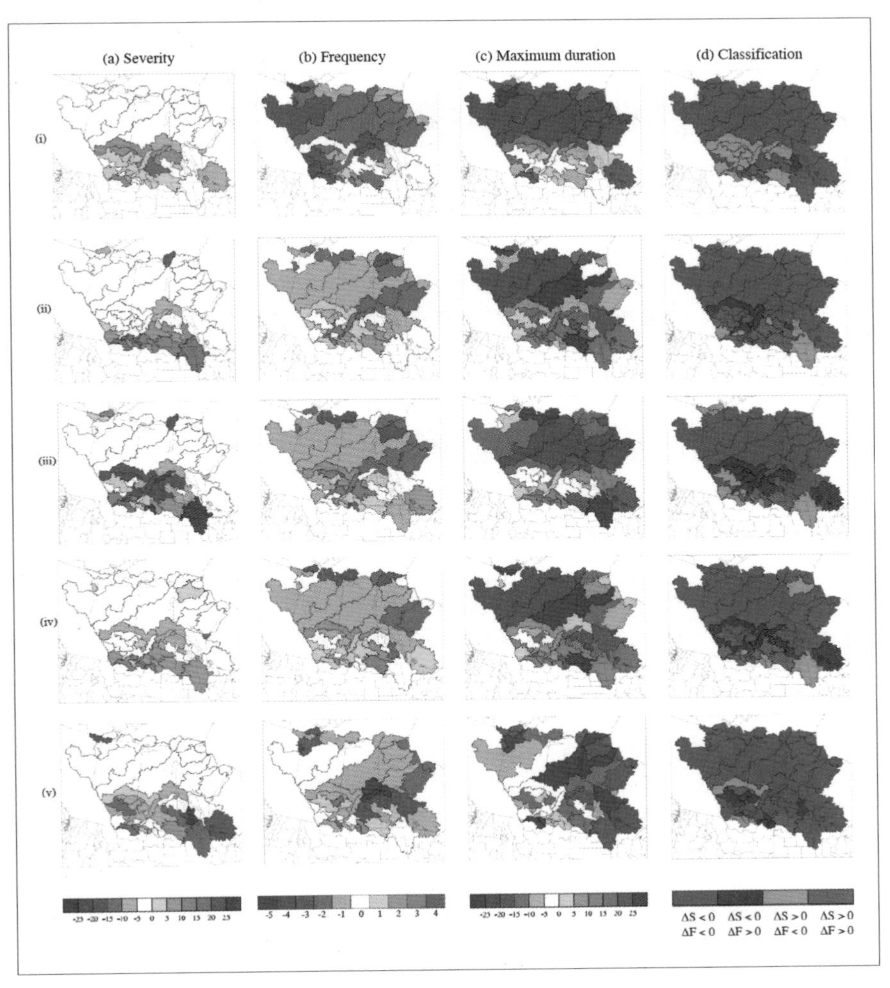

Figure 3. Projected changes to a) severity (%), b) frequency, and c) maximum duration (months) for 10-month drought events at the watershed scale, and d) classification of watersheds based on projected changes to the severity and frequency of 10-month events for the 47 watersheds in the Canadian Prairies for five pairs (i–v) of Canadian Regional Climate Model simulations
(Source: PaiMuzumber et al. 2012: Figure 12)

some increases in inter-annual variation, which indicate that multi-year droughts will become more common and more intense, especially with higher emission scenarios by 2100.

PaiMazumber et al. (2012) estimated future durations of drought severity; their results show that 6- and 10-month long droughts will become more severe over southern Saskatchewan and Manitoba in the 2050s compared with the 1971–2000 baseline. The 10-month droughts are expected to increase in frequency by as many as four events in the 2050s. Maximum durations of long-term droughts are projected to increase for a large part of the southern Prairies, and the largest increases are expected for droughts lasting 10 months or longer. The most vulnerable watersheds were found to have future possible increases in both severity and frequency of 10-month droughts for five pairs (GCM/RCM) of climate simulations for the 2050s (Figure 3). The CRCM and the high-emission scenario (A2) were used to develop climate scenarios, and monthly precipitation deficits were used to measure drought severity. Limitations of the research include the CRCM's ability to simulate precipitation, the use of only one model, and the use of precipitation alone to describe drought.

Bonsal et al. (2013) produced one of the most comprehensive descriptions of future possible drought for the Canadian Prairies and were the first to use three time periods—pre-instrumental period, instrumental (or observational) period, and future—spanning the years 1365–2100. Their study area was Alberta and western Saskatchewan from the US border north to past Edmonton (i.e., 54° North). They used five climate scenarios downscaled from two versions of the CGCM and the UK Hadley climate model (HadCM3), as well as the baseline period of 1961–90. Summer (June, July, August) self-calibrated PDSI and SPI values were averaged over the study area, and time series were produced from 1900 to 2100. They examined the time series of the areal averaged PDSI and SPI for the 1901–2099 period for the five GCMs, their means, and the nine-year running means (Figure 4).

Bonsal et al.'s (2013) results indicate that the pattern of the future mean PDSI values shows drying from the present to 2020, followed by a slight improvement with much variability to 2040. After 2040, persistently negative values occur with a downward trend, reflecting drier to drought conditions. The authors suggest this trend indicates a permanent regime shift to a more arid climate. In contrast, the SPI time series for the future period reveal no strong change compared with the instrumental period to about 2040; however, a higher persistence of multi-year droughts is found in the central and southern portion of the study area. This result occurs

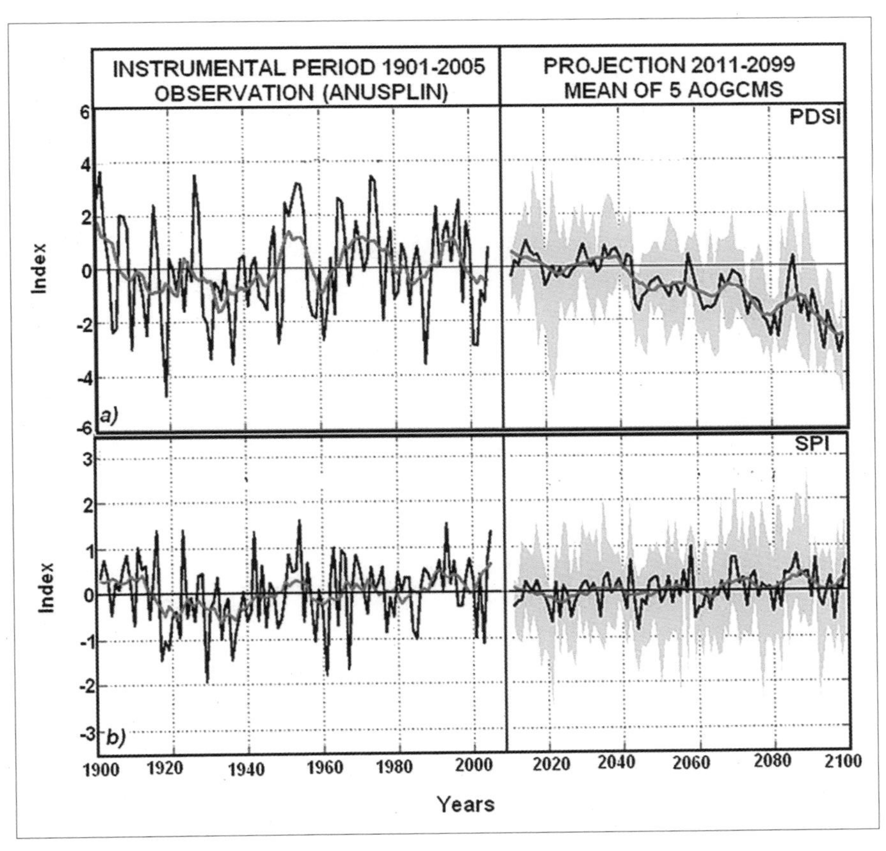

Figure 4. Summer a) Palmer Drought Severity Index (PDSI) and b) Standardized Precipitation Index (SPI) area-averaged values for the instrumental period (1901–2005) and the future (2011–2099). The black lines are the future ensemble-mean values from the five climate model runs, and the red lines are the nine-year running means. The minimum and maximum climate projections for each summer are shown in grey. (Source: Bonsal et al. 2013: Figure 9)

because SPI is calculated using only precipitation and not temperature. Drought indicators that consider precipitation alone are insufficient to determine future drought characteristics (e.g. Bonsal et al. 2013).

Drought area was also estimated by Bonsal et al. (2013); they found a substantial increase in the area and frequency of severe drought and worse (i.e., PDSI of -3 or less) in the area of the Canadian Prairies they studied. Even SPI shows that most future summers have severe drought conditions in some portion of the study area. These patterns suggest that severe droughts will become a more permanent feature in some areas of the southwestern Canadian Prairies in terms of characteristics such as occurrence, duration, and/or severity.

Multi-year droughts were also investigated by Bonsal et al. (2013) using the PDSI and found to be more frequent in the future period compared with the instrumental period (105 years). The length of a drought was considered to be the average number of consecutive summers with a negative value. Summer droughts of five years and longer have a frequency of 1.9 occurrences per 100 years during the instrumental period. This frequency is expected to more than double to 4.2 per 100 years in the future. The frequency of droughts of 10 years or longer increases to 3.1 per 100 years in the future. This result is even worse than the paleo record frequency of 3.0 per 100 years (see Chapter 2 by Sauchyn and Kerr in this volume). A worst-case situation is for increased frequency of drought of 10 years with consecutive summer droughts (i.e., negative PDSI values).

Although the general climate is projected to become drier, substantial variability could occur. The IPCC (2012) has identified areas of expected changes for the return period of intense daily rainfall events globally. For central North America, including the agricultural prairies, a decrease of 5 to 10 years is projected for the return period of a maximum 20-year rainfall event. This means that an extreme daily rainfall event could occur as much as twice as often as during 1981 to 2000. This result is for the middle 50% of models for the medium (A1B) to extreme (A2) emission scenarios. Dai (2010) also reports that the type of rainfall is expected to change with continued warming to more intense rainfall events and fewer light rainfalls. This pattern would tend to exacerbate drought, because intense rainfalls do not recharge soil moisture as well as more gentle rainfalls. Drier soils also tend to increase the risk of drought, because heating is used to warm soils to higher temperatures instead of evaporating water. This effect is similar to the cooling effect of water on one's skin as compared with dry skin that stays warmer.

Wheaton et al. (2013) reviewed projections of extreme precipitation globally and for the Canadian Prairies and found consistent estimates from several sources of increases in future extreme rainfall. Therefore, it seems that long periods of dry to drought conditions would be punctuated by periods of extreme rainfall. Some of the mechanisms behind this trend

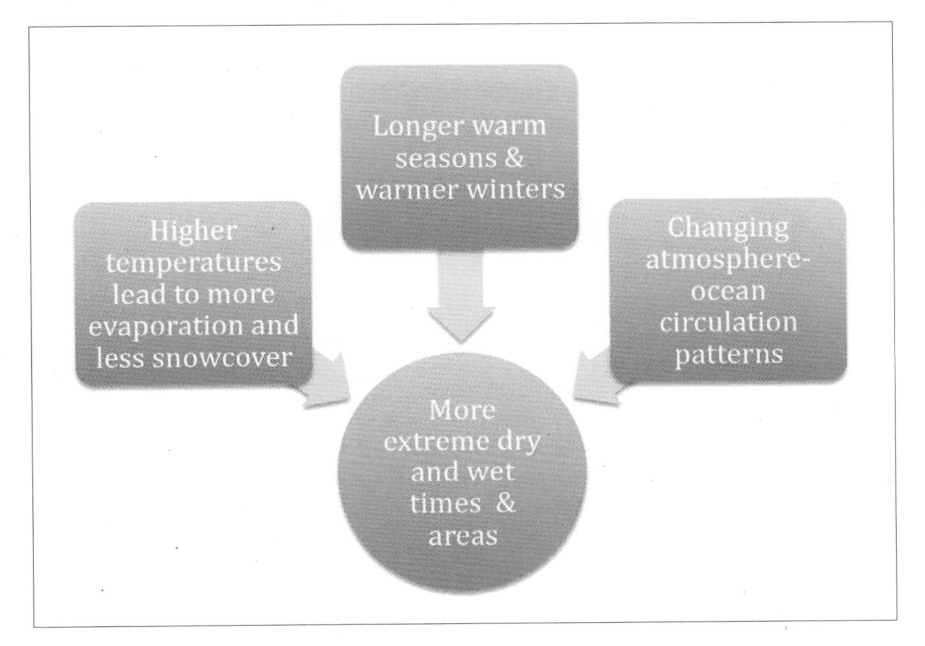

Figure 5. Dry times become drier and wet times become wetter (Source: Wheaton 2013)

include higher temperatures, shorter snow-cover seasons, longer warm seasons, and changing atmospheric and oceanic circulation patterns, as shown conceptually in Figure 5.

Based on research estimating future drought on a regional or global basis, there is a clear evidence for increasing risk of more common and persistent severe future droughts, including on the Canadian Prairies. A summary of projections of probable future drought characteristics emphasizes the consistency in projections of dry times and places becoming drier (Table 1). The Canadian Prairies will not be the only area facing more severe future drought. The IPCC (2012) states that drought will intensify in the twenty-first century in some seasons and areas. These regions include central North America, southern and central Europe, the Mediterranean, Central America and Mexico, northeastern Brazil, and southern Africa.

Table 1. Future possible drought characteristics, Canadian Prairie agricultural region

Drought indices	Projections	Time period	Spatial pattern	Climate models	Comments	References
P-PET	Decrease of about 200 mm and more farther south; driest area expands considerably	2050s mean annual water-year	Greater increase farther south, larger increase in SK	CRCM with CGCM3 HRM3	PET increases in intensity and area with time for all simulations; risk of moisture deficit becomes more severe.	Barrow 2010
AMI	Increase of about 1 degree- day/mm from current 3.5	2050s	Greater increase or dryness farther east	56 climate scenarios with range selected by AMI	Increased dryness occurs for all sites and time periods in the grassland region.	Barrow 2009
Precipitation deficits (monthly)	6- and 10-month droughts worsen	2050s	Southern SK and MB	CRCM with A2	The longer, 10-month drought adds about 4 events in the future.	PaiMazumber et al. 2012

PDSI negative for consecutive summers	Frequency of 3.1/100y for 10y+ droughts; more than 3× current rate	2080s		Statistically downscaled, 5 climate model runs	Frequency also increases for 5y+ droughts compared with instrumental and pre-instrumental.	Bonsal et al. 2013
Summer PDSI (areal average)	Permanent drought occurs after about 2040.	2011–2100 time series	Conditions are worse in the eastern Prairies.	Statistically downscaled with 5 climate model runs	Most statistics show worse droughts, including area, frequency, and intensity.	Bonsal et al. 2013
Multi-year droughts PDSI	Frequency more than doubles to 4.2/100y of 5y or longer droughts.	2011–2100 time series	Time series	Five scenarios ensemble-mean	Frequency of 10y+ droughts or longer triples.	Bonsal et al. 2013

Source: Wheaton et al. 2013.

Notes: PET = potential evapotranspiration; P = precipitation; PDSI = the Palmer Drought Severity Index; SK = Saskatchewan; MB = Manitoba; HRM3 = UK Hadley Regional Climate Model driven by HadCM3; CRCM = Canadian Regional Climate Model driven by CGCM 3 T47.

The annual moisture index (AMI) is GDD/P, the annual growing degree-days (base 5°C) divided by the total annual precipitation; multi-year drought mean duration is the average number of consecutive summers with a negative value of PDSI.

Lessons for the Future from Paleoclimates

At the end of this chapter, we conclude that current and past droughts may seem mild compared with future possible droughts. By "past" droughts we mean those that have affected the Canadian Prairies since agriculture was introduced, that is, those occurring after the settlement of the Prairies by Euro-Canadians. Also, nearly all planning and resource management that involves weather and water is based on direct observations and information collected from water gauges and weather instruments. This direct observation of weather and water began soon after the railroad was built and settlers arrived, so these records appear to be long, but they actually are very short compared to the age of the Prairie landscape and stream network that formed with the retreat of the continental ice sheet between 12,000 and 18,000 years ago. Climate varies over a large range of temporal scales, spanning seasons to climatic cycles that may last for tens of thousands of years. A weather record that spans decades to at most about 100 years will reveal only the shorter cycles. These short weather records are embedded in longer cycles that can be detected only from indirect study or inference of climate from geological and biological indicators (proxies) of climate variability and change.

The past climate or paleoclimate of the Canadian Prairies has been reconstructed from various climate proxies, including trees growing at the margins of Prairie grasslands and in island forests like the Cypress Hills, and the types and relative abundance of certain minerals, plant remains (e.g., pollen, spores, seeds), and aquatic organisms (e.g., diatoms, ostracodes) found in buried soils and lake sediments. The sampling and analysis of these remnants of prior ecosystems has revealed shifts in climate, in some cases abruptly, over the past 10,000–12,000 years of relative landscape stability. For example, the paleoecology of the Peace–Athabasca Delta (Wolfe et al. 2012) and the paleolimnology of Humboldt Lake, Saskatchewan (Michels et al. 2007) show systematic shifts in moisture regime, including extended dry periods (megadroughts) during the Medieval Climate Anomaly in the ninth to eleventh centuries. These past periods of higher temperature and aridity have been used as temporal analogues of the warmer climate emerging as a result of anthropogenic effects.

In Chapter 2, Sauchyn and Kerr look in detail at the nature of these various climate proxies and how they are used to infer past climate and water conditions. Here we are interested in what the paleoclimate of the Prairies can reveal about the climate to expect in coming decades. In the future, our climate will be increasingly influenced by human modifications of the atmosphere and Earth's surface. Anthropogenic emissions of greenhouse gases have been apparent only since the mid-nineteenth century and have become a major factor affecting climate only in recent decades. Knowledge of the regional climate regime is extremely important to detect an anthropogenic signal and to separate natural climate variation from what is human-induced. Future climate will be affected by both, although at some point the distinction between natural and anthropogenic will become irrelevant because the "natural" drives of climate (excluding volcanic eruptions), notably ocean-atmosphere circulation anomalies, are part of an increasingly artificial climate system. The paleoclimate record gives us a baseline; it shows the climate cycles as they exist in a mostly natural climate regime. Climate scientists expect that, for at least the next few decades, regional climate fluctuations will mostly consist of natural climate variability (Deser et al. 2012). This scenario applies, in particular, to regions like the Canadian Prairies that have a high degree of climate variability, and thus where the anthropogenic signal is more difficult to detect against the background of extreme inter-annual and decadal variability.

Based on the research above, we can expect that prolonged and severe droughts, similar to those that are evident in the paleoclimate record and discussed in the previous chapter, will reoccur in the coming decades. These droughts were of longer duration and, in some cases, greater severity than the worst droughts of the post-settlement period—those recorded by weather and water gauges. In the absence of global warming, we would expect unprecedented drought conditions. Global warming only amplifies the probability that future droughts will be more severe than those that have produced much of the adaptation of our communities and economy to a dry climate.

Summary and Conclusions

This chapter reviews recent literature regarding characteristics of future drought in the Canadian Prairies. Overall, research results, especially for the Prairies, indicate that dry times are expected to become much drier, and wet times wetter. Probable future droughts in the Canadian Prairies are likely to be drought types that, although perhaps not catastrophic, have the power to slowly erode adaptive capacity of both human and natural capital. Alternatively, the worst-case scenarios for future droughts may have low probability but could be catastrophic.

Current and past droughts may seem mild compared with future possible droughts, and the disruption of the climate by increasing greenhouse gases might result in some additional surprising effects on climate. The nature of future drought is particularly concerning because of insufficient water for increasing atmospheric demands and increasing (and even stable) societal demands. Much-improved adaptation to extremes, such as drought, is needed.

Estimating future droughts and extreme precipitation has several limitations, but projections using several different indicators, climate models, and emission scenarios provide compelling evidence of the risk of increased intensity, duration, frequency, and area of future droughts and extreme precipitation.

References

Barrow, E. 2009. *Climate Scenarios for Saskatchewan.* Prepared for the Prairie Adaptation Research Collaborative (PARC). Regina, SK: Climate Research Services.

Barrow, E. 2010. *Hydroclimate Data for the Prairies: An Analysis of Possibilities.* Regina, SK: Climate Research Services.

Bonsal, B., R. Aider, P. Gachon, and S. Lapp. 2013. "An Assessment of Canadian Prairie Drought: Past, Present, and Future." *Climate Dynamics* 41, no. 2: 501–16. doi: 10.1007/s00382-012-1422-0.

Bonsal, B., E. Wheaton, A. Chipanshi, C. Lin, D. Sauchyn, and L. Wen. 2011. "Drought Research in Canada: A Review." *Atmosphere-Ocean* 49, no. 9: 303–19.

Bonsal, B., E. Wheaton, A. Meinert, and E. Siemens. 2011. "Characterizing the Surface Features of the 1999–2005 Canadian Prairie Drought in Relation to Previous Severe Twentieth Century Events." *Atmosphere-Ocean* 49, no. 9: 320–38.

Dai, A. 2010. "Drought under Global Warming: A Review." WIREs *Climate Change* 2, no. 1: 45–65. doi: 10.1002/wcc.81.

Deser, C., R. Knutti, S. Solomon, and A.S. Phillips. 2012. "Communication of the Role of Natural Variability in Future North American Climate." *Nature Climate Change* 2: 775–79.

IPCC (Intergovernmental Panel on Climate Change). 2012. "Summary for Policymakers. Pp. 1–19 in C.B. Field, V. Barros, T.F. Stocker, Q. Dahe, D.J. Dokken, K.L. Ebi, M.D. Mastrandrea, K.J. Mach, G.-K. Plattner,

S.K. Allen, M. Tignor, and P.M. Midgley (eds.), *Managing the Risks of Extreme Events and Disasters to Advance Climate Change Adaptation*. Cambridge, UK: Cambridge University Press.

Meteorological Service of Canada (MSC) Drought Study Group. 1986. *An Applied Climatology of Drought in the Canadian Prairie Provinces*. Meteorological Service of Canada, Canadian Climate Centre, Report no. 86-4. Downsview, ON: MSC.

Michels, A., K.R. Laird, S.E. Wilson, D. Thomson, P.R. Leavitt, R.J. Oglesby, and B.F. Cumming. 2007. "Multi-decadal to Millennial-scale Shifts in Drought Conditions on the Canadian Prairies over the Past Six Millennia: Implications for Future Drought Assessment." *Global Change Biology* 13: 1295–1307.

PaiMazumber, D., L. Sushama, R. Laprise, M. Khaliq, and D. Sauchyn. 2012. "Canadian RCM Projected Changes to Short- and Long-term Drought Characteristics over the Canadian Prairies." *International Journal of Climatology* 33, no. 6: 1409–23. doi:10.1002/joc.3521.

Price, D., D. McKenney, L. Joyce, R. Siltanen, P. Papadopol, and K. Lawrence. 2011. *High-resolution Interpolation of Climate Scenarios for Canada Derived from General Circulation Model Simulations*. Information Report NOR-X-421. Edmonton, AB: Northern Forestry Centre, Canadian Forest Service.

Sauchyn, D., S. Kennedy, and J. Stroich. 2005. "Drought, Climate Change, and the Risk of Desertification on the Canadian Plains." *Prairie Forum* 30, no. 1: 143–56.

Sushama, L., N. Khaliq, and R. Laprise. 2010. "Dry Spell Characteristics over Canada in a Changing Climate as Simulated by the Canadian RCM." *Global and Planetary Change* 74: 1–14.

Thorpe, J. 2011. *Vulnerability of Prairie Grasslands to Climate Change*. Prepared for the Prairies Regional Adaptation Collaborative. Saskatoon, SK: Saskatchewan Research Council.

Wheaton, E. 2013. "Future Risks of Wet and Dry Climate Extremes in Saskatchewan, Canada." Presentation to the Climate Extremes and Mining Advisory Group. Regina, SK, April 2013.

Wheaton, E., B. Bonsal, and V. Wittrock. 2013. *Possible Future Dry and Wet Extremes in Saskatchewan, Canada*. Prepared for the Water Security Agency. Saskatoon, SK: Saskatchewan Research Council.

Wheaton, E., V. Wittrock, G. Koshida, E. Siemens, and D. Smeh. 2011. *What are the Benefits of Drought? Concepts and Examples to Promote Effecive Climate Adaptation Information Services*. Prepared for Adaptation and Impacts Research Section, Environment Canada. Saskatoon, SK: Saskatchewan Research Council.

Wheaton, E., S. Kulshreshtha, V. Wittrock, and G. Koshida. 2008. "Dry Times: Lessons from the Canadian Drought of 2001 and 2002." *The Canadian Geographer* 52, no. 2: 241–62.

Wilhite, D. 2000. "Drought as a Natural Hazard: Concepts and Definitions." Pp. 3–18 in D. Wilhite (ed.), *Drought: A Global Assessment*. Vol. 1. New York, NY: Routledge Press.

Williams, D., and E. Wheaton. 1998. "Estimating Biomass and Wind Erosion Impacts for Several Climatic Scenarios: A Saskatchewan Case Study." *Prairie Forum* 23, no. 1: 49–66.

Wittrock, V., E. Wheaton, and E. Siemens. 2010. *More than a Close Call: A Preliminary Assessment of the Characteristics, Impacts of and Adaptations to the Drought of 2008–2010 in the Canadian Prairies*. Prepared for Environment Canada. Saskatoon, SK: Saskatchewan Research Council.

Wolfe, B.B., R.I. Hall, T.W.D. Edwards, and J.W. Johnston. 2012. "Developing Temporal Hydroecological Perspectives to Inform Stewardship of a Northern Floodplain Landscape Subject to Multiple Stressors: Paleolimnological Investigations of the Peace-Athabasca Delta." *Environmental Reviews* 20: 1–20.

PART 3

DEALING WITH PRAIRIE DROUGHTS: CRISES AND ADAPTIVE RESPONSES

CHAPTER 4

THE IMPACTS OF THE 2001–2 DROUGHT IN RURAL ALBERTA AND SASKATCHEWAN, AND CANADA

Suren Kulshreshtha, Elaine Wheaton, and Virginia Wittrock

Droughts are a recurring event in the Canadian Prairie provinces. The paleoclimatic data indicate that severe droughts (of long duration) have been observed in the nineteenth century.[1] Although droughts occur in many regions of North America, the Prairie region is the most susceptible. Droughts also occur in all seasons as part of normal climate variability; however, the effects are most severe during the warmer seasons because of the increased demand for water due to higher temperatures.

This chapter focuses on droughts in the Prairie region, which includes the agricultural portion of Alberta, Saskatchewan, and Manitoba. Its primary objective is to synthesize and discuss information on impacts of droughts and adaptation to them in the Prairie provinces. In particular, this chapter focuses on the 2001–2 drought, with emphasis on the following: first, to illustrate the conceptual economic and social impacts of droughts on various economic sectors; second, to link the sectoral impacts to economic and social impacts on rural communities in Saskatchewan and Alberta; third, to identify adaptation measures undertaken by producers and communities in response to droughts; and finally, to present knowledge gaps in drought impacts and adaptation to droughts and implications for policy makers for future drought planning and policy formulation. Addressing these knowledge gaps is key to advancing future drought impacts and adaptations research. In addition, the chapter briefly reviews the major characteristics of droughts in the region as well as some conceptual issues regarding drought impacts. The nature of these droughts is compared with the most recent intense and extensive drought, which occurred in 2001–2 and is used here as a case study for discussing impacts on communities.

The empirical results presented in this chapter are based on earlier studies[2] dealing with economic and social impacts of the 2001–2 drought on various economic sectors and on government institutions.[3] These results are complemented by community-level research under the auspices of two projects—Institutional Adaptation to Climate Change (IACC) and Rural Communities Adaptation to Drought (RCAD).[4]

Background on Droughts and Their Impacts

An improved understanding of droughts is needed so that the information can be used to enhance adaptation to droughts and thereby lessen vulnerability. Improved adaptation to drought is required because drought causes severe and extensive socio-economic and ecological disruption and damage. Drought is more costly than any other form of natural disaster (Wilhite 2000). This is especially true of the Canadian Prairies where drought is more frequent, intense, extensive, and damaging than in other parts of Canada (Wheaton et al. 2008).

In terms of frequency and chronology, several extensive, multi-year droughts on the Prairies have been identified, including those of the 1890s, 1910s, 1930s, 1980s, and 1999–2005 (with 2001–2 being the peak of the drought) (Bonsal et al. 2011). A comparative spatial incidence of these droughts is shown in Figure 1. These droughts represent five major episodes during 120 years. Shorter but severe droughts have also occurred during this period (e.g., 1961). During the recent drought of 2001–2, which affected not only the Prairies but also other areas of Canada, parts of the Canadian Prairies experienced their most severe dry conditions for the last 100 years. Some locations, such as Saskatoon, had their lowest annual precipitation on record in 2001, and others had their lowest Palmer

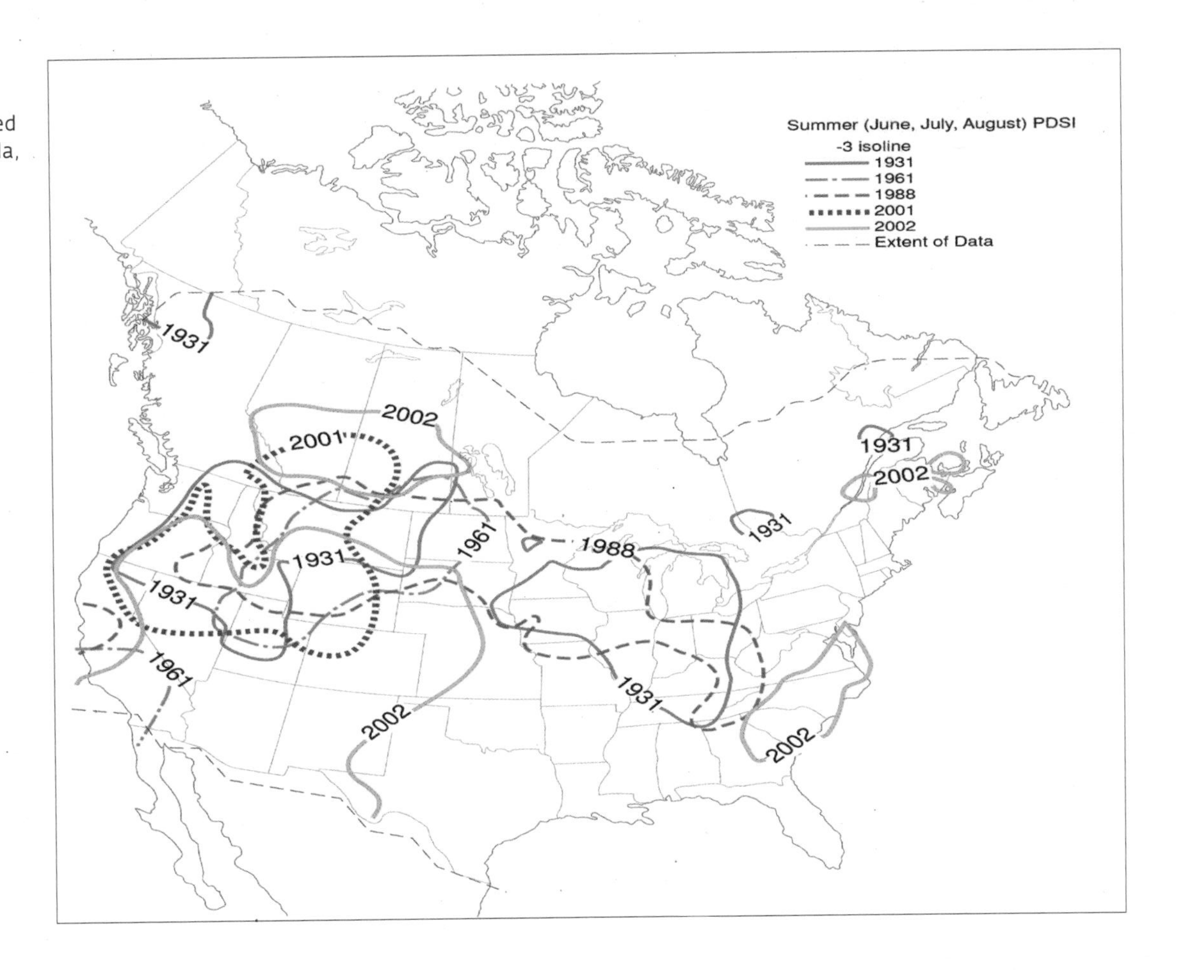

Figure 1. Spatial patterns of selected droughts in Canada, 1931–2002 (Source: Wheaton et al. 2008)

Drought Severity Index (PDSI) scores (indicating worst drought) on record (Wheaton et al. 2008). Hanesiak et al. (2011) examined many databases for several drought characteristics and found that the 1999–2005 drought on the Prairies was one of the driest meteorological and hydrological events on record.

Not only is drought an intrinsic part of the variable climate of the Prairies, but the potential for future droughts is increasing because of human-induced climate change and increasing water demand (see Chapter 3 by Wheaton et al. in this volume). Therefore, it is critical that adaptation measures, strategies, and policies consider increasing droughts and their impacts in future years.

Droughts, particularly those lasting over a period of a few years, can completely devastate a region, in terms of biophysical changes as well as economic and social impacts. Some parts of the world are more prone to droughts than others. In Canada, the southern part of the Prairie provinces (Manitoba, Saskatchewan, and Alberta) belongs to this group of regions. Here, droughts have represented a major natural disaster. Of the top 11 most costly natural disasters in Canada, 7 of them were Prairie droughts (Table 1). In fact, the most costly natural disaster in Canada was the 2001–2 drought, which had a direct impact of $5.8 billion (Wheaton et al. 2008).

Significant changes in the hydrological cycle have the biggest impact on agricultural production, but these changes also have other social and economic impacts. Among these are health effects; as Stern (2007: 89) indicated, "droughts (and floods) are harbingers of diseases, as well as causing death from dehydration."

Conceptual Framework to Describe Impacts of Agricultural Droughts

As discussed in Chapter 1, there are different perspectives on droughts—biophysical perspectives and socio-economic–political perspectives. The biophysical perspective includes studies of drought patterns, their severity and frequency, and their impacts on the physical environment, while the socio-economic–political perspective focuses on identifying the effects of precipitation deficiencies on people and their institutions. Initially, drought impacts are felt in terms of biophysical changes and experienced

Table 1. Estimated economic cost of Canadian droughts compared with other hazards

Date of occurrence	Event	Location	Estimated total cost (billion*)
2001–2	Drought	Prairies, Ontario, Nova Scotia, Prince Edward Island	$5.8
1980	Drought		$5.8
	Freezing rain	Ontario to New Brunswick	$5.4
1988	Drought	Prairies	$4.1
1979	Drought	Prairies	$3.4
1984	Drought	Prairies	$1.9
	Flood	Québec	$1.6
May 1950	Flood	Manitoba	$1.1
	Hurricane Hazel	Toronto and southern Ontario	$1.1
1931–38	Drought	Prairies	$1.0
1989	Drought	Prairies	$1.0

* For comparison purposes, all values from various studies were converted into constant dollars using 2000 as the base.

Source: Koshida 2010.

by local people and their communities, but over time, their impacts are exacerbated and extend to the larger regional, national, and even international settings. These impacts have two dimensions, a sectoral/spatial dimension and a temporal dimension, both of which are discussed below.

Parry and Carter (1987) distinguish between two types of approaches to study drought impacts:[5] the impact approach and the interaction approach. The impact approach is based on the assumption of direct cause and effect. Here an activity (economic or social) is exposed to a climatic event (such as a drought) and then experiences an impact. This approach could be unrealistic (and perhaps misleading), since many other factors affect the socio-economic activities. The interaction approach assumes that a drought (or other climate-related event) is just one of many processes that may affect the exposure unit. Furthermore, the impact may be multi-dimensional through various interaction processes.

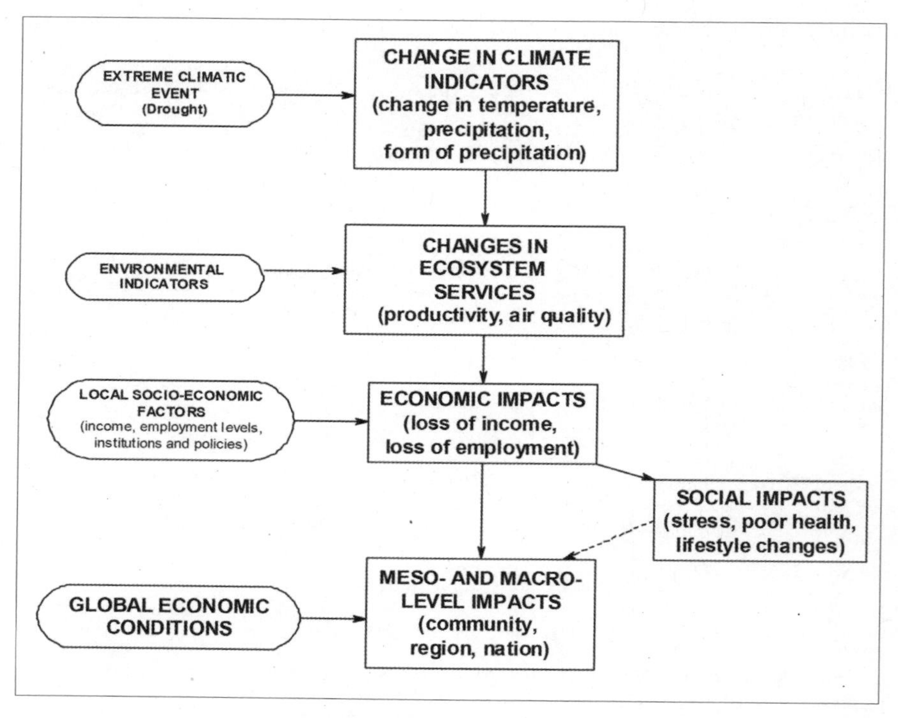

Figure 2. Interaction between drought and various sectors

For a hypothetical drought, an interactive process is shown in Figure 2. Here, three-level impacts are hypothesized. The initial order of impact during the drought period is biophysical in nature—temperature and precipitation regimes change both in terms of amount and timing. These biological changes would have an impact first on the ecosystem services (level of productivity of natural resources) and, through these changes, on the socio-economic system. These impacts are called second-order impacts. These impacts would vary according to the type of socio-economic activities that are present in a region. These second-order impacts may also lead to third-order impacts (such as changes in regional-level productivity of resources and level of income of people, as well as changes at the national economy level). For a country with open borders (such as Canada), economic impacts of the drought (negative or positive) will also be felt at the international level.

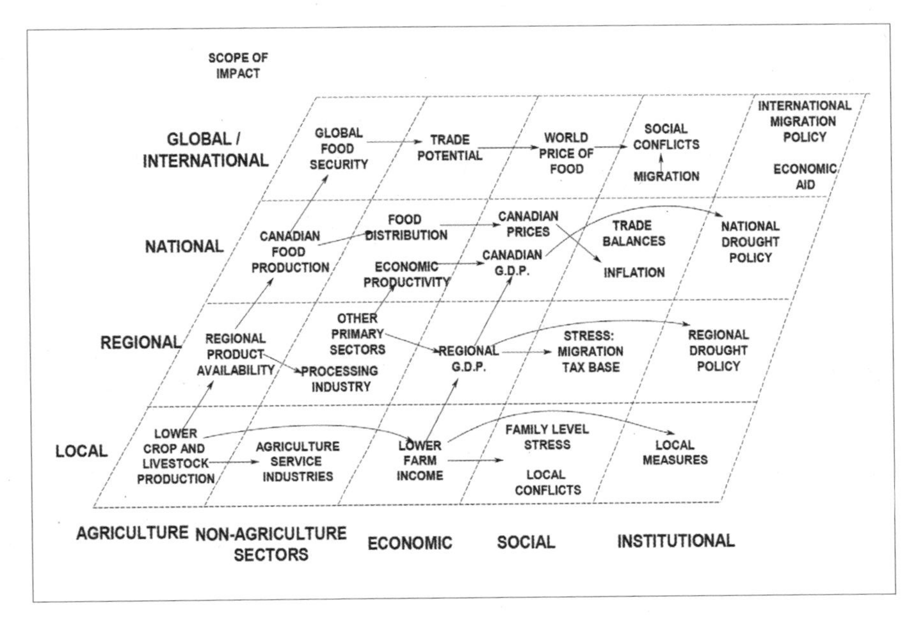

Figure 3. Hypothesized pathways of a drought incident
(Source: Adapted from Parry and Carter 1987)

A case in point is the recent experiences of the Australian drought of 2006 and the US drought of 2010. During these periods, farm-level wheat price in Saskatchewan rose from a five-year average (2002–6) of $142.20 per tonne to $301 per tonne in 2007 (more than doubling) and to $256 per tonne in 2010 (an increase of 80%).[6] Both of these situations illustrate the international connections in commodity markets. Price booms like this are welcomed by exporting nations, but they may create social hardships in other parts of the world and may initiate a series of changes leading even to inter-regional or international migration of people. A more detailed account of these changes is presented in Figure 3.

When an agricultural drought occurs, more than just agriculture suffers from the lack of water. Rural communities, municipalities, industries, and processors are also affected. In some local regions, rationing may be required, but unless the water source becomes completely depleted, the right to the use of the water is relatively secure for the users. Overall, the economic conditions during the drought period and immediately after that would be adverse, either through economic losses or through impacts on ecosystem services. Impacts can be hypothesized to occur in two dimensions—sectoral and spatial.

For the purposes of studying droughts, the economy should be segregated into two sectors: agricultural and non-agricultural. Some non-agricultural sectors may experience two types of impacts—direct impacts of droughts and indirect impacts induced by losses in agricultural production. These impacts would lead to several other types of impacts within the local region, culminating in regional and national (as well as international) impacts. For example, loss of agricultural production (e.g., livestock production) would affect agricultural processing industries and then affect the rest of the food supply chain. Some of these industries would suffer from higher processing costs and would also need to import their required raw material from other regions. Social impacts might be experienced from lower economic conditions for some people, communities, and businesses, which might lead to higher stress levels and might even culminate in health impacts.

Although droughts are typically confined to a certain period of time, their impacts are not necessarily limited only to that time period. For example, Figure 4 depicts a hypothetical region that has been experiencing economic growth over the past few time periods (as shown by line *Oa* in the figure). The region suddenly experiences a severe drought[7] in time period t_1. If the region did not experience that drought, it would have moved along line *Oac* to time period t_2. The direct (one period) impact of the drought is measured by the vertical line *ab*. However, the actual cost of the drought would depend on the path of recovery taken by the economy. If the economy reaches the same point where it would have been without that drought occurring, then the cost of the drought is approximated by the area *abc*. However, if the growth rate is sluggish and the economy needs more time to recover, the cost would likely be higher than approximated by that area.

In addition to changes in economic activities, drought may also affect ecosystem goods and services. Changes in land productivity resulting from drought, or loss of vegetation and wildlife resulting from drought, would also affect many other socio-economic activities in the future (such as recreation, hunting, and tourism). A true total cost of a drought must therefore sum all economic and environmentally induced costs over a

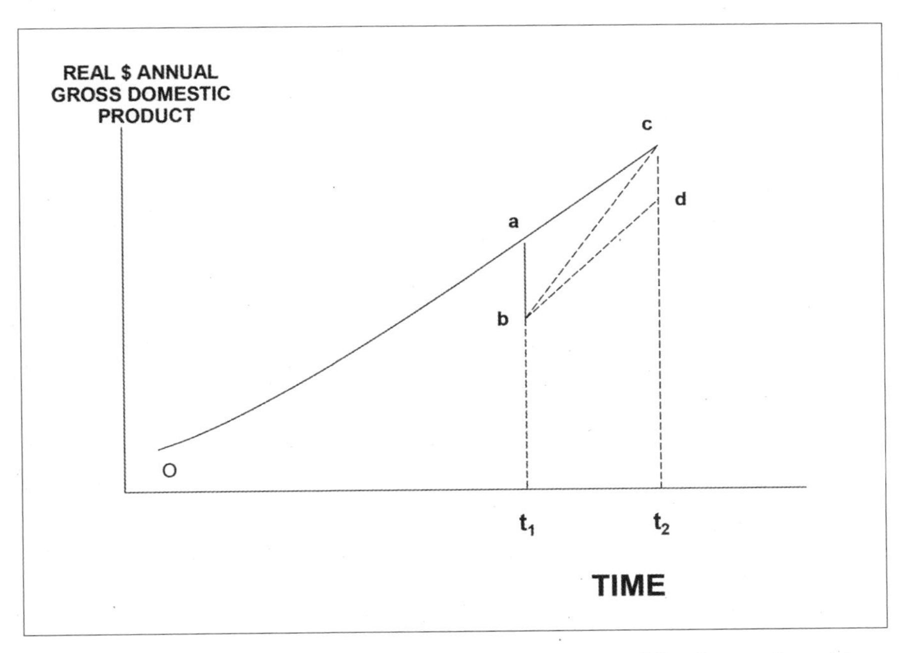

Figure 4. Time path of adjustment in regional economy resulting from a drought (Source: Adapted from Dore and Etkin 2000)

period of time. However, such a study has yet to be undertaken for the Prairie provinces.

The 2001-2 Drought Impacts in Canada

Past droughts in Canada have been more spatially fragmented, less intense, and shorter than what was witnessed in Canada in 2001-2. This drought was exceptional by many measures: it was unusually large in area, severe, and embedded in a long dry period (Wheaton et al. 2008). As a result, it affected many sectors and people residing in a large part of Canada. The two Prairie provinces—Saskatchewan and Alberta—were particularly hard hit by these back-to-back droughts. In 2001, Canada experienced one of the worst droughts on record by many standards, including its coverage across Canada and its intensity. Further details on its impacts are provided below.

The genesis of the 2001–2 drought was in the autumn of 2000 in southern to southeastern Alberta. The drought then spread across into central-western Saskatchewan, but the province of Manitoba had near normal temperature and precipitation conditions. The drought intensified in spring and summer 2001 in Alberta and Saskatchewan. Only the northwestern agricultural portion of Manitoba was dry in spring and summer 2001. The warm, dry trend continued into the autumn and winter of 2001–2. Conditions changed in spring 2002, but only in temperature, resulting in an unusually dry and severely cold spring across western Canada. The 2001 drought was confined to a smaller region—primarily located in the southern and east-central parts of the province (Wheaton et al. 2008). The 2002 drought in Alberta covered most of the province at some point in time during the agricultural season.

The higher temperatures accompanied by lack of precipitation resulted in several biophysical impacts, such as wind erosion, reduced streamflows, dry dugouts, and groundwater reductions. More prominent impacts in the region included the following:

- The areas of most frequent wind erosion were estimated to have occurred in the drought areas of southern Alberta and in Saskatchewan, particularly in the central area along the provincial border. The month of peak wind erosion occurred in May for both 2001 and 2002, but was nearly as high in April 2002. Alberta had the most wind erosion events during 2001, while Saskatchewan had more in 2002.

- Many rivers and streams in Alberta and Saskatchewan had well-below-average flows in 2000, 2001, and 2002.

- Many of the 19 groundwater observation wells examined in the Canadian Prairies (7 in Alberta, 8 in Saskatchewan, and 4 in Manitoba) recorded declining water level trends, depending on location of the observation well.

- Dry dugouts were first reported in the fall of 2000, with the area of dry to one–quarter-full dugouts expanding through 2001. In 2002, the area of dry dugouts shifted northward (Wheaton et al. 2008).

Table 2. Impact of the 2001–2 drought on agricultural production in Saskatchewan and Alberta

Particulars	Alberta		Saskatchewan	
	2001	2002	2001	2002
Reduction in value of production before government payments (millions)	$412.90	$1,400.70	$925.30	$1,520.10
Reduction in value of production after government payments and other adjustments (millions)	$271.16	$1,008.50	$654.90	$1,001.00
Drought losses as a percentage of average 1998–2000 value of production	5.97	20.26	16.14	26.52

Source: Wheaton et al. 2004.

These biophysical impacts led to other second-order impacts on the socio-economic activities in the two provinces, such as adverse impacts on agricultural production in Alberta and Saskatchewan. In both provinces, crop yields and harvested areas were below average for 2001 and 2002. This led to reduced farm cash income in both years. The overall impact of the drought was a loss in gross farm cash receipts of $413 million in 2001 and $1,401 million in 2002 for Alberta and $925 million in 2001 and $1,520 million in 2002 for Saskatchewan (Table 2). These losses included changes in crop production and in livestock production.

Producers also reduced input costs in response to drought conditions. A reduction in fertilizer application occurred in 2002 because the 2001 crop did not use the nutrients that were applied to it. Fuel purchases were down in 2002 because of reduced harvested area. Adjusting for the reduction in cost of production (through reduced farm input costs) and for payments received under various safety-net programs (mainly crop insurance), the net effect of the drought on crop production was estimated. Adjusting for losses in livestock production and adding them to adjusted crop production effects, net losses to Alberta producers were estimated at $271 million in 2001 and $1,009 million in 2002. Similar estimates for Saskatchewan producers were $655 million in 2001 and $1,001 million in 2002. Total losses of producers in the region were therefore around $926 million in 2001 and $2,010 million in 2002. In both provinces, these losses were over 16% of average 1998–2000 net farm income.

The 2001–2 drought had profound impacts on the water supply in some parts of the Prairie provinces. At the farm level, dugouts were affected the most, although domestic water supplies were also at risk. The hardest-hit regions were southern Alberta (in 2001) and central Saskatchewan (during 2001 and 2002). Producers used various methods to supplement water, including hauling, drilling new wells, and sourcing new water supplies, such as pipelines from distant secure sources.

As a direct consequence of loss in production and lower farm incomes, non-agriculture sectors were also affected. In Alberta, major changes on non-agricultural industries included the following:

- New investment in 2001 was down by 4.6% in agriculture, forestry, fishing, and hunting activities.
- Some negative impacts of the drought were noted on sales of new farm machinery and equipment in these areas.
- Agricultural processing firms reported no change in their sales, but they faced higher prices for their raw materials, thereby affecting their profit margin.
- Some firms had to find new suppliers for their raw materials.
- Forest-fire occurrences were five times higher than the previous 10-year average during 2002 in Alberta.
- Some recreational areas were affected due to low water levels in water bodies and open-fire restrictions in some areas (Wheaton et al. 2008).

Table 3. Reduction in gross domestic product and employment resulting from the 2001–2 drought in the Prairies

Particulars	Unit	2001*	2002*
Loss of gross domestic product	Millions of dollars	$1,434.62	$3,108.33
Loss of employment	No. of workers	10,083	17,803

* These estimates include data for Manitoba; however, direct impacts in Manitoba were relatively small and accounted for only 0.7% of total impacts on the Prairie region in 2001 and 1.3% in 2002. Thus, these estimates for the Prairie region largely reflect impacts for Alberta and Saskatchewan. Source: Wheaton et al. 2004.

In Saskatchewan, impacts of the drought were very similar to those noted above for Alberta, but there was also a reduction in the amount of hydroelectric power generated, requiring the Saskatchewan Power Corporation to purchase additional power from other sources.

The Canadian economy (and within that the economy of the Prairie provinces) represents an integrated system of activities. Regions depend on each other for raw materials as well as for markets for the good produced. Loss of production in Alberta and Saskatchewan therefore had consequences for other sectors in other parts of Canada. Using an input-output model, total loss for the region was estimated. Results for the Prairies are summarized in Table 3. The region lost a total of $1.4 billion in 2001 and $3.1 billion in 2002. These losses also culminated in loss of employment. About 10,000–17,000 jobs were lost in the region.

Droughts and Rural Communities

As previously mentioned, community-level research was carried out through two main projects: the IACC project and the RCAD project. Under the umbrella of the IACC, studies examined five rural communities in Saskatchewan and Alberta, as well as one First Nation reserve in Alberta (Figure 5). The five rural communities were Taber (Taber Municipal District [MD]), Hanna (Special Area No. 2), Cabri (Riverside Rural Municipality [RM]), Stewart Valley (Saskatchewan Landing RM), and Outlook (Rudy RM). The First Nation reserve in Alberta was the Kainai Blood Indian Reserve (KBIR). The RCAD studies occurred in six different

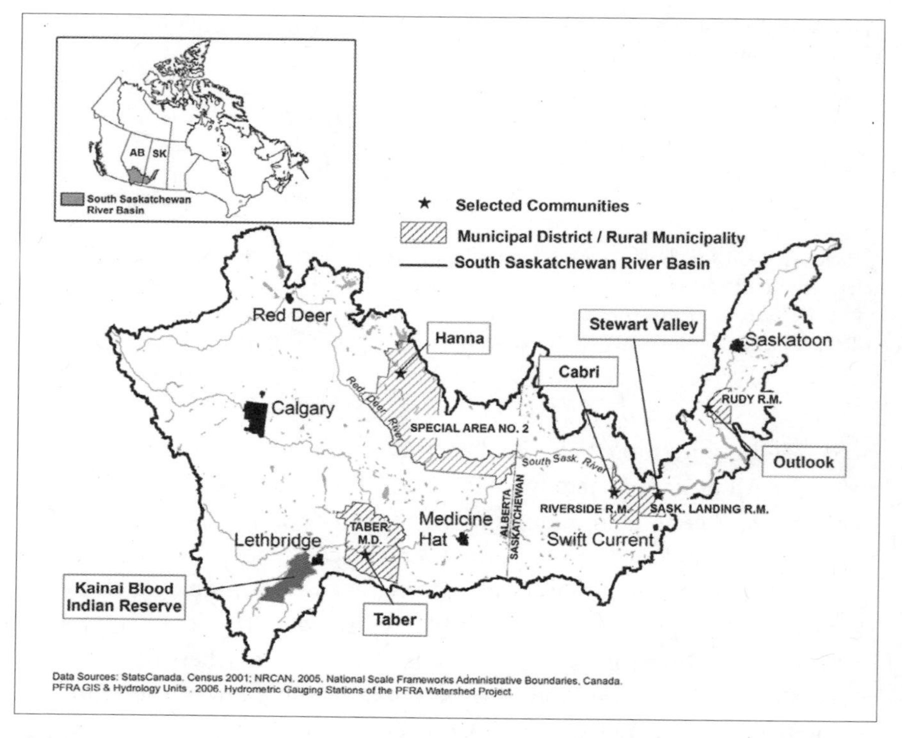

Figure 5. Communities in the Institutional Adaptation to Climate Change project (Source: Adapted from Patino 2011)

communities in Saskatchewan: five were located in the Palliser Triangle area, a well-documented location of reoccurring drought, and one was located just south of the North Saskatchewan River. Communities in the Palliser Triangle included Shaunavon (Grassy Creek RM and Arlington RM), Coronach (Hart Butte RM), Gravelbourg (Gravelbourg RM), Kindersley (Kindersley RM), and Maple Creek (Maple Creek RM). The other community was Maidstone (Eldon RM) (Figure 6).

This section details how the 2001–2 drought impacted the case-study communities both economically and socially. Because the IACC and RCAD projects used different methods, non-standard information, with an attempt at standardization, is provided here.

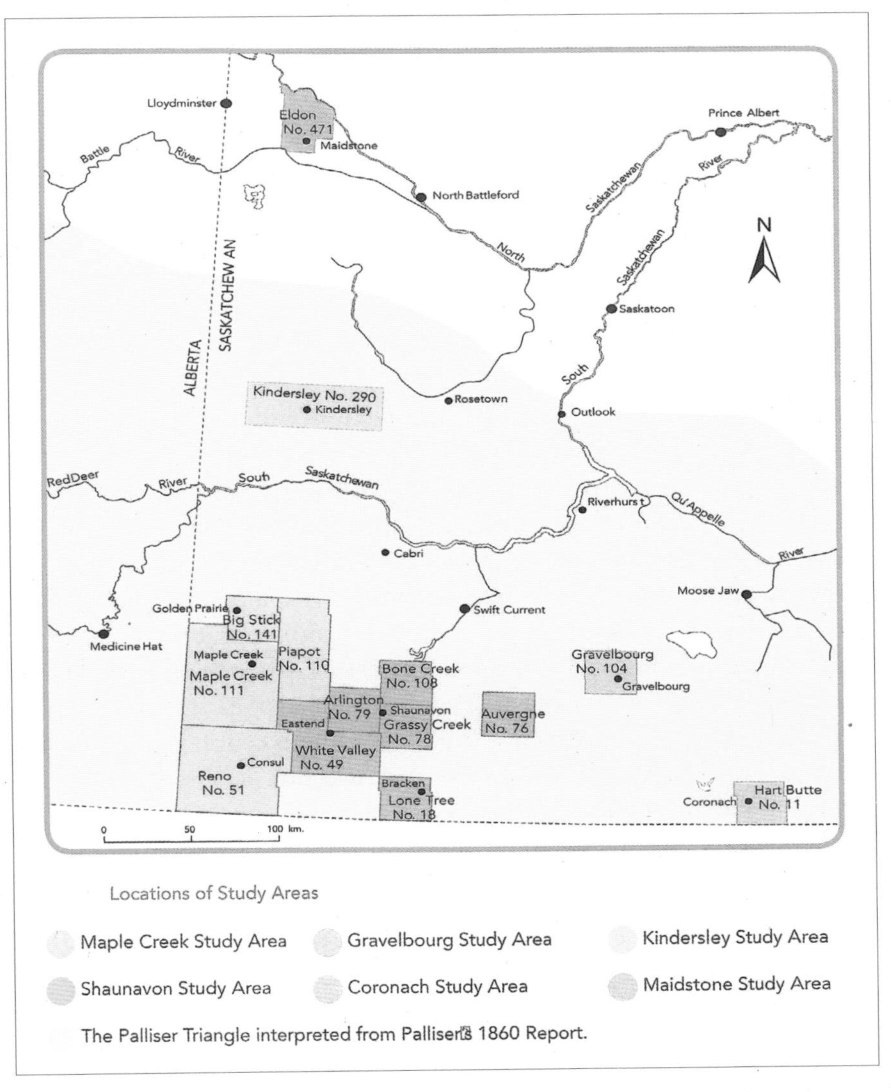

Figure 6. Communities in the Rural Communities Adaptation to Drought project (Source: Perrick 2012)

Table 4. Cost of the 2001–2 drought through lost crop production

Year	Value in dollars per hectare				
	Taber MD	Special Area No. 2	RM of Rudy	RM of Riverside	RM of Saskatchewan Landing
2001	-38.38	-87.93	-76.61	-78.03	-63.48
2002	-35.12	-171.08	-62.91	-20.44	-7.09

Sources: Wittrock et al. 2012; Wittrock et al. 2007.

Economic Costs to Communities

The cost of the 2001–2 drought for all of Canada was estimated to be nearly $6 billion (Table 1). The economic cost breakdown was carried out for the local study areas by the IACC project. The largest crop production losses of the 2001–2 drought were in Special Area No. 2 (in central-west Alberta), with a nearly $88 per hectare loss in 2001 and nearly double that amount in 2002 (Table 4). The Special Areas in Alberta were established under the auspices of the Special Areas Board in response to previous negative impacts from droughts in the early twentieth century (see Marchildon et al. 2008; see also Chapter 8 by Marchildon in this volume). The second-highest crop production losses occurred in the RM of Rudy when both 2001 and 2002 are examined together. While 2001 losses in the RM of Rudy were not as extreme (crop production loss of $76.61 per hectare or $4.23 million) as for the RM of Riverside, the drought conditions continued to plague the RM of Rudy in 2002, resulting in a continued loss of crop production by nearly $63 per hectare ($3.48 million). Taber MD also suffered crop losses but not as extreme as those in these other areas. In 2001, the loss was about $38 per hectare ($7.48 million) and in 2002 was about $35 per hectare ($6.84 million) (Wittrock et al. 2012; Wittrock et al. 2007).

A large rainstorm went through southern Alberta and southwestern Saskatchewan on 8–11 June 2002 (Szeto et al. 2011). Partly because of this event, crops recovered somewhat and the negative financial impact of crop loss was reduced in the RM of Riverside and the RM of Saskatchewan

Landing in southwestern Saskatchewan. The crop production loss in 2001 was about $78 per hectare in 2001 (or $8.57 million) in the RM of Riverside, but this value greatly improved to a loss of just over $20 per hectare in 2002 ($4.60 million). The RM of Saskatchewan Landing's financial situation improved between 2001 and 2002; loss in crop production was more than $63 per hectare in 2001 ($5.55 million), but the RM had near normal production in 2002 (Table 4).

Other economic costs were incurred by the communities but were not quantified. These included reduced fertilizer sales (Taber), reduced advertising in local newspapers (Taber), increased costs for market-garden operations (RM of Rudy), reduced new farm-machinery sales (Outlook), and increased water costs for the oil and gas industry (Special Area No. 2). Some industries that were more severely impacted by the drought moved out of the regions, such as grain brokers (RM of Rudy). Other sectors benefited from the drought, such as financial institutions, which profited because demand for money rose with the drought (Taber) (Wittrock et al. 2012; Pittman et al. 2010; Wittrock et al. 2007). The economic impact on the livestock industry could not be estimated due to a lack of data (Wittrock et al. 2012).

The RCAD project examined economic impacts though loss of crop production in the RMs surrounding the communities. Kindersley suffered decreases of more than 50% in 2001 and almost 100% crop loss in 2002 for wheat and canola (Abbasi 2014). The RM of Eldon (Maidstone) had crop losses of more than 70% in both 2001 and 2002 (Abbasi 2014). Prior to 2002, the RM of Eldon had not been severely impacted by an extreme drought, because this region in northwestern Saskatchewan's agricultural region generally has crop yields above the provincial average (Warren 2013). Producers in the RM of Gravelbourg did not perceive any major impact due to the 2001–2 drought conditions. They found that the adaptation measures they implemented due to the drought conditions in the late 1980s lessened their vulnerability to the 2001–2 drought (Luk 2011).

Other economic challenges emerged, in part, because of the 2001–2 drought, including those associated with upgrading water supply systems at various locations after the drought. Such improvements took place at Maple Creek (at a cost of $3.7 million) (Warren 2013), at Cabri (Wittrock et al. 2006), Maidstone (Abbasi 2014; Warren 2013) and Kindersley (Abbasi 2014; Warren 2013).

The KBIR stands out as a special case of drought impact because of its own style of governance, including that related to property rights. It is one of the largest reserves in Canada, with a population between 4,000 and 10,000 people. Agriculture is the predominant land use with some irrigation. However, much of the irrigated lands are leased out to non–First Nations people. The KBIR also has a beef cattle operation.

During the 2001–2 drought, the KBIR was affected in several ways (Kulshreshtha et al. 2011): i) local government costs increased from delivering water to homes on the reserve; ii) some road maintenance equipment was damaged due to extremely dry road conditions; iii) the livestock operation had higher feed costs resulting in some cattle being culled; and iv) residents on the reserve faced increased costs and time to obtain water.

Social Impacts of Droughts on Communities

Social and economic vulnerabilities to communities co-exist and tend to be accentuated by exposure to extreme climatic events including drought (Diaz et al. 2009). All the communities examined in the IACC and RCAD projects are relatively small, ranging in population from just over 100 (Stewart Valley) to 6,000 (Taber). Many of the communities have similar social issues, including depopulation—particularly of the younger generation—and centralization of services, which make the communities more vulnerable to external stressors and reduce their adaptive capacity (Diaz et al. 2009). Drought is one of these added stressors and creates additional impacts on the communities. A common impact of drought and resulting stressors throughout most of the communities was the lack of water (Wittrock et al. 2011). The meteorological and hydrological drought of 2001–2 resulted in low water supplies affecting available water for activities for some farmers and in some towns and villages across the Canadian Prairies. Low water supplies resulted in water use restrictions for some towns as well as restrictions on the agricultural community's access to town water. These restrictions resulted in agricultural producers having to find alternative water sources and the government (both federal and provincial) providing some assistance to farmers/ranchers to find adequate quality water for their livestock (Wittrock et al. 2012; Wittrock et al. 2011; Wheaton et al. 2008). This scenario played out in many of the communities examined in this chapter. For example, the towns of Taber and Cabri imposed water rationing (Wittrock et al. 2007; Wittrock et al. 2006). The town of Cabri took the additional step of not allowing agricultural producers to access the town's potable water supply (Diaz et al. 2009; Wittrock et al. 2006). The water rationing in Taber may have negatively impacted production by some industries or resulted in them having to invest in water conservation technology (Wittrock et al. 2006). The communities of Kindersley, Maidstone, Maple Creek, Gravelbourg, and Coronach all had water supply issues due to the drought (Abbasi 2014; Warren 2013; Luk 2011). The town of Outlook had easier access to water through the development of Lake Diefenbaker. Dry conditions increased demand for domestic water use and increased the revenues of the water utility of Outlook. This situation improved the town's financial position (Wittrock et al. 2007). In addition to impacts on communities as a whole, the drought of 2001–2 impacted individuals' well-being. For example, community officials had difficulty coping at the personal level with the cumulative effects of the drought and the associated secondary and tertiary impacts (Maple Creek) (Warren 2013).

Adaptation to Droughts: Overview

Extreme climatic events can have devastating consequences for agriculture as well as the accompanying community. Adapting to these extreme climatic events is critical in reducing vulnerability and decreasing the recovery time. An adaptation framework was formulated in Wittrock and Wheaton (2007) and is used here to assess the various strategies implemented.[8]

In general, two types of adaptation strategies exist: short term and long term. These strategies can also be subdivided into subcategories, including technology/research, government programs, farm management, farm and agriculture financial management, and community support—for crops, livestock, and water. These subcategories can then be assessed based on key topics. For example, some key topics for cropping adaptation strategies may include weed control, pest control, or crop rotation (see also Chapter 5 by Warren on minimum till in this volume). Many secondary impacts to both agricultural producers and/or communities may also require adaptation strategies.

Adaptation by Producers

Canadian Prairie agricultural producers have always been impacted by droughts. Some of the historic droughts have been short, such as the drought in 1961, while other droughts have lasted for extended periods, such as the droughts in the 1920s and 1930s (Marchildon et al. 2008) and more recently in 1999–2005 (Bonsal et al. 2011). Consequently, many adaptation measures have been implemented, resulting in a moderately proactive response leading to lower vulnerability and fewer or less negative impacts. Other portions of the study region (such as northern Saskatchewan and Alberta) have not experienced many severe droughts, resulting in lower implementation of adaptation strategies and thus higher vulnerability to droughts.

Southern Alberta and western Saskatchewan have a history of droughts. This portion of the Canadian Prairies is in the Palliser Triangle, where droughts are frequent. This vulnerability has resulted in many adaptation measures being implemented over several decades, and thus the most recent drought event of 2001–2 had lower negative impacts than might have resulted without this experience.

The agricultural industry reduced its vulnerability to the 2001–2 drought by implementing short- and long-term adaptation strategies. Many adaptation strategies are initially reactive in nature, but turn into proactive strategies when used for long periods of time. Examples of short-term adaptation strategies used by the agricultural community are listed in Table 5.

The long-term adaptation strategies apply to crops, livestock, water, and land use, and include three different groups of adaptations—technology/research, government programs, and farm and agriculture financial management. These strategies have a longer time frame either through implementation (e.g., research into drought-resistant crops and forage crops) and/or usage (e.g., minimum till expansion, conservation cover program). However, even with these extensive adaptation measures, harsh droughts such as the 2001–2 event, can stress coping levels, as indicated earlier, and result in large losses and difficult recoveries at the community to national levels.

Table 5. Examples of short-term adaptation strategies for the agricultural sector

Particulars	Technology / research	Government programs	Farm management	Farm and agriculture financial management	Community support
Crops	Modify equipment to deal with shortened crops	Crop insurance Net Income Stabilization Account Low interest rates	Use cropping strategies (crop rotation, seeding times, crop diversification, drought-tolerant species)	Sell crops when higher commodity prices are available Take off-farm jobs	
Livestock	Use Web to buy and sell livestock and forage	Forage/hay insurance Tax deferral from livestock sales Farm Income Disaster Program	Import feed from non-drought-stricken regions Reduce stocking rates in pastures Use annual feeds	Cull older cattle Increase cow/calf sales	HayWest program, Ducks Unlimited opened its property to livestock producers
Water		Partial funding for installing temporary water pipelines National Water Supply Expansion Initiative	Conserve water in agriculture, urban areas, and industry Ration water Haul water and/or install temporary remote water systems	Temporarily trade or sell water rights to other producers or industry	

Source: Adapted from Wittrock and Wheaton 2007.

Table 6. Examples of longer-term adaptation strategies for the agricultural sector

Particulars	Technology / research	Government programs	Farm and agriculture financial management
Crops	Drought-resistant crop development Long-range weather forecasts Extreme climatic events research to reduce vulnerability	Assessment of future government assistance programs Agriculture Policy Framework	Minimum tillage expansion Increased use of high-efficiency irrigation systems Increase crop diversification
Livestock	Research into drought-resistant forage crops	Conservation Cover Program	Different grass and pasture management strategies Purchase land in different parts of the Prairies Change to different livestock breeds that survive better in drought situations
Water	Hydrologic modelling to assist with planning and operational design Examination of expansion of water storage and irrigation	Modifications in water allocation (Alberta) Moratorium on new water licenses in fully allocated river basins (Alberta) Assistance with building of dugouts and/or groundwater wells	
Land use			Increase acreage under irrigation Increase value-added commodities with more farm level processing activities

Source: Adapted from Wittrock and Wheaton 2007.

Adaptation by Communities

The level of community adaptation to drought varies by length, timing, and intensity of drought; location of the community; and the community's level of adaptive capacity. A community's level of vulnerability is determined by its exposure to environmental and societal stresses and its capacity to adapt to those stressors (Brklacich and Woodrow 2007; see also Chapter 1 in this volume).

Two assessments were undertaken for the IACC project to determine the level of vulnerability through adaptation measures. Diaz et al. (2009) examined how successful various portions/sectors of the community were in responding to drought and the reasons behind their success or failure. Wittrock et al. (2011) examined the adaptive capacity of the communities and rated them based on the method by Brklacich and Woodrow (2007).

The town of Outlook was assessed as the least vulnerable community to the 2001–2 drought mainly due to its secure potable water supply. The community also has an income close to the provincial average and has a higher-than-average formal education base. The community of Cabri was rated as the most vulnerable to the 2001–2 drought mainly due to its inadequate water supply (Wittrock et al. 2011). Because of its inadequate water supply, citizens implemented adaptation measures, including water conservation and use of grey water (e.g., clothes' washing water) to water gardens. The local government implemented additional measures to combat the low potable water supply, including restricting lawn watering and restricting agricultural producers from accessing the town's limited water supply (Diaz et al. 2009).

The drought of 2001–2 triggered initial reactive adaptation strategies in many of the communities mainly due to the lack of potable water. For example, Kindersley had a historic adaptation to limited potable water supply by installing a water pipeline from the South Saskatchewan River in the 1960s. This infrastructure required an upgrade to maintain a feasible level of potable water for the community. Maidstone was perhaps most severely impacted by the drought due to the extreme negative effect on its potable water supply. This may also have been an effect of a reactive adaptation strategy used by the town. This strategy was to drill more groundwater wells and install a potable water pipeline, thus decreasing the vulnerability of the community to future extreme drought events.

Areas for Further Research

This overview of research on drought impacts and of adaptation strategies to reduce these impacts was based on available data, which was sometimes limited. This section provides several suggestions for planning and undertaking future research on drought impacts to provide more comprehensive information and understanding.

The timeline of impacts was not included in past studies. This timeline would likely illustrate the cumulative impacts that occurred due to the drought. These impacts could have a dampening effect on the economy in the future, particularly for livestock production. Results on livestock production were based on provincial-level data. Regional data, particularly on the drought regions, were not available, thus limiting the analysis of regional level drought impacts.

In addition to the agricultural sector, drought may affect other sectors (e.g., forestry; hydroelectric power generation; transportation industries, including water transportation; tourism and recreation; food processing industries; and farm input industries). Attempts should be made to collect more information on these sectors to enable a more comprehensive analysis of the estimated impacts of the drought.

A concern that needs to be more fully explored in the future relates to the impacts of drought on the environment. Various aspects of the environment can be impacted by prolonged droughts (such as soil quality, air quality, water quality). Such changes could affect the sustainability of Prairie agriculture and the associated economy. Another major concern is the looming possibility of future droughts that will make past droughts appear mild in comparison. These more severe droughts would make adequate adaptation much more difficult and would push the limits of adaptation.

Summary

Droughts are frequently experienced in the southern part of the Prairie provinces. Although paleoclimatic data suggest past droughts were of longer duration, recent droughts have been mostly single-year or consecutive-year events. The drought of 1999–2005, which peaked in 2001–2, was a longer event. It created havoc for the agriculture industry and for people associated with it. In addition, many non-agricultural sectors were either directly or indirectly affected by the drought conditions. Overall, the provinces of Saskatchewan and Alberta were the hardest hit in Canada. Drought affected central-west and southwestern Saskatchewan, and central-east Alberta. In the region, total gross domestic product declined by $1.4 billion in 2001 and by $3.1 billion in 2002. These economic losses were associated with employment losses in the agricultural sector and associated industries.

Rural communities in the drought region suffered as a result of losses in the agriculture industry and shortage of water. Many of these communities, as well as agricultural producers, undertook adaptation measures in response to the droughts. In some cases, new sources of water were found, while in other cases, existing sources were improved to secure water. Adaptation to climate change (particularly drought events) represents a challenge for the Prairie economy; however, adaptation can reduce vulnerability to future events, within limits. Although humans have always adapted to changing climate and to non-climatic changes, more can be done to help people to prepare for these conditions.

NOTES

1 In southwestern Saskatchewan and southeastern Alberta, decade-long droughts have been estimated during the early and late 1800s (see Chapter 2 by Sauchyn and Kerr in this volume; see also Sauchyn 2002).

2 Details on these studies can be found in Wheaton et al. (2008, 2004), Wittrock et al. (2012), and Kulshreshtha et al. (2011).

3 For details on historical development of institutions in response to drought, see Marchildon et al. (2008). See also Chapters 9 and 10 by Hurlbert in this volume.

4 Details on the IACC project are reported by Wittrock et al. (2012, 2011, 2007, 2006), Pittman et al. (2010), and Kulshreshtha et al. (2011). Similarly, the RCAD project results are summarized by Diaz and Warren (2012), Abbasi (2014), Luk (2011), and Warren (2013).

5 These approaches were originally suggested by Kates (1985).

6 These data are from the Government of Saskatchewan (2013).

7 Although in this example, we have assumed a drought, any other climate-related natural disaster may have similar impacts.

8 Data regarding the communities and agricultural sector are from the IACC and RCAD projects, as well as from Wheaton et al. (2008).

References

Abbasi, S. 2014. "Rural Communities Adaptation to Drought: Case Studies of Kindersley and Maidstone, Saskatchewan." MES dissertation, University of Saskatchewan.

Bonsal, B.R., E.E. Wheaton, A. Meinert, and E. Siemens. 2011. "Characterizing the Surface Features of the 1999–2005 Canadian Prairie Drought in Relation to Previous Severe Twentieth Century Events." *Atmosphere-Ocean* 49: 320–38.

Brklacich, M., and M. Woodrow. 2007. *A Comparative Assessment of the Capacity of Canadian Rural Resource-based Communities to Adapt to Uncertain Futures.* Ottawa: Carleton University.

Diaz, H., S. Kulshreshtha, B. Matlock, E. Wheaton, and V. Wittrock. 2009. "Community Case Studies of Vulnerability to Climate Change: Cabri and Stewart Valley, Saskatchewan." *Prairie Forum* 34: 261–88.

Diaz, H., and J. Warren. 2012. *RCAD Rural Communities Adaptation to Drought Research Report.* Regina: Canadian Plains Research Center Press.

Dore, M., and D. Etkin. 2000. "The Importance of Measuring the Social Costs of Natural Disasters at a Time of Climate Change." *Australian Journal of Emergency Management* 15, no. 3 (Spring): 46–51.

Government of Saskatchewan. 2013. "Agriculture." http://www.agriculture.gov.sk.ca/agriculture_statistics/HBV5_Result.asp. Accessed 6 June 2013.

Hanesiak, J., R. Stewart, B. Bonsal, P. Harder, R. Lawford, R. Aider, B. Amiro, E. Atallah, A. Barr, T. Black, P. Bullock, J. Brimelow, R. Brown, H. Carmichael, C. Derksen, L. Flanagan, P. Gachon, H. Greene, J. Gyakum, W. Henson, E. Hogg, B. Kochtubajda, H. Leighton, C. Lin, Y. Luo, J. McCaughey, A. Meinert, A. Shabbar, K. Snelgrove, K. Szeto, A. Trishchenko, G. van der Kamp, S. Wang, L. Wen, E. Wheaton, C. Wielki, Y. Yang, S. Yirdaw, and T. Zha. 2011. "Characterization and Summary of the 1999–2005 Canadian Prairie Drought." *Atmosphere-Ocean* 49, no. 4 (December): 421–52.

Kates, R.W. 1985. "The Interaction of Climate and Society." Pp. 3–36 in J. Ausubel and M. Berbarian (eds.), *Climate Impact Assessment: Studies of the Interaction of Climate and Society.* SCOPE 27. Chinchester, England: Wiley.

Koshida, G. 2010. "Disasters through History." Pp. 2–6 in D. Etkin (ed.), *Canadians at Risk: Our Exposure to Natural Hazards.* Research Paper Series No. 48. Toronto: Institute for Catastrophic Loss Reduction.

Kulshreshtha, S.N., E. Wheaton, and V. Wittrock. 2011. "Natural Hazards and First Nations Community Setting: Challenges for Adaptation." Pp. 277–88 in C.A. Brebbia and S.S. Zubir (eds.), *Management of Natural Resources, Sustainable Development and Ecological Hazards.* Ashhurst, UK: WIT Press.

Luk, K.Y. 2011. "Vulnerability Assessment of Rural Communities in Southern Saskatchewan." MES dissertation, University of Waterloo.

Marchildon, G., S.N. Kulshreshtha, E. Wheaton, and D. Sauchyn. 2008. "Drought and Institutional Adaptation in the Great Plains of Alberta and Saskatchewan, 1914–1939." *Natural Hazards* 45: 391–411.

Parry, M.L., and T.R. Carter. 1987. "Climate Impact Assessment: A Review of Some Approaches." Pp. 165–87 in D.A. Wilhite and W. Easterling (with D. Woods) (eds.), *Planning for Droughts—Toward a Reduction of Social Vulnerability.* Boulder: Westview Press.

Patino, L. 2011, "Participatory Mapping and the Integration of Knowledge in Climate Change Adaptation and Vulnerability: Rural Communities of the South Saskatchewan River Basin." PhD dissertation, University of Regina.

Perrick, D. 2012. "Location of Study Areas Map." In H. Diaz and J. Warren, *RCAD Rural Communities Adaptation to Drought Research Report.* Regina: Canadian Plains Research Center Press.

Pittman, J., V. Wittrock, S.N. Kulshreshtha, and E. Wheaton. 2010. "Vulnerability to Climate Change in Rural Saskatchewan: Case Study of the Rural Municipality of Rudy No. 284." *The Journal of Rural Studies* 27: 83–94.

Sauchyn. D. 2002. "Role of Prairie Adaptation Research Collaborative in Climate Change Impacts and Adaptation Research." Pp. 12–21 in S. Kulshreshtha, R. Herrington, and D. Sauchyn (eds.), *Climate Change and Water Resources in the South Saskatchewan River Basin.* Saskatoon: Department of Agricultural Economics, University of Saskatchewan.

Stern, N. 2007. *The Economics of Climate Change—The Stern Review.* Cambridge: Cambridge University Press.

Szeto, K., W. Henson, R. Stewart, and G. Gascon. 2011. "The Catastrophic June 2002 Prairie Rainstorm." *Atmosphere-Ocean* 49: 380–95.

Warren, J. 2013. "Rural Water Governance in the Saskatchewan Portion of the Palliser Triangle: An Assessment of the Applicability of the Predominant Paradigms." PhD dissertation, University of Regina.

Wheaton, E., S. Kulshreshtha, and V. Wittrock (eds.). 2004. *Canadian Droughts of 2001 and 2002: Climatology, Impacts and Adaptations.* Report prepared for Agriculture and Agri-Food Canada—PFRA. Publication No. 11602-1E03. Saskatoon: Saskatchewan Research Council.

Wheaton, E., S. Kulshreshtha, V. Wittrock, and G. Koshida. 2008. "Dry Times: Hard Lessons from the Canadian Drought of 2001 and 2002." *Canadian Geographer* 52: 241–62.

Wilhite, D. 2000. "Drought as a Natural Hazard: Concepts and Definitions." Pp. 3–18 in D. Wilhite (ed.), *Drought: A Global Assessment.* Vol. 1. New York, NY: Routledge Press.

Wittrock, V., D. Dery, S. Kulshreshtha, and E. Wheaton. 2006. *Vulnerability of Prairie Communities' Water Supply during the 2001 and 2002 Droughts: A Case Study of Cabri and Stewart Valley, Saskatchewan.* SRC Publication No. 11899-2E06. Saskatoon: Saskatchewan Research Council.

Wittrock, V., S. Kulshreshtha, and E. Wheaton. 2011. "Canadian Prairie Rural Communities: Their Vulnerabilities and Adaptive Capacities to Drought." *Mitigation and Adaptation Strategies to Global Climate Change* 16: 267–90.

———. 2012. "Bio-Physical and Socio-economic Vulnerabilities of Selected Prairie Communities in South Saskatchewan River Basin Facing Droughts." Pp. 267–87 in D.F. Neves and J.D. Sanz (eds.), *The Droughts: New Research*. New York: Nova Science Publishers.

Wittrock, V., S. Kulshreshtha, E. Wheaton, and M. Khakpour. 2007. *Vulnerability of Prairie Communities' Water Supply during the 2001 and 2002 Droughts: A Case Study of Taber and Hanna, Alberta, and Outlook, Saskatchewan*. Saskatoon: Saskatchewan Research Council.

Wittrock, V., and E. Wheaton. 2007. *Towards Understanding the Adaptation Process for Drought in the Canadian Prairie Provinces: The Case of the 2001 to 2002 Drought and Agriculture*. Prepared for Government of Canada's Impact and Adaptation Program. SRC Pub #11927-2E07. Saskatoon: Saskatchewan Research Council.

CHAPTER 5

THE "MIN TILL" REVOLUTION AND THE CULTURE OF INNOVATION

Jim Warren

Introduction

Over the course of the agricultural period on the Canadian Prairies, extending from the mid-1880s until today, the region has experienced several periods of severe region-wide drought, along with numerous localized episodes (Wheaton et al. 2005; Wheaton 2007; Lemmen et al. 1997; see also Chapter 8 by Marchildon in this volume). As noted in other chapters in this volume, major droughts affecting the region have at times been followed by significant adaptation efforts, including the creation of new institutions such as the Prairie Farm Rehabilitation Administration (PFRA) and Alberta's Special Areas Board (see Chapter 8 by Marchildon on the history of drought in the region). This chapter focuses on the efforts of agricultural producers and local machinery manufacturers to enhance drought resilience through the invention and adoption of new machine technology and land management practices.

The chapter contends that the propensity of dryland agricultural producers in the region to adopt new farming practices and machinery in response to drought has helped reduce their vulnerability. It also proposes that the adoption of innovative practices which enhance resilience to drought has become an institutionalized social value for dryland farmers in the Palliser Triangle. These arguments are supported by an assessment of historical literature on the evolution of farming practices and equipment used in the region (Shepard 2011; Ward 2011; Bruneau et al. 2009; Hall 2003; Dale-Burnett 2002; Wetherell and Corbet 1993; Archer 1980) and by ethnographic fieldwork data obtained in the Rural Communities Adaptation to Drought (RCAD) project, a major study of adaptation to drought in the region (RCAD 2012; see the introduction to this volume for a discussion of the RCAD project). The chapter makes frequent reference to Warren and Diaz (2012), a book which assesses research conducted for the RCAD project as well as the final report of the RCAD project itself (RCAD 2012). The principal task of this chapter is to view the RCAD data through the lens of the diffusion of innovations theory developed by Rogers (1962).

Min Till

A recent manifestation of widely embraced adaptation in response to agricultural drought on the Canadian Prairies has been the near universal adoption of a family of farming practices collectively referred to as "min till"—an abbreviation for minimum tillage, also referred to as conservation tillage and less accurately as zero till (Bruneau et al. 2009; Hall 2003).

Min till describes a set of technological innovations that reduce soil disturbance and conserve soil moisture, often without the need for mechanical summer fallowing. Reducing mechanical summer fallowing and seedbed disturbance prior to and during planting (referred to as direct seeding) helps retain moisture and reduce wind-driven soil erosion typically associated with severe drought. Min till relies on specialized farming equipment and chemical applications that reduce soil disturbance as well as the frequency of field operations.

Min till methods also include continuous cropping practices, which have significantly reduced the amount of land formerly dedicated annually to summer fallow. Where soil and climate conditions are considered inappropriate for continuous cropping, the application of chemical herbicides (chem fallow) has replaced mechanical weed control methods.

Continuous cropping has increased the need to apply chemical fertilizer. However, crop rotations also help replace lost nutrients and control crop-specific pathogens. Increased use of crop rotations has been facilitated by the adoption of a host of new crops and crop varieties, primarily in the 1990s. Crop diversification in the Palliser Triangle is represented by a significant increase in the acreage devoted to heat-resistant canola varieties and nitrogen-fixing legume crops (referred to as pulse crops), such as field peas, chickpeas, and lentils—crops that were relatively unknown in the region prior to the 1990s.

Min till's advocates contend that leaving standing stubble and trash (crop residue) on the soil surface helps capture winter snow, reduces wind-driven soil erosion, and provides an insulating mulch, which helps reduce evapotranspiration and thereby conserves soil moisture. The mulch is eventually incorporated back into the soil, contributing to soil nutrient and fibre levels.

The adoption of straw spreading attachments for combine harvesters has facilitated the retention of trash. Prior to widespread adoption of this innovation, straw was deposited in windrows, which allowed for the baling for livestock feed and bedding. Particularly thick windrows that were not baled could make spring field work difficult, and they were often burned. Straw spreading has reduced stubble burning as well as the availability of straw for the livestock industry.

The adoption of soil conservation and drought mitigation practices has a long history in the Palliser Triangle, extending back to the early decades of agricultural settlement on the Prairies, the period from the mid-1880s up to World War I (Ward 2011; Shepard 2011; Wetherell and Corbet 1993; Archer 1980). However, the adoption of the collection of min till practices currently in use began in the late 1980s and became widespread over the course of the late 1990s, partly in response to a series of severe drought years in the second half of the 1980s.

Crop yield data and other agronomic observations suggest that min till practices have enhanced the drought resilience of dryland agriculture in the Palliser Triangle. The vast majority of RCAD respondents, including farmers and agrologists, reported that when severe region-wide drought conditions returned to the area in 2001–2, the impacts on crop yields and soil conditions were relatively less severe in some areas than conditions experienced in the 1980s. Many respondents indicated that dust storms, while they did occur, were less common and severe during the drought of 2001–2 compared with the dry years of the late 1980s and the 1930s (see also Luk 2011 and Bruneau et al. 2009: 142–43).

Notwithstanding the contribution of min till to drought resilience and soil conservation, it is not a panacea. In the second consecutive year of a severe drought, yields on min till fields can be significantly reduced. By the second or third consecutive year of severe drought, crop failures can occur on min till fields. Most RCAD respondents reckoned that after two to three consecutive years of severe drought most of the farmers and ranchers in the Palliser Triangle would be experiencing considerable economic hardship. They predicted that three years of severe back-to-back droughts would force many producers to exit agriculture. This grim forecast was thought to apply to producers in general, including those employing min till practices, but with the possible exception of irrigators. Nevertheless, most RCAD respondents also attested to the ability of min till to reduce wind-driven soil erosion and conserve moisture in the early stages of a prolonged drought better than would typically be the case for methods used prior to the 1990s.

It is also noteworthy that min till practices are suited to a particular agricultural production model in a particular climatic environment—dryland annual crop production in a semi-arid climate region that experiences accumulations of snow over winter and periodic drought. Min till practices are not as well-suited to irrigation agriculture and are somewhat less popular among dryland farmers operating in the moister regions of the Prairies outside the Palliser Triangle. And, as will be discussed later in the chapter, min till has detractors who contend that while it may reduce soil erosion and conserve moisture, those benefits come at the cost of increased dependency on fertilizer and herbicide price levels. Critics of min till also maintain that it generates chemical and nutrient pollution, which is harmful to ecosystems and human health.

Furthermore, while min till may indeed produce economically beneficial yield improvements and input cost reductions, there are many other factors besides crop yield that affect the survival of individual agriculture units, including the cost-price squeeze described in Chapter 7 by Fletcher and Knuttila in this volume. Despite widespread adoption of min till, there has been a significant reduction in the number of farms on the Canadian Prairies. In the early 1990s, when min till was in the initial stages of widespread adoption, there were approximately 60,000 farm units in Saskatchewan; in 2015, there were less than 37,000 farms (Saskatchewan Ministry of Agriculture 2015). Diverse factors such as commodity price fluctuations or changes to government farm support programs can have at least as much impact on the survival of a farm as the yield and input cost benefits attributed to min till.

Notwithstanding the qualifications just noted, the principal purpose of this chapter is not to assess the economic and agronomic benefits of min till in a precise way, but rather to describe how and why it emerged as a widespread adaptation to drought on the Canadian Prairies.

Imagining Innovation as a Cultural Value

Interview data collected by the RCAD project and by Warren and Diaz (2012) provide examples of the socio-economic conditions and decision-making processes that supported the widespread adoption of min till technology. That research shows that farmer adoption of min till on the Canadian Prairies reflects the influence of many of the factors contributing to innovation identified under the diffusion of innovation theory famously described by Rogers (1962).

Rogers assesses the processes through which innovation in agricultural technology occurs and provides a list of socio-cultural conditions that can contribute to or detract from the diffusion of innovations. The propensity to innovate is described along a temporal continuum that begins with "the innovator." Innovators are individuals or groups of individuals who are "the first to adopt new ideas in their social system" (Rogers 1962: 193). The affinity of others for the innovations adopted by innovators, which Rogers refers to as "innovativeness," is ordered along the time continuum, beginning with early adopters, followed by the early majority and the late majority adopters, and finally, by laggards who may never adopt the innovation (Rogers 1962: 19).

According to Rogers, the adoption of technological innovations by farmers usually depends on the relative advantage of the innovation over existing practices—measured primarily in economic terms (Rogers 1962: 312). He adds that relative advantage can be emphasized by crises such as drought-induced crop failure. Clearly, the desire to capture potential economic advantages is a facet of innovation that is especially applicable in the Palliser Triangle. Notwithstanding the foregoing, the apparent economic utility (or relative advantage) of any particular agricultural innovation, while important, can by itself be insufficient to generate widespread adoption. No less important in fostering diffusion are embedded cultural factors (Rogers 1962: 57–75). For example, Rogers contends that the "innovativeness of individuals is related to a modern rather than traditional orientation" and that "an individual's innovativeness varies directly with the norms of his social system on innovativeness" (Rogers 1962: 311).

The RCAD research and the literature on Prairie farm technology shows that the adoption of min till on the Canadian Prairies reflects each of the characteristics just noted. Min till practices offered practical economic and agronomic advantages. Severe drought in the late 1980s made innovation more desirable, and there were important socio-cultural conditions on the Prairies that facilitated its adoption. For example, not only is there a population of active innovators on the Prairies, but hundreds of them have been both inventors and manufacturers of farming equipment. In addition, a pattern of historical learning combined with the utility of numerous previous innovations has fostered a propensity for abandoning traditional practices in favour of new ideas that make economic and agronomic sense.

The data compiled in association with the RCAD project suggest that the min till adoption process was facilitated by cultural values supporting innovativeness, which extend across a wide section of the agricultural population of the Palliser Triangle. Over the past century, innovativeness has become institutionalized—a recognized and valued social characteristic relevant to achieving socially important goals. Dryland farmers in the region understand that being adaptive is a key contributor to the long-term, typically multi-generational, survival of agricultural production units on the Prairies. In other words, adaptive capacity resides within a reflexive process whereby agricultural producers recognize the value of being innovative and understand themselves to be innovators and enthusiastic adopters of ideas they perceive will enhance their resilience. This encourages ongoing innovation and adaptation, further reinforcing the value of the "innovative norm."

The propensity of dryland farmers in the Palliser Triangle to adopt innovations stands as an important dimension of the human capital available to enhance resilience to drought in a dry land. Human capital has been described by the Intergovernmental Panel on Climate Change as one of the determinants of adaptive capacity (IPCC 2001: 893). It includes the knowledge, skills, and expertise available to people dealing with adversity:

> This [human] capital includes not only knowledge obtained in the formal education system, but also local knowledge and experiences that could be used to employ, modify and develop other types of resources. Important in this context of human capital are the capacities to wisely manage materials and human resources, learning from experience, as well as the ability to gain access to and process information. (Warren and Diaz 2012: xviii)

In the context of drought in the Palliser Triangle, the propensity to innovate constitutes a key component of the human capital available for reducing vulnerability and enhancing the sustainability of dryland agriculture.

Historical Learning and Innovation

The discussion that follows in this section describes the evolution of tillage practices on the Canadian Prairies from the 1880s to the present. Table 1 presents a timeline of the adoption of new farming practices and machinery from the innovator to early and late majority stages of diffusion.

The first few decades of the agriculture settlement period in the Palliser Triangle, extending from the mid-1880s until the early 1920s, were relatively drought-free. Prior to the 1920s, one of the few more notable incidences of severe region-wide drought in the settled portion of the Prairies occurred in 1886 (Archer 1980: 102). Nonetheless, a number of influential pioneer farmers and government researchers recognized that farming methods and crop varieties developed in the settled regions of North America and Europe would need to be adjusted to account for the relatively dry average conditions and short growing season typical of the Palliser Triangle.

Archer (1980: 99, 102) writes that during the early phase of the settlement period, "the agricultural potential and limitations of the physical environment were not yet understood, with the result that settlers groped

Table 1. The evolution of tillage technology on the Canadian Prairies

Type of innovation	Approximate date for adoption by innovators	Climate and economic conditions at the innovator stage	Approximate period of early and late majority adoption stages	Climate and economic conditions at the early and late majority adoption stages
Mechanical summer fallowing	Late 1880s	Severe drought in 1885 and dry average conditions/early stages of settlement period	Over the course of the settlement period 1886–1913, as new arrivals became familiar with local practices	The period 1886–1913 had no exceptional episodes of widespread severe drought.
Duck-foot cultivators, one-way disc plows, chisel plows, and hoe drills	1920s	Severe droughts and soil drifting in certain regions, and a decline in grain prices following a peak in 1919	Lengthy adoption period from the 1920s to the early 1950s	Adoption was hampered by drought and low farm incomes in the 1930s and later by shortages of steel due to World War II.
Combination of reduced tillage, seeding, and fertilizer application implements	1950s	Increasing farm size and post-war increase in implement manufacturing	1950s	Increasing farm size and post-war increase in implement manufacturing encourage adaptation

Mega-sized tillage implements and tractors	1970s	The increase in farm size continues.	1970s–1980s	Low farm commodity prices encourage farmers to seek economies of scale through farming more acres.
Continuous cropping and trash conservation	1970s	Severe widespread drought in late 1980s hampers adoption until moister conditions return.	1990s	Low farm commodity prices encourage farmers to seek economies of scale through farming more acres.
New crops and varieties facilitate continuous cropping via rotations)	1970s	Severe widespread drought in late 1980s hampers adoption in drier regions until moister conditions return.	1990s	Low farm commodity prices encourage farmers to seek economies of scale through farming more acres. Marketing companies begin offering contracts for specialty crops.
Min till air seeder technology along with advanced minimum tillage and packing tools	1980s	Prairie manufacturers master the technology, but severe drought in late 1980s affects rate of adoption.	1990s	Equipment purchased in the 1960s and 1970s is exceeding useful life spans, and air seeders allow for combined operations with large minimum tillage type equipment.
Chem fallow, trash conservation, and larger chemical applicators	1970s	The high cost of herbicides and low farm commodity prices retard adoption.	1980s–1990s	A 50% decline in herbicide prices and high diesel prices makes chem fallow economically attractive.

toward a suitable agricultural technology." Archer adds that the conditions settlers encountered on the Prairies required them "to adapt or leave."

That initial phase of adaptation involved collaboration between inventive farmers and agronomists working for the federal government. Farmer-agronomist collaboration is reflected in the adoption of regular summer fallowing as a method for conserving moisture, controlling weeds, and enhancing crop yields in a dry country. One of the early experimenters was Angus MacKay, whose fields left fallow in 1885 produced relatively good wheat yields despite drought conditions in 1886. MacKay's innovativeness was recognized by the federal government, which placed him in charge of one of the first agricultural research stations established on the Prairies in 1887. Similarly, Marquis Wheat—a quicker-ripening variety suited to dry conditions and the short growing season on the Prairies—was developed through the combined efforts of farmers (the Saunders family) and the Dominion Experimental Farms (Archer 1980: 102, 121).

According to Archer (1980: 102), notwithstanding the subsequent adoption of locally developed innovations, agricultural practices in western Canada during the settlement period "were largely an extension of traditional [eastern and mid-western North American] methods of wheat cultivation" (see also Ward 2011: Dale-Burnett 2002; Wetherell and Corbet 1993). Imported moldboard plows and peg and disc harrows were the principal tillage tools during the settlement period (Ward 2011: 149; Wetherell and Corbet 1993: 121). By the early 1920s, duck-foot cultivators and chisel plows were beginning to replace moldboard plows and disc harrows for use in summer fallowing and seedbed preparation. Experience with dry conditions suggested that plowing followed by excessive harrowing dried and pulverized the soil, making it subject to wind erosion and moisture loss. A series of droughts during the 1920s in southern Alberta and southwestern Saskatchewan had confirmed this for a growing number of producers. The nine dry years of the 1930s made the observation apparent to many more.

Growing interest among farmers in new tillage implements and practices was supplemented by the efforts of government and university extension agrologists. The PFRA, established by the federal government in 1935, promoted the use of strip farming and the establishment of treed shelterbelts to reduce wind erosion. The PFRA also developed irrigation projects in the handful of neighbourhoods where reasonably dependable surface water supplies were available. The PFRA also took thousands of acres of lighter land (presumed to be unsuited to annual field crop agriculture) out of crop production altogether, reseeding it to grass and establishing community pastures (Gray 1967; see also Chapter 8 by Marchildon in this volume).

A number of locally designed innovative tillage implements were developed on the Canadian Prairies in response to drought conditions in the 1920s and 1930s. Prominent innovations included the Noble blade, the one-way disc plow, the rod weeder, and a variety of high-clearance cultivators (including duck-foot cultivators and chisel plows). Nearly all of these new implements were being designed and manufactured by innovative farmers and machine-shop operators located on the Canadian Prairies (Wetherell and Corbet 1993: 120–121). The development of these implements reflected the beginnings of a shift in practice away from "black summer fallowing," whereby fields were tilled and harrowed to the point that weeds and crop residues were no longer visible on a smooth, clean soil surface. The new thinking supported tillage methods that retained trash (stubble and crop residue) on, or at least near, the soil surface—and left an irregular as opposed to smooth soil surface (Wetherell and Corbet 1993: 118). An important goal of these innovations was to reduce wind-driven soil erosion—a particularly serious problem during drought years. However, the adoption of these implements throughout the farming community was delayed by adverse on-farm economic conditions during the Depression of the 1930s and by limits on the availability of steel for farm implement manufacturing during World War II (Warren and Diaz 2012: 43; Dale-Burnett 2002; Wetherell and Corbet 1993). The first post-war decades coincided with a return to relative prosperity on the farm, enabling producers to take full advantage of innovations such as the combine harvester and improved tillage equipment, which had been invented as far back as the 1920s.

A farmer who participated in the RCAD project described how learning based on experience with severe drought prompted the development of new approaches to soil management. In this instance, the drought which encouraged adaptation occurred in 1961—a year of severe widespread drought in southern portions of the Palliser Triangle:

> There have been some important changes in farming since I started and a number of them were prompted by drought. We used to summer fallow 50–50 around here. In 1961 there was a serious drought. The ground dried out and the wind blew the dirt away right down to the hard pan in places. It blew out whole 40-acre strips in places. In some places dirt drifted up over the top wire on fences. To this day you can still see the effects. I can still show you which fields were in summer fallow that year. And once the topsoil is gone, it's gone. Oh sure, it is starting to come back in places, but it will never be back to what it was in my lifetime. . . That's the sort of experience that led people to come up with solutions like minimum tillage and continuous cropping. Adaptations like those were borne out of necessity. . . Years like 1961 taught my dad that summer fallowing just so you could watch your topsoil blow away afterwards was a good way to go broke. (Warren and Diaz 2012: 5)

Local Innovation and Local Farm Equipment Manufacturing

The adoption of new tillage technologies was supported by the development of a regional farm equipment manufacturing industry on the Canadian Prairies. Local manufacturers understood their neighbours' needs and produced equipment suited to the region's climate and soil conditions. Wetherell and Corbet (1993) indicate that the growth of the local implement manufacturing industry was spurred in part by the reluctance of most major farm machinery manufacturers based in central North America to develop equipment specifically suited to dryland farming on the Canadian Prairies (and the northern plains of the United States). Major manufacturers apparently did not consider the northern plains to be a large enough market to warrant investment in new regionally specialized lines of implements. Farmers and repair shop operators on the Canadian Prairies perceived the value of new types of tillage equipment and were well positioned to cost-effectively service local markets. The region's harsh winters had an influence as well. With several months of downtime, when field work was impossible, innovative farmers had the time to think and tinker.

Rogers (1962: 196) reports that well-equipped farm shops and a population of mechanically adept farmers contributed to the pace of innovation and adaptation in North American farming communities. This was clearly the case on the Canadian Prairies, where mechanical aptitude and the availability of shop equipment, particularly welding equipment, contributed to on-farm modification of existing machinery and the invention of new implements.

One of the RCAD project respondents epitomized the level of mechanical and welding skills resident in the farm population of the region. In 1955, this respondent and his neighbour purchased a dilapidated antique well drilling rig, refurbished it, and dug hundreds of water wells in their neighbourhood. In addition to being able to repair and modify his own farm equipment, this respondent put his technical skills to work for his community.

> I suppose . . . having the ability to meet our own well drilling needs here in the neighbourhood says something about our ability to respond to different challenges. We've done a lot of that sort of thing in this area. Back when I was Reeve. . . we decided we needed a new fire truck for the RM [rural municipality]. Buying one was too expensive so we got together, modified a used truck and had ourselves a fire engine. I was on the rink board when we decided we should get a Zamboni. Well, as usual, money was tight so we got an old Volkswagen car and converted it into a Zamboni. When I was on the hospital board we found ourselves in need of an ambulance. For some time we'd been borrowing the hearse from the local funeral home and that wasn't always the best situation. So, we built our own ambulance by modifying a van. (Warren and Diaz 2012: 45)

An early adopter of min till practices interviewed for the RCAD project described how the capacity to develop and modify machinery on the farm contributed to adaptation:

> We got into continuous cropping on this farm by the mid-70s. In fact my dad put together a little invention of his own to help us do it. We were having trouble running our hoe drills through stubble. The disturbed stubble was piling up and plugging up the works. It was like you were pulling a rake. He rigged up a cycle mower blade run by the power take-off that rode ahead of the drills and cut the stubble off so it would be reduced enough to pass easily through the drills. . . Some years later I was looking over the new inventions on display at the *Farm Progress Show* in Regina [Canada's largest annual farm machinery exhibition] (I try to get over there to see that when I can). There was a guy there with the exact same deal on display—a mower blade that travelled ahead of the drills. I told him he was behind the times. (Warren and Diaz 2012: 5)

As noted above, over the course of the twentieth century, a growing population of farmer-inventors and repair shop operators began supplementing their incomes by building and marketing farm equipment. By the early 1990s, 267 farm equipment manufacturers were reported to be in business on the Canadian Prairies (Wetherell and Corbet 1993: 231–52).

Difficult times in agriculture resulting from drought and low commodity prices, combined with the relative hardships of rural versus urban life, contributed to a significant reduction in the number of farmers in the Palliser Triangle region. The number of people living on farms in Saskatchewan, for example, peaked at 573,894 in 1936. By 1951, only 398,279 people were living on 119,451 farms in Saskatchewan. The number of farms in Saskatchewan declined to 60,000 by the close of the 1980s, and, as of 2011, the number of farms in Saskatchewan was 36,952 (Saskatchewan Ministry of Agriculture 2015; Shepard 2011: 182, 183).

Those farmers who remained in business in the immediate post–World War II period were typically farming more land. It was assumed that economies of scale could improve the profitability of farms. The relative dearth of farm labourers during World War II and into the post-war period stimulated the adoption of labour-saving technology. These pressures prompted the invention and diffusion of new tillage and seeding machinery that combined two or more functions into a single implement and field operation.

For example, seeding equipment was attached to minimal soil disturbance tillage implements, such as the one-way disc plows already coming into widespread use. Saskatchewan-based Canadian Co-operative Implements began manufacturing discers with attached seed boxes in 1950—the first major manufacturer in North America to do so. Mounting seeding and packing attachments to discers and cultivators allowed farmers to combine pre-planting tillage, seeding, and seedbed packing into a single operation—saving person-hours (always an important consideration on the Prairies given the short growing season) and diesel fuel. The hoe drill was the second most popular seeding implement in use on the Prairies prior to the 1990s (after disc seeders). By the 1980s, farmers were experimenting with tillage tools and soil packers that allowed them to seed with hoe drills without having to pre- or post-till the seedbed (a min till practice referred to as direct seeding). While disc drills kept trash close to the soil surface, appropriately modified hoe drills left more residue directly on the surface.

The need to cover more acres within the short growing season available on the northern plains prompted local manufacturers such as Olaf Friggstad of Frontier, Saskatchewan, to manufacture and market huge tillage implements, including one of the largest field cultivators (80 feet) ever marketed in North America. Larger implements required larger tractors, and manufacturers on the US and Canadian northern plains responded in the 1970s and 1980s by building large, articulated four-wheel drive tractors—years ahead of the major full line equipment manufacturers (e.g., Versatile Manufacturing of Winnipeg, Manitoba; Steiger Tractor of Fargo, North Dakota; and Big Bud Tractors of Havre, Montana).

An RCAD project respondent recalled the move to larger tillage machinery that occurred on the Prairies in the 1970s and 1980s:

> I can't recall exactly who started the minimal till thing around here. I remember that just before minimal till caught on, the race was on to buy bigger cultivators. Buy as many feet of cultivator as you can, that was good management then. We've got one of the biggest cultivators ever made, we've got an 80 footer. But then it turned out that it was better to summer fallow with chemicals instead of cultivators. I can't say precisely when that was we began to use chemical summer

> fallow, but it was back in the Glean [a brand name herbicide] days, maybe the early 90s. (Warren and Diaz 2012: 35)

By the close of the 1980s, Prairie equipment manufacturers had made considerable strides in developing air seeder technology. Companies including Ezee-On Manufacturing of Vegreville, Alberta; Bourgault Industries of St. Brieux, Saskatchewan; and Saskatoon-based Flexi-Coil, among others, had developed implements that combined high-capacity seed/fertilizer tanks, pneumatic seed delivery systems, and large tillage equipment. New tillage and packing tools were developed in conjunction with pneumatic seed delivery, allowing for minimal disturbance of trash and soil and precision application of fertilizer and seed in a single operation.

A parallel development in the post-war period was growth in the use of chemical fertilizers, herbicides, and pesticides (Argue et al. 2003). After decades of farming, which included periods of drought-induced soil erosion, farmers in the post–World War II period increasingly relied on fertilizer to replace depleted soil nutrients. New chemical herbicides and pesticides capable of controlling weeds and pathogens in growing crops and on summer fallow were becoming available and were marketed to farmers. Not surprisingly, new implements were developed for applying fertilizer, herbicides, and pesticides on increasingly larger farms. A number of manufacturers on the Canadian Prairies specialized in manufacturing large-capacity field sprayers, and as noted above, tillage implements were adapted to combine seeding and fertilizer application operations (Wetherell and Corbet 1993: 152–57).

An initially controversial innovation receiving attention during the post-war period was continuous cropping. A minority of farmers and agrologists had begun to challenge long-standing conventional wisdom regarding the need to leave land fallow every other year or every third year. A farmer from southern Saskatchewan described how his family became early adopters of continuous cropping and other min till practices in response to drought conditions in the 1960s:

> Summer fallow was supposed to be a great moisture conservation measure. But it didn't help you much if your soil blew away. The best Dad did when summer fallowing, the best crop I think he ever grew, was probably about 35 bushels an

> acre. Okay, but it took him two years to grow that. When you divide that by two it gives you 17½ bushels an acre. So with continuous cropping I'm getting 20, 24 bushels an acre. Sure, you'd maybe get more out of a summer fallow crop. But I still get my 20–24 bushels per acre and I get it every year. So the summer fallow guy, he's getting his 35 once every two years. I'm getting my 40 or 50 when you take it over two years. You don't have to be a rocket scientist or mathematician to figure that one out. (Warren and Diaz 2012: 4, 5)

By the late 1980s, government researchers were conducting studies that questioned the benefits of tilled summer fallow. In its 1987 *Guide to Farm Practice in Saskatchewan*, Saskatchewan's Department of Agriculture was reporting on studies from the federal research station at Swift Current, Saskatchewan, which suggested that leaving standing stubble on fields over the winter was possibly a more effective method for retaining moisture than leaving land idle for a year as tilled summer fallow (Saskatchewan Agriculture 1987: 100). Researchers had begun to speculate that the increased yield effects associated with summer fallowing were more likely due to the nitrogen-accumulating effects of tilled summer fallow than to the long-held assumption that it was entirely the result of moisture retention. If this was indeed the case, it could prove more cost-effective to forego summer fallowing in favour of continuous cropping combined with increased applications of nitrogen fertilizer. As we have seen, summer fallowing had been among the first innovations adopted by farmers on the Prairies during the early days of settlement. Now it appeared that it was a traditional practice that should be abandoned in the face of new and better information—and that is precisely what would happen on a large scale in the 1990s.

The series of dry years experienced in the 1980s frustrated proponents of continuous cropping. However, evidence was mounting that, under average moisture conditions or even moderate drought, continuous cropping could out-produce summer fallow farming—particularly in moister areas of the Palliser Triangle.

An RCAD respondent reflected upon the diffusion of continuous cropping in his neighbourhood:

> I knew a guy who was an early adopter of continuous cropping, but he was trying it in the 80s and it wasn't working. The idea was right but it was just too dry. Everybody was looking and saying, "see it doesn't work." But on further reflection people started to say, "I think it would have worked but we needed a little bit more rain." (Warren and Diaz 2012: 92)

The Convergence of Forces in the Early 1990s

By the mid-1980s, the technological ingredients required to support the family of minimum tillage technologies in use today were essentially in place. Nonetheless, the explosion of widespread adoption, typical of the early and late majority phases described by Rogers (1962: 11), did not occur until the 1990s. Farmers interviewed for the RCAD (2012) project and by Warren and Diaz (2012: 5–6, 35, 60–61) attributed the rapid pace of change in the 1990s to the convergence of several key factors, including

- heightened interest in increasing drought resilience in the aftermath of the severe droughts of the 1980s;
- availability and awareness of locally manufactured, specialized minimum tillage and seeding equipment and chemical applicators suited to the large farm sizes typical of the Palliser Triangle region;
- a significant reduction in the cost of glyphosate herbicides in the early 1990s (i.e., glyphosate dropped in price from approximately $25 per litre in the 1980s to $10 per litre in the 1990s), which made chemical summer fallowing for weed control more cost-competitive with mechanical summer fallowing practices reliant on higher diesel fuel consumption and more labour;
- research and promotional activities, including on-farm field days, of farmer-operated soil conservation associations (sometimes supported by government extension agrologists, local manufacturers, and herbicide marketers), which en-

couraged min till practices including greater use of continuous cropping and chemical summer fallowing;

- development and promotion of new crop varieties such as pulses (annual legumes such as peas, beans, and lentils) that facilitated continuous cropping through crop rotations; and
- a population of innovative farmers and ready adopters who were amenable to developing and implementing new farming practices.

One of the notable differences in the pattern of diffusion associated with min till in the 1990s and previous phases of agricultural adaptation on the Prairies was the relative increase in the influence of farmer innovators as opposed to innovation co-led by extension agrologists from universities and government. While government agencies contributed funds toward the field testing of min till techniques and new crop varieties, government-backed crop insurance programs initially penalized producers who experimented with continuous cropping. Also of significant importance was the role played by Prairie manufacturers who built and marketed the necessary equipment and by chemical manufacturers and distributors who encouraged the shift to more chemical-intensive agriculture.

Min Till as the Product of an Adaptive Culture

The converging factors described above correspond to characteristics that Rogers (1962: 124–33) attributes to innovations that are likely to be widely adopted. These characteristics include the relative advantage offered by the innovation, often measured in terms of its ability to enhance economic profitability. Min till practices met this criterion by virtue of their capacity to conserve moisture, sustain yields, and protect soil from erosion more cost-effectively than conventional practices.

Another characteristic identified by Rogers (1962: 57–75) is the compatibility of an innovation with the values and past experiences of the adopters. This characteristic is reflected in the historical pattern of adaptation and the wide acceptance of inventiveness and adaptability as positive social attributes on the Canadian Prairies. An important contributor to the adaptive culture is the fact that most farms operating on the Prairies are second- or third-generation operations. Intergenerational learning within families and communities has contributed to an appreciation of adaptation as an iterative process that has helped enable succeeding generations to survive in agriculture. The valuable lessons provided by previous generations are not so much the particular innovations they adopted, but that they were flexible enough to adapt.

A producer from Wardlow, Alberta, reported that the experience of earlier generations was valuable because it demonstrated that being prepared to do things differently than one's antecedents was integral to survival—and indeed it was that attitude which enabled subsequent generations of survivors in agriculture to succeed:

> There are plenty of things that the older generations of ranchers and farmers learned about how to survive in this country and you have to respect that. But you don't want to get into that mindset where you start to think their way is the only way. It's tough to make a buck in this industry, and it doesn't seem to be getting any easier . . . The point is, you need to keep adapting if you want to survive. A fellow told me one time that if you run into one of these guys who says, "If it was good enough for grandpa, and it was good enough for dad, it is good enough for me," you can bet if he carries on like that, before too long there will be a "For Sale" sign on his gate. (Warren and Diaz 2012: 249)

An early adopter of min till technology characterized the reflexive mindset required for survival in family farm agriculture as *planning that accommodates flexibility*:

> You need to spend some time on your butt thinking . . . A lot of fellows get into trouble because they fly out into the field and go to work without thinking. Another thing is to always have a plan B. Don't go down the road there, with hard and fast rules that this is what's going to be done come hell or high water. You've got to stay flexible and roll with the punches. You have to stay flexible or you're history. (Warren and Diaz 2012: 5)

Interestingly, farmers interviewed for the RCAD project sometimes employed the language and concepts used by academics to describe the diffusion of innovations. Echoing Rogers' (1962: 196) characterizations, inventors, innovators, and early adopters from the Palliser Triangle understand that they march to a different drummer and are somewhat deviant—but that theirs is a socially beneficial form of deviance. A farmer who was active in the promotion of min till practices in southwest Saskatchewan in the 1980s and 1990s describes how innovators appreciate that their early efforts can be met with skepticism but nonetheless proceed:

> And of course there are those bright, eccentric, inventive farmers that you find here and there around the country who aren't afraid to be criticized by their neighbours for trying something radically different. I recall talking to one of the first direct seeders in the country, a fellow who farmed up near Biggar. He told me how at first people thought his new methods were pretty goofy, but within a short time virtually everyone was into direct seeding. He said, "I went from wing nut to innovator in about five years." (Warren and Diaz 2012: 60)

Many of the dozens of producers interviewed in connection with the RCAD project understood the importance of innovation and adaptation to survival in Prairie agriculture. They also demonstrated an understanding of how the process works. The following comments are not untypical:

> That's what happens, out of necessity somebody comes up with a new idea. His neighbours watch him for a while to see if it really works, and if it does, before long they're doing it too. That's what's happened with lots of equipment. I remember the first time I saw an air seeder. A fellow had one in at the *Farm Progress Show* one year and before long all sorts of companies like Flexi-Coil were making them. It was the same with Friggstad's from Frontier [Saskatchewan], they came up with a better header [a harvest machinery attachment] and pretty soon other people wanted them too. Some farmers are good at doing that in this country, not all the good ideas come from the universities or government research stations—we

come up with a lot of them on the farm; especially new machinery. (Warren and Diaz 2012: 5)

Currently, min till practices have been adopted by most dryland farmers in the Palliser Triangle. Bruneau et al. (2009: 143) report that as of the first decade of the twenty-first century, conventional tillage was used on just 18% of the cropland in Saskatchewan and 25% in Alberta. Given that there are moister cropland areas outside the boundaries of the Palliser Triangle in both provinces, it is reasonable to assume that within the drier regions the proportion of farmers using min till is higher than the proportions reflected in the provincial averages.

Laggards as Innovators

Rogers (1962) contends that innovators, opinion leaders, and early adopters tend to be more cosmopolitan and modern in their thinking compared with those who are especially slow to adopt a new idea. There are indeed dryland farmers operating in the Palliser Triangle who have not fully embraced min till technology. That being said, many of these producers do not consider themselves to be atavistic Luddites but rather as innovators in their own right. They reject the heavy use of herbicides and chemical fertilizer associated with min till, preferring to farm organically. Organic producers interviewed in association with the RCAD project argued that chem fallow and continuous cropping were inimical to soil health—the very thing that those methods were intended to protect. Also influential are the human and ecological health concerns that organic farmers and many consumers associate with agricultural chemicals. Another detriment identified by RCAD respondents and the literature is that the cost-effective implementation of min till technology depends on the prices of fertilizers and chemicals, which are largely beyond the control of individual farmers (Argue et al. 2003). A spike in these prices has the potential to reduce the economic advantages of min till relative to mechanical weed control and summer fallowing.

The fact that min till mitigates wind-driven soil erosion during droughts is generally considered to be an environmental benefit. However, organic producers contend that this advantage needs to be considered within the context of the environmental problems it exacerbates.

For example, the increased application of fertilizer required under min till farming has been identified as a factor that contributes to eutrophication (nutrient pollution) in prairie lakes (Environment Canada 2014; Carpenter et al. 1998).

Organic farmers argue that their products can obtain premium prices from health-conscious and environmentally conscious consumers, which offset differences in yield. They have, indeed, developed niche markets throughout North America and in Europe. One might reasonably assert that their ingenuity as marketers is equivalent to that of conventional min till producers. Notwithstanding the relative strength of the arguments advanced by organic agriculture over chemically supported agriculture, organic producers remain a minority of the farming population in the Palliser Triangle. For example, as of 2013, approximately 2,000 certified organic farms were operating in Saskatchewan, out of a total of about 37,000 farms (Saskatchewan Ministry of Agriculture 2013).

Organic farmers do not view themselves as backward-thinking but rather as innovators striving to avoid widespread maladaptation. Indeed, some RCAD respondents wondered whether the success of min till might encourage a sort of drought-defying hubris whereby overconfident farmers break lighter, erosion-prone land that had been seeded to grass in the wake of the droughts of the 1930s (RCAD 2012). Shifting land from permanent grass cover to cultivation was reportedly occurring at some locations in southwest Saskatchewan. The suspected danger is that crops and soil resources on this type of land could be vulnerable to erosion under severe drought conditions that exceed recent experience on the Canadian Prairies. Min till is assumed to have enhanced drought resilience since the 1980s. However, none of the droughts occurring since the 1980s have lasted as long as the drought of the 1930s or the megadroughts identified in paleoclimatic records by Sauchyn and Kerr (see Chapter 2 in this volume).

The min till versus organic debate suggests that Rogers' characterization of laggards may not apply to everyone who fails to innovate. Rogers' classification casts laggards as less cosmopolitan and economically astute than early adopters. These aspersions would be difficult to apply across the board with respect to organic producers in the Prairie provinces. That being said, during a long and severe drought the ongoing use of mechanical summer fallowing in organic farming on the Prairies would contribute to soil erosion and a reduction in drought resilience.

Conclusion

Agricultural producers in the Palliser Triangle have been adapting to dry conditions and drought for over a century. Farmers in the region understand that survival in agriculture under dry climate conditions, drought, and frequently unfavourable markets has benefited from adopting a series of technological innovations. Within the dryland farming community, innovation and adaptation are well understood and valued processes. Multi-generational survival of farming units is a matter of some pride, especially since tens of thousands of family operations have failed to survive. The governments of Alberta, Saskatchewan, and Manitoba honour "century farms"—operations that have remained under the ownership of the same family for 100 years. Multi-generational survival suggests that intergenerational transmission of adaptive capacity has occurred.

The capacity to innovate is a matter of considerable pride as well. The contributions of farmer innovators, including those who developed local machinery manufacturing concerns such as Charles Noble, George Morris, Olaf Friggstad, and many others, are widely recognized. Innovation and the diffusion of innovations have been enhanced by the recognition of flexibility and adaptability as important values, as well as by the technological and mechanical proficiency available in dryland farming communities in the region. These are important facets of the human capital available in the region that have facilitated drought resilience, reducing vulnerability under the range of climate conditions experienced over the past century. Indeed, the adaptation-enhancing features of the culture of agriculture on the Canadian Prairies are largely consistent with Rogers' contention that innovativeness "varies directly with the norms of his social system on innovativeness" (Rogers 1962: 311). Whether these cultural assets will prove sufficient in providing the resilience required to adapt to the climate conditions projected for the upcoming century is unclear (see Chapter 3 by Wheaton et al. in this volume)—but they should help.

References

Archer, J. 1980. *Saskatchewan: A History.* Saskatoon: Western Producer Prairie Books.

Argue, G., B. Stirling, and H. Diaz. 2003. "Agricultural Chemicals and Agribusiness." Pp. 207–22 in H.P. Diaz, J. Jaffe, and R. Stirling (eds.), *Farm Communities at the Crossroads: Challenge and Resistance.* Regina: Canadian Plains Research Center Press.

Bruneau, J., D.R. Corkal, E. Pietroniro, B. Toth, and G. van der Kamp. 2009. "Human Activities and Water Use in the South Saskatchewan River Basin." Pp. 129–52 in G. Marchildon (ed.), *A Dry Oasis: Institutional Adaptations to Climate on the Canadian Plains.* Regina: Canadian Plains Research Center Press.

Carpenter, S.R., N.F. Caraco, D.L. Correll, R.W. Howarth, A.N. Sharpley, and V.H. Smith. 1998. "Nonpoint Source Pollution of Surface Waters with Phosphorous and Nitrogen." *Ecological Applications* 8: 559–68.

Dale-Burnett, L. 2002. "Agricultural Change and Farmer Adaptation in the Palliser Triangle, Saskatchewan, 1900–1960." PhD dissertation, University of Regina.

Environment Canada. 2014. "Phosphorus and Nitrogen Levels in Lake Winnipeg." Environment Canada website. http://www.ec.gc.ca/indicateurs-indicators/default.asp?lang=en&n=55379785-1. Accessed 15 April 2014.

Gray, J. 1967. *Men Against the Desert.* Saskatoon: Western Producer Prairie Books.

Hall, A. 2003. "The Adoption of Conservation Tillage: An Understanding of the Social Context." Pp 267–88 in H.P Diaz, J. Jaffe, and R. Stirling (eds.), *Farm Communities at the Crossroads: Challenge and Resistance.* Regina: Canadian Plains Research Center Press.

IPCC (Intergovernmental Panel on Climate Change). 2001. *Climate Change 2001: Impacts, Adaptation, and Vulnerability. Contribution of Working Group II to the Third Assessment Report of the Intergovernmental Panel on Climate Change.* Cambridge: Cambridge University Press.

Lemmen, D.S., R.E. Vance, S.A. Wolfe, and W.M. Last. 1997. "Impacts of Future Climate Change on the Southern Canadian Prairies: A Paleoenvironmental Perspective." *Geoscience Canada* 24(3): 121–33.

Luk, K.Y. 2011. "Vulnerability Assessment of Rural Communities in Southern Saskatchewan." MES dissertation, University of Waterloo.

Marchildon, G. (ed.). 2011. *Agricultural History.* Regina: Canadian Plains Research Center Press.

RCAD (Rural Communities Adaptation to Drought Project). 2012. *RCAD Research Report.* Regina: Canadian Plains Research Center Press.

Rogers, E. 1962. *Diffusion of Innovations.* New York: The Free Press.

Saskatchewan Agriculture. 1987. *Guide to Farm Practice in Saskatchewan.* Regina: Saskatchewan Department of Agriculture.

Saskatchewan Ministry of Agriculture. 2013. "Agricultural Statistics Fact Sheet." SMA website. http://www.agriculture.gov.sk.ca/Saskatchewan_Agriculture_Statistics_Fact_Sheet. Accessed 12 July 2013.

———. 2015. "Census Farms." SMA website. http://www.agriculture.gov.sk.ca/Number_Census_Farms. Accessed 31 May 2015.

Shepard, B. 2011. "Tractors and Combines in the Second Stage of Mechanization on the Canadian Plains." Pp. 167–86 in G. Marchildon (ed.), *Agricultural History.* Regina: Canadian Plains Research Center Press.

Ward, T. 2011. "Farming Technology on Early Prairie Farms." Pp. 145–65 in G. Marchildon (ed.), *Agricultural History.* Regina: Canadian Plains Research Center Press.

Warren, J., and H. Diaz. 2012. *Defying Palliser: Stories of Resilience from the Driest Region of the Canadian Prairies.* Regina: Canadian Plains Research Center Press.

Wetherell, D., and E. Corbet. 1993. *Breaking New Ground: A Century of Farm Equipment Manufacturing on the Prairies.* Saskatoon: Fifth House Publishers.

Wheaton, E. 2007. "Drought." Pp. 40–52 in B.D. Thraves, M.L. Lewry, and J. Dale (eds.), *Saskatchewan: Geographic Perspectives.* Regina: Canadian Plains Research Center Press.

Wheaton, E., S. Kulshreshtha, and V. Wittrock (eds.). 2005. *Canadian Droughts of 2001 and 2002: Climatology, Impacts and Adaptations.* Saskatoon: Saskatchewan Research Council.

CHAPTER 6

THE TROUBLED STATE OF IRRIGATION IN SOUTHWESTERN SASKATCHEWAN: THE EFFECTS OF CLIMATE VARIABILITY AND GOVERNMENT OFFLOADING ON A VULNERABLE COMMUNITY

Jim Warren

Introduction

Irrigation has facilitated the development of agriculture in many of the world's drier regions. It is often associated with areas where agricultural production would be difficult or impossible without the water resources and infrastructure that allow for the delivery of water to land that receives inadequate precipitation to support crops. Similarly, in areas such as the Palliser Triangle, the driest region in the Canadian Prairies, where precipitation can be unreliable, irrigation purportedly allows for crop production in those years when rainfall is scarce. The ability to irrigate in a region that experiences periodic severe droughts might reasonably be considered to be the consummate adaptation to drought. However, the experience of farmers and ranchers in the southwest corner of Saskatchewan demonstrates that investment in irrigation infrastructure alone does not always ensure drought resilience.

The history of irrigation in this region underlines the importance of context when considering the utility of various strategies for enhancing drought resilience. In Chapter 5 on min till in this volume, we observe how the adoption of new farming practices and machinery has reduced the impact of drought on crop yields and soil health in the context of dryland annual field crop production. The adoption of min till practices was attributed in part to the adaptive proclivities of dryland farmers and local machinery manufacturers. However, climate and soil conditions in portions of the southwest corner of Saskatchewan are frequently different from those areas where min till farming is predominant. In the southwest, cattle ranching is the dominant agricultural activity, partly because local conditions are frequently considered too dry to facilitate dryland crop production. Thus, the adoption of grazing-based agriculture stands as one of the principal long-term adaptations to drought in the region. Dry conditions also limit the ability of ranchers to produce dryland hay to feed their cattle over winter. Irrigation is attractive since irrigated hay land typically produces yields that are 200% or more above those available from dryland hay production. Furthermore, irrigation supposedly ensures that hay crops will not fail due to the moisture deficits normally associated with agricultural drought.

Notwithstanding the purported drought resilience available through irrigation, the fortunes of irrigation agriculture in Saskatchewan's dry southwest have been frustrated by three consecutive decades of hydrological drought—reflected in low streamflows and reservoir levels. From 1979 until 2010, there were several years when irrigation farmers in the southwest of the province struggled with reductions in the amount of water available for irrigation and, in some years, had to contend with a total lack of water (RCAD 2012: 23–29; Warren and Diaz 2012: 124–49, 322–30). The changing availability of water illustrates the impact of a significant reduction in the value of natural capital available to ranchers in the region.

Infrastructure improvements, which promised to compensate for reduced water availability in the 1990s and 2000s, have not been made (Warren and Diaz 2012; PFRA 1992). The effects of this irrigation infrastructure deficit have been exacerbated by the Canadian government's decision to end its six-decade history of financial and technical support for irrigation infrastructure in southwest Saskatchewan. It is uncertain whether the necessary investments in infrastructure enhancement can be made without support from senior government. This situation reflects a significant decline in the institutional capital available to producers in their efforts to deal with drought.

The experience of irrigators in Saskatchewan's southwest presented in this chapter also demonstrates how understanding drought through the lens of appropriate definitions (as discussed in Chapter 1 of this volume) is beneficial in appreciating its impacts on communities. The previous chapter on min till shows how changes in tillage technology moderated the impacts of agricultural drought under dryland farming conditions. However, we find that irrigated forage production in southwest Saskatchewan is primarily vulnerable to hazards associated with hydrological drought, such as low streamflows and reservoir levels. Indeed, there have been many years in which well-timed precipitation allowed for normal grazing and average dryland farming yields in the region, yet at the same time irrigation activity was reduced. Later in this chapter we will see how the failure of a government support program to extend assistance to producers under a 2010 drought support program was in part a failure to adequately consider the effects of hydrological drought on forage production.

The previous chapter emphasized the resilience-enhancing benefits of a culture of innovation on the capacity of dryland farmers to adapt to drought. That process demonstrated the importance of human capital for communities adapting to drought. The innovations associated with min till were attributed in large part to local farmer innovators and machinery manufacturers on the Prairies. These innovations were not driven primarily by the institutions and agencies of government. While governments occasionally supported the adoption of min till, they also put barriers in the path of innovators in the form of crop insurance penalties.

In the case of irrigation in southwest Saskatchewan, government agencies assumed responsibility for most infrastructure development after the 1930s (SIPA 2008a, 2008b; Saskatchewan AgriVision 2004). Systems that had been developed by farmers and ranchers without government assistance in the first decades of the twentieth century were largely absorbed into the government-managed irrigation projects. The assessment presented in this chapter suggests that government involvement, while necessary for the creation of many projects, did not require producers to engage in self-reliant innovation to the same degree as their dryland counterparts. Some observers suggest that this may have contributed to an unhealthy dependency, which is now inhibiting the development of producer-driven solutions to the region's irrigation problems. Producer reliance on government support for irrigation is especially troublesome today, as Canada's governments reduce financial support for primary agricultural production.

In summary, this chapter describes how drought resilience of irrigation farmers in southwestern Saskatchewan has declined due to a combination of forces, including hydrological drought, infrastructure deficits, poor system management, low cattle prices, rising input costs, and the unwillingness of senior governments to provide ongoing support for irrigation. It also suggests measures that could enhance the coping capacity of irrigators in the region. Furthermore, it contends that the mixed success of irrigation in the region underlines the importance of incorporating long-range climate records and forecasts into adaptation planning.

The chapter relies on a substantial store of ethnographic field research data produced in association with the Rural Communities Adaptation to Drought (RCAD) project (RCAD 2012) and the Institutional Adaptation to Climate Change project (IACC 2009), and collected by Warren and Diaz (2012).

Historical and Climatic Context

When the Canadian federal government responded to a succession of severe droughts on the Prairies during the 1930s, the development of irrigation infrastructure was one of the pathways it took to increasing the drought resilience of farmers. In 1935, a new federal government agency, the Prairie Farm Rehabilitation Administration (PFRA), was established to ameliorate the combined effects of a succession of years that featured low prices for farm commodities and crop failures due to drought (Gray 1967; see also Chapter 8 by Marchildon in this volume). Over the course of the next six decades, the PFRA developed and operated 11 irrigation projects in southwestern Saskatchewan, providing irrigation opportunities for hundreds of farmers and ranchers. However, in 2007, the PFRA informed the producers who rely on these projects that it intended to abandon its irrigation responsibilities and turn the project infrastructure over to the irrigators (Warren and Diaz 2012: 322–30). The impacts of that departure are described later in this chapter, following a brief overview of the development of irrigation agriculture in Saskatchewan and Alberta.

Irrigation before PFRA

Prior to the megadroughts of the 1930s, farmers, ranchers, and farmland speculators had developed a number of individual (single farm) irrigation systems as well as larger, multiple-user projects (Warren and Diaz 2012: 245; SIPA 2008a, 2008b; Saskatchewan AgriVision 2004). Two natural conditions prompted development of irrigation systems on the Canadian Prairies. First, these systems tended to emerge where conditions were driest—areas such as the Palliser Triangle, where even in years when moisture conditions were average, crop yields were low compared to less dry portions of the Prairies (see Chapter 8 by Marchildon in this volume). Second, development relied on the availability of readily accessible source water. An area with an especially dry climate adjacent to a reliable stream was the most likely sort of neighbourhood to acquire an irrigation system (SIPA 2008a, 2008b; Saskatchewan AgriVision 2004).

Agricultural pioneers on the Prairies of what would become southern Alberta had access to several reliable streams originating in the Rocky Mountains and their foothills. In southern Alberta, farmers and ranchers choosing to irrigate sometimes developed individual private systems, but more often, they partnered with neighbours to share the cost of constructing and maintaining the necessary "works"—the infrastructure required for irrigation such as dams, reservoirs, canals, and ditches. Real-estate speculators, including the Canadian Pacific Railway (CPR), also invested in the development of multiple-user irrigation projects. The CPR anticipated that the availability of irrigation would attract immigrants and traffic to some of the drier regions traversed by its rail lines. As the surrounding communities became settled, the railway's irrigation infrastructure was transferred to producer-operated district irrigation associations (Warren and Diaz 2012: 245; Brownsey 2008; SIPA 2008a, 2008b, 2000; Saskatchewan AgriVision 2004).

Irrigation developed at a much slower and more erratic pace in the section of the Prairies that became the province of Saskatchewan. In the drier regions of Saskatchewan, where investment in irrigation made the most agronomic sense, reliable supplies of source water were far less abundant than was the case in southwestern Alberta (SIPA 2008a, 2008b; Saskatchewan AgriVision 2004). Exceptions included lands transected by the South Saskatchewan River and streams originating in the Cypress Hills. One of the first multiple-user irrigation projects in Saskatchewan was developed in 1903 by the Richardson and MacKinnon families along Battle Creek between the Cypress Hills, where the creek originates, and the boundary with the United States (SIPA 2000: 8, 9).

The creation of the PFRA facilitated a significant increase in irrigation development on the Canadian Prairies. Federal funds and engineering expertise were directed at expanding and intensifying irrigation activity in Alberta and Saskatchewan. In Alberta, new project areas were brought on-stream with PFRA support, including those near the communities of Rolling Hills and Brooks (Gray 1967: 199; see also Chapter 8 by Marchildon in this volume). From the late 1930s on through the 1940s and 1950s, the PFRA built dozens of dams and reservoirs in Alberta and Saskatchewan, along with 11 flood irrigation projects in the dry southwestern corner of Saskatchewan (SIPA 2008a; 2000).

Most of the PFRA's irrigation projects in southwestern Saskatchewan were supplied by streams originating in the Cypress Hills. Flows on these streams were assumed to be reliable enough to support irrigation, and until 1979, they essentially were (Warren and Diaz 2012: 124–49, 322–30). Streamflows in the southwestern corner of Saskatchewan are considerably smaller than those that supplied irrigators in southwestern Alberta. Consequently, the total area irrigated in southwestern Saskatchewan was much smaller than in Alberta. The difference in the scope of irrigation activity in the two provinces has persisted to the present. Approximately 1.3 million acres of land is irrigated in Alberta compared with just 350,000 acres in Saskatchewan (SIPA 2008a, 2008b; Saskatchewan AgriVision 2004).

As far back as the 1930s, it was widely assumed that the amount of irrigated land in Saskatchewan would expand exponentially if infrastructure was developed to provide farmers with access to flows on the South Saskatchewan River. Since that river originates in the Rockies, its flows are less vulnerable to drought than streams that originate in the Palliser Triangle. PFRA planners supported by powerful political champions of Prairie agriculture envisioned a massive dam and reservoir project on the South Saskatchewan River as a means to launch much larger irrigation projects in Saskatchewan (Herriot 2000; Archer 1980). Construction work on the South Saskatchewan River Dam project (now Gardiner Dam and Lake Diefenbaker Reservoir) began in 1959, and by the early 1970s, infrastructure was in place to facilitate the development of Alberta-size irrigation projects in Saskatchewan. By the close of the 1980s, over 20,000 acres of land were under irrigation in the Outlook area (SIPA 2008a).

Nonetheless, irrigation proponents were disappointed in the rate of irrigation uptake by farmers who had access to Lake Diefenbaker water. There was far more water and irrigation infrastructure available for use in the Lake Diefenbaker area than farmers willing to use it. Low uptake was attributed, in part, to the fact that irrigation works associated with Lake Diefenbaker were located in an area that straddled the northern boundary of the Palliser Triangle, where moisture conditions allowed for the production of acceptable crops using less capital-intensive dryland methods (Suderman 1966).

Notwithstanding the disappointing growth in the number of farmers irrigating, the Lake Diefenbaker projects provided irrigators with highly reliable water supplies. Producers were essentially able to apply as much water as they wanted whenever they wanted it. This was not the case for the PFRA projects in the southwestern corner of Saskatchewan (RCAD 2012; Warren and Diaz 2012).

The PFRA Projects in Southwestern Saskatchewan

Following construction of the infrastructure supporting the PFRA's projects in southwestern Saskatchewan, individual producers were encouraged to purchase flood irrigation plots on approximately 10,000 available acres. Purchasers would be required to pay an annual fee for the delivery of water to their plots. Under normal operations, water would be delivered twice annually—allowing participants to harvest two irrigated hay crops per year. The PFRA retained responsibility for the maintenance of system infrastructure, including dams, reservoirs, ditches, and gates. As well, PFRA employees performed "ditch riding" functions—managing the canals, ditches, and gates that distributed water to each participant's plot.

As of 2010, approximately 300 producers had plots on the PFRA projects or on projects reliant on PFRA infrastructure—there are approximately 1,200 irrigators in the whole of the province (Warren 2013;

Map 1. Irrigation districts in the Palliser Triangle

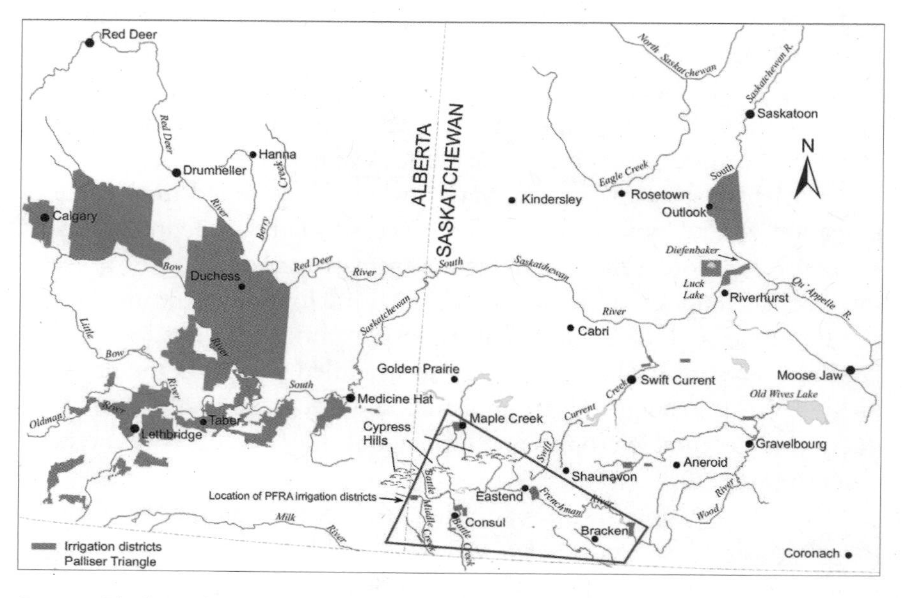

Sources: The boundaries of the Palliser Triangle were derived from Spry (1995). The location of irrigation districts in Saskatchewan were derived from Thraves et al. (2007: Plate 39). The location of irrigation districts in Alberta were derived from Alberta Agriculture and Rural Development (2013).

SIPA 2008a, 2008b; Saskatchewan AgriVision 2004). Individual plot sizes on the various projects supplied by PFRA infrastructure in the southwest range in size from as little as 20 acres to 320 acres.

Irrigation was a welcome development in the dry southwestern corner of Saskatchewan. During drought years, such as those experienced in the 1920s and 1930s, it had been virtually impossible to grow winter feed for the region's beef cattle herds using dryland methods. Not surprisingly, interest in irrigation was relatively high; there were more producers requesting access to water for irrigation than there was water to allocate. As of today, streamflows in southwestern Saskatchewan are deemed fully allocated, whereas only 11% of the water available for use in Lake Diefenbaker is being used (Warren 2013: 214). In the southwest, irrigation appeared to provide a welcome assurance that feed could be grown locally, even during especially dry years. And from the early 1950s until 1979, the projects generally met producers' expectations.

Map 1 locates the multiple-user irrigation projects and districts in the Palliser Triangle described above.

The Impact of Three Dry Decades

The year 1979 marked a major turning point in the operation of irrigation agriculture in the southwestern corner of Saskatchewan. It was the first in a succession of years extending to 2009 when the availability of water for irrigation became unreliable (RCAD 2012; Warren and Diaz 2012). It was the last year that irrigators on three of the largest projects in the southwest (the Consul, Vidora, and Eastend projects) could count on two full water allocations per year. PFRA managers determined that there simply was not enough water available in streams and reservoirs to flood fields twice per season. Making matters worse, in some years there was only enough water available to irrigate half the available land once per year on some of the projects. Worse yet, there were some years during the three-decade period when water supplies were too low to allow for any irrigation at all. The inability of producers to irrigate all of their land twice per year and the complete lack of water in some years had significant economic implications for the region's ranchers.

Comments provided by a rancher who relies on irrigated hay for overwintering her cattle reflect those of most RCAD interviewees who rely on PFRA irrigation infrastructure:

> When Cecil and I were first married, we used to get two full irrigations. And we've been married since 1977. Now we're lucky if we get to irrigate half our land once a year. Last year we had no irrigation at all and we had just a single half-irrigation during each of the four years prior to that. The problem is there's just been no water. Cypress Lake [reservoir] was drained down to where we couldn't pump from it last year. It used to have a lot of water but the weather has become drier and the lake doesn't provide adequate storage, because it is so shallow. There has been less snow and less runoff—we haven't had two full irrigations since 1979 . . . So the end

result of reduced irrigation is that we've been buying feed. And we're pretty well at the end of our rope with that option. You can't sustain a cow-calf and backgrounding [feeding calves over their first winter] operation down here if you are buying feed. The prices you pay for the feed and having it hauled don't match what you get at the market for your cattle . . . We probably spent $30,000 on feed last year. If there was any profit to be made with the cattle, that pretty well used it up. And we're not alone. I mean, everybody who counts on irrigation down here is using every strategy they can think of to make ends meet. They're trying to get the banks to increase their operating loans—just to try and get through to next year. (Warren and Diaz 2012: 126)

The principal culprit identified by irrigators and government water managers was a multi-decadal decline in the depth of the region's annual winter snowpack, which resulted in reduced spring runoff and streamflows. However, irrigators and water managers also identified system design and management flaws as contributing factors. By the mid-1980s, irrigators and PFRA officials had concluded that the development of additional reservoir capacity could solve much of the problem produced by drier conditions. During the 1980s, PFRA engineers worked on plans for three major infrastructure projects to increase water storage capacity. These included a new dam and reservoir on Battle Creek, a new dam at Cypress Lake Reservoir, and the enlargement of the Eastend Reservoir on the Frenchman River. As of 2013, the only project to proceed to completion was an enhancement of the Eastend Reservoir. However, that project fell short of the original design specifications, resulting in minimal water supply benefit for irrigators (Warren and Diaz 2012: 328).

The seriousness of the problem varies between projects and the various streams with which they are associated. RCAD respondents reported that a few individual irrigators along the Frenchman River were able to irrigate at least once per year every year between 1979 and 2010. At the other extreme, some irrigation systems on the south slope of the Cypress Hills rarely provided water for two irrigations over the three dry decades. One of the RCAD respondents with a privately owned and operated irrigation system described this situation:

> We have a 70-acre parcel of land . . . that is irrigated in theory. In the 20 years that Dad had it he only missed irrigating one year—just one year without water. After Lou and I took it over in 1974 there have only been about four years when we have had enough water to irrigate it . . . Yes, there were only four or five years out of 35 years that we got enough water to irrigate that piece. That is no kidding. That is how things have changed. There hasn't been any snowpack to speak of for a lot of winters. It is as simple as that. (Warren and Diaz 2012: 115)

Irrigators were also critical of ineffective management of streamflows on the part of the Saskatchewan Watershed Authority (SWA) and PFRA water managers. RCAD respondents reported that allocation management was haphazard. There were people who regularly irrigated without having an official allocation or having to pay for the water they used, while at the same time, there were irrigators who received no water at all despite making annual infrastructure upkeep payments (Warren and Diaz 2012: 135–37). Irrigators questioned the competence and capacity of senior water managers located in distant cities to supervise the water management decisions of PFRA employees located onsite at the irrigation projects. Producers also suspected that provincial and federal water managers were far more committed to meeting treaty obligations requiring that 50% of the flows on transboundary streams be available to the United States than they were to supporting Canadian irrigators (Warren and Diaz 2012: 26–30, 124–141, 322–30).

The Challenge of Uncertainty

Irrigators on the PFRA projects were presented with additional frustration and uncertainty in 2007 when the PFRA announced it intended to transfer responsibility for its irrigation projects to patrons effective 2017. Patrons were concerned that a number of infrastructure components were in need of upgrading and wondered if they could afford to make the necessary enhancements. They also wondered how they would be able to finance system improvements, such as the three new dams that had been planned by the PFRA in the 1980s and 1990s (Warren and Diaz 2012: 124–49, 322–30; PFRA 1992). In addition, the PFRA informed patrons that when they assumed ownership they would become legally responsible for any environmental cleanup associated with the projects that provincial or federal environmental authorities might require. This prospect was troubling given that many of the irrigation works built by the PFRA incorporated creosote-treated timbers. When the works were constructed, creosote was not deemed as environmentally harmful as it is today. Some irrigators worry that the cost of environmental rehabilitation could exceed the actual value of the existing works. Furthermore, the economic condition of the area's agricultural producers has been compromised by successive years when they have been required to purchase feed due to irrigation restrictions. Problems with the irrigation systems coincided with perennially increasing input costs and years of depressed cattle prices, including the price collapse associated with the 2003–7 bovine spongiform encephalopathy (BSE) crisis.

According to several RCAD respondents from Saskatchewan's southwest, insult was added to injury in 2010 when they discovered they would not be eligible for assistance under a federal-provincial drought relief program. The program was intended to assist producers affected by drought in parts of southeastern Alberta and southwest Saskatchewan in 2008–9. Apparently, government officials applied standard agricultural drought indices to determine which municipalities had suffered drought. As was discussed in Chapter 1 of this volume, agricultural drought is closely associated with low soil moisture conditions, especially at the time of seeding and early crop development. Apparently most dryland farmers in the southwest had received adequate moisture at the right time. Irrigation agriculture in southwest Saskatchewan, on the other hand, principally depends on accumulations of snow over winter and a well-timed runoff. Since hydrological conditions in the southwest did not enter into the program eligibility equation, irrigators did not receive drought support despite the fact that many of them were unable to irrigate in 2009.

An RCAD respondent commented on the apparent irrationality of program eligibility requirements and the frustrations felt by producers:

> I can tell you that we've had problems getting senior levels of government to recognize that we've been affected by drought in this area. This latest drought assistance program [2010 Canada-Saskatchewan Pasture Recovery Initiative] didn't include

> producers from RMs 51 and 111 [the RMs in the southwest corner of the province]. After four years of restricted irrigation you think we'd have been included within the drought disaster area. Going into this year they didn't think we were going to be able to irrigate anything again. We had no water whatsoever [in 2009] and still we weren't included. (Warren and Diaz 2012: 127)

This situation underlines the importance of applying the appropriate definitions of drought and understanding how drought affects producers using different production models. There is probably no one-size-fits-all model that will improve drought resilience in all contexts. Indeed, the economic success of agricultural producers using different production models on the Canadian Prairies is often at cross purposes. For example, when grain prices are higher than average, dryland farmers can generate higher-than-average gross incomes. However, higher grain prices often translate into lower calf prices for ranchers since the cost of finishing cattle to slaughter weight (by feeding them grain) increases. Similarly, there are occasions when dryland farmers may experience drought (perhaps due to low springtime precipitation), yet at the same time, irrigation agriculture can operate at near optimal levels. Again, reservoir levels often depend on factors such as winter snow accumulations or precipitation occurring outside the Palliser Triangle along the east slope of the Rockies. If dryland producers across a wide portion of the Palliser Triangle experience drought-induced crop failures due to low early growing season precipitation, there can be shortfalls in the supplies of commodities such as livestock forage and feed grain and a corresponding spike in prices. If irrigators are still able to water their crops during such a drought, they can take advantage of the associated price increases.

Chapters 13 and 14 in this volume, which discuss drought in Chile and Argentina, similarly describe how drought can sometimes produce both winners and losers depending on local conditions and the production models involved. In Chile, below-average moisture conditions can improve some fruit qualities desired by certain grape growers while causing harmful yield reductions for others. In Argentina, improved irrigation has mitigated the impact of drought for irrigators at higher elevations while at the same time causing adverse effects for downstream goat ranchers.

Another frustration for irrigators in the southwest is criticism of the agronomic and hydrological practices used on the projects. Government agrologists and water managers, as well as irrigators, operating on the larger districts associated with Lake Diefenbaker are generally critical of flood systems, seeing them as wasteful of water. Furthermore, the projects in the southwest had been situated on land that was suited to gravity-flow flood irrigation (the principal irrigation method available at the time) but did not necessarily have soils that were optimally suited to irrigation. Consequently, hay yields on the PFRA projects tend to be lower than yields obtained by irrigators in other areas where mechanical pivots apply water on better-suited soils. In conjunction with the widespread adoption of sprinkler pivot irrigation in most of western Canada in the 1970s and 1980s, agronomic best practices evolved to suggest water should be conveyed from its source to the most appropriate soils available.

Irrigators from the southwest have responded to critics of their flood systems and soil conditions:

> Sure, we don't have the best land in the world, but we're still getting a lot more production out of it than we would if it wasn't irrigated. We feed a lot of cows in the wintertime out of this project [Eastend project]. People count on it. Just think about how much dryland it would it take to replace all that irrigation production in a dry year . . . Right now, our project works off gravity. Mother Nature's doing all the work. So I wonder if they take that into account when they say flood irrigation isn't environmentally friendly. If we went to pivots we'd be using a whole lot of energy and that leaves a footprint too. Currently, the energy footprint for this system is about zero. I think that's a reasonable trade-off; we might be using more water than we would with pivots, but we're not consuming any electricity. (Warren and Diaz 2012: 325, 327)

One of the most daunting issues for irrigators on the PFRA projects is that the SWA has not yet agreed to transfer the water allocation currently awarded to the PFRA to the producers should they agree to assume ownership of the projects (Warren and Diaz 2012: 324). Producers face the prospect of taking over projects in need of costly infrastructure upgrades (and potential environmental cleanup costs), with no assurance that they will be allocated the water required to irrigate. Irrigators are uncertain about precisely why the allocation has not been guaranteed. Some suspect it is because provincial water managers are reluctant to endorse flood irrigation on suboptimal land. Others imagine that provincial authorities are reluctant to become involved because the province does not want to incur additional financial responsibilities. The uncertainty has led some producers to argue that the PFRA should simply buy out existing irrigators, compensating them for paying premium prices for irrigated land that the PFRA is no longer prepared to irrigate (Warren and Diaz 2012: 138–139).

The Return of Snow and Rain

Three decades of relatively dry years were followed by a year of record flooding in the Cypress Hills region in 2010. Indeed, since the spring of 2010, snowfall and runoff levels have increased, and area reservoirs are full. Unfortunately, the 2010 flood damaged irrigation works supplying irrigators on the northern slope of the Cypress Hills. A PFRA weir essential to the operation of the Maple Creek irrigation project washed out and, as of 2015, has not been repaired. No flood irrigation has occurred on the Maple Creek flats portion of the project since 2010, although above-average precipitation from 2011 to 2014 has allowed ranchers to produce hay using dryland methods. The PFRA was disbanded in 2010, and officials from other sections of Agriculture and Agri-Food Canada now have responsibility for the projects. Officials from Agriculture and Agri-Food Canada have not indicated whether the government will replace the weir prior to transfer of the system to the producers. Irrigators who offered to hire their own contractor to repair the weir were informed that this would not be allowed since Agriculture and Agri-Food Canada and Saskatchewan's watershed authority would still require that any repairs would have to be managed according to government engineering parameters, which the locals were apparently deemed unable to meet (SSFIG 2013). Yet the government agencies concerned have not, as of November 2015, offered to provide the necessary engineering support. This is an instance in which the irrigators endeavoured to develop their own self-reliant, innovative response to their circumstances but have been prevented from doing so by government.

The Changing Role of Governments

From the mid-1930s until just recently, Canada's federal government subsidized the development and maintenance of irrigation on the Canadian Prairies. That being said, the PFRA's longstanding involvement in irrigation in southwestern Saskatchewan was something of an anomaly. The management of major projects in Alberta, and the Lake Diefenbaker projects, which initially received considerable federal financial and engineering support, were turned over to provincial authorities and irrigator associations decades ago. As the level of financial and technical support available from the federal government waned, provincial government support was stepped up—particularly in Alberta. The Government of Alberta has entered into long-term funding agreements with irrigation district associations. Under the current agreement, the irrigation districts receive 50% or more of the funds required to upgrade system components from the province (Warren and Diaz 2012: 279; Saskatchewan AgriVision 2004: 10–11).

In Saskatchewan, support from the province has been less generous and less dependable. A number of prominent observers contend that the lack of consistent financial support from the province has retarded irrigation development in Saskatchewan, particularly in the Lake Diefenbaker area (SIPA 2008a, 2008b; Saskatchewan AgriVision 2004). On the other hand, as noted above, some observers attribute the slower than anticipated pace of irrigation development in the Lake Diefenbaker area to the fact that moisture conditions in that region generally allow for the production of acceptable crop yields under dryland methods.

Proponents of irrigation enhancement and expansion in Saskatchewan maintain that the economic benefits associated with increased crop production and the development of added-value food processing and livestock feeding associated with more densely concentrated irrigation agriculture far outweigh the initial investment in infrastructure made by governments (SIPA 2008a, 2008b; Saskatchewan AgriVision 2004). The Alberta government apparently agrees and continues to make significant investments in irrigation infrastructure. The high level of added-value processing and employment associated with Alberta's irrigation districts is interpreted as evidence of the economic multiplier effect that concentrated irrigation agriculture can generate. Saskatchewan's governments have behaved more erratically. Some governments have actively promoted the expansion of irrigation, only to be followed by new administrations that were less enthusiastic (SIPA 2008a: 4; Saskatchewan AgriVision 2004: 125).

Supporters of expanded irrigation in Saskatchewan, including the Saskatchewan Irrigation Projects Association (SIPA) and Saskatchewan AgriVision Corporation—an agri-business think tank—contend that increasing the amount of irrigated land in the province from the current 350,000 aces to 500,000 acres (with the increase occurring primarily in the Lake Diefenbaker area) would generate a benefit-cost ratio of 14:1. It is assumed that a government investment of $2.9 billion in new infrastructure would generate direct and indirect benefits totalling approximately $60 billion (SIPA 2008a: ii). SIPA contends that the multiplier effect of additional irrigation would generate tax revenues that would more than offset the government's investment.

A 1991 PFRA study by Kulshreshtha (1991) assessed the economic return on the federal government's investment in the irrigation projects in southwestern Saskatchewan and indicated that the projects have not generated the revenues required to fully offset costs. Kulshreshtha (1991: i) reports that even when ancillary benefits such as the use of water by urban municipalities and for recreation are taken into account, the projects have not paid for themselves. He identifies a benefit-cost ratio of 0.85. That being said, Kulshreshtha's analysis does not consider the possibility that low economic productivity on the projects could be significantly enhanced by additional investment in infrastructure. According to the irrigators, increasing reservoir capacity would have resulted in higher yields and revenues over the relatively dry 1970–2010 period. Notwithstanding the modest net deficit he identified, Kulshreshtha underlined the importance of the PFRA projects to the sustainability of cow-calf ranching and communities in southwestern Saskatchewan. This argument is echoed by participants in the projects.

The Impacts of Offloading on the Irrigation Community

The decision by the PFRA to abandon its irrigation responsibilities is just one example of a wider pattern of declining federal government support for Prairie agriculture and the offloading of responsibilities onto provincial governments and producers. In 2010, just three years after the PFRA announced that it was giving up its irrigation responsibilities, the PFRA itself was disbanded. And, in 2011, the federal government informed the provinces that it would cease operating the community pastures and tree nursery formerly managed by the PFRA. Federal officials have indicated that the provinces of Manitoba, Saskatchewan, and Alberta and/or pasture patrons (as is the case for irrigation project patrons) have the opportunity to operate the pastures if they wish but without financial support from the federal government. A concern for pasture patrons and irrigators in Saskatchewan is that the provincial government has been reluctant to assume responsibility for the ongoing operations of the pastures and irrigation projects. This reluctance is evidenced by the SWA's (since renamed Saskatchewan Water Security Agency) failure to promise that the water allocations will be awarded to the patrons.

Irrigation project patrons are worried that without greater government support the projects could cease to operate. The loss of access to irrigation is a daunting prospect for the irrigators who rely on it for winter feed. It could also accelerate the decline of the few urban communities that survive in the southwestern corner of Saskatchewan. One of the irrigators interviewed by RCAD project researchers expressed concern over the potential loss of access to irrigation:

> Without irrigation many of us would simply not be able to survive as cow-calf ranchers. Maybe we could survive by switching to straight grazing operations. Sell off our cows and run calves raised by someone else on our land as yearlings but that would involve a big reduction in a rancher's income. Irrigation is one of the few things that has kept Consul [the only remaining village in Rural Municipality #51] going... I don't know how they expect us to survive down here. There aren't

> that many of us left and if we can't irrigate there will be even fewer of us. (Warren 2013: 241)

Journalist Sheri Monk, who frequently writes on issues affecting agriculture in southwestern Saskatchewan, recently posted the following comments regarding the PFRA's demise:

> Whether intentional or merely the inevitable result of catastrophic policy decisions, the area [southwestern Saskatchewan] is being depopulated. Piece by piece, all the pillars of economic sustainability are being removed. Sure the federal government may be motivated by their economic ideology, but it's the people who are going to suffer for it. Even the staunchest libertarians will admit that maybe it wasn't the government's place so many decades ago to create the framework and infrastructure for the PFRA projects, but now that it's here, ripping it away from the people who have built generations of lives around it is criminal. (Monk 2013)

Some government water managers suggest that the PFRA's operation of irrigation projects in southwestern Saskatchewan resulted in counterproductive dependency and complacency on the part of patrons. For example, patrons did not incorporate their own district irrigation associations until after the PFRA announced its plans to transfer the projects. However, there are irrigators participating in several multiple-user projects in the southwest who have been operating without PFRA management and operational support for decades (these systems are referred to as provincial projects). The producers on provincial projects have always had their own district associations, and with the exception of some major works such as dams and reservoirs (which are managed by the province or PFRA), they look after the full cost of system maintenance and operations. It was not until 2008 that irrigators on the PFRA projects hired their own "ditch riders" (the technicians who manage water distribution on the projects).

Federal officials have reported that the government has been operating the projects in the southwest at a loss (Warren and Diaz 2012: 167). The fees that producers are charged for water and system maintenance do not cover actual costs. Irrigators counter that costs incurred by the PFRA include unnecessarily high head-office staff costs, redundant local employees, and gold-plated engineering and construction costs. Furthermore, the inconsistent delivery of water and the relatively low yields achieved on some projects warrant fees that are somewhat lower than those paid by irrigators on more reliable projects. Indeed, some irrigators interviewed held that some of the PFRA's charges were inordinately excessive. One respondent noted that the PFRA charged participants on the Middle Fork project a system maintenance fee in years when no water was available for irrigation, and no one from the PFRA appeared to have even visited the project over the course of those years (Warren and Diaz 2012: 144, 145).

Some producers speculate that if patrons were required to invest more of their own money in system improvements, they might recognize the value of making yield-increasing improvements on their plots. Indeed, PFRA and SWA officials, as well as some producers interviewed in association with the RCAD project, held that many patrons on the PFRA projects were not regularly renewing hay stands (by reseeding) or applying fertilizer in conformity with widely recognized best management practices.

Notwithstanding the potential benefits of increased producer investment, patrons find themselves locked into a classic "Catch-22" scenario. Given decades of low yields and restricted irrigation, their bottom lines have been stressed. Yields and water availability would probably improve if they invested in new infrastructure and more intensive plot management. But given their experience under existing economic and hydrological conditions, they lack the financial resources required to improve their situation. For example, the construction of a dam across the Cypress Lake Reservoir would allow for deeper, more drought-resistant water containment and would most likely allow for more efficient conveyance of reservoir water to the Consul, Vidora, and Govenlock irrigation projects (Warren and Diaz 2012: 30). Current cost estimates run at up to $4 million. Shared among the approximately 100 irrigators on these projects, the investment per irrigator would be $40,000. Given that some patrons are incurring significant costs to purchase feed when irrigation is restricted (over $30,000 annually for some producers), $40,000 to ensure more regular irrigation might appear to be a good investment. However, this conclusion assumes that producers have access to the required capital. Based on the RCAD interviews, it is apparent that some lack either the savings or access to credit that would be required. And borrowing money to enhance irrigation infrastructure in the absence of a guaranteed water allocation is something that both lenders and borrowers would no doubt finding troubling.

Conclusions

The research suggests a number of preconditions need to be in effect before significant system improvements can be entertained. First, patrons need assurance that the water allocations currently held by Agriculture and Agri-Food Canada will be transferred to the district associations. Second, the transfer of assets to the irrigators should be free of pre-existing environmental cleanup liabilities. If the federal government is prepared to consider its investment in infrastructure a sunk cost, it seems reasonable to treat environmental costs similarly. Third, to make the sort of infrastructure improvements that could enhance the drought resilience of the projects, the producer associations will require access to government grants and/or the sort of patient financing that would forego significant upfront cash contributions by producers (many simply do not have the cash or borrowing capacity today), allowing for the repayment of loans over an extended period of time. However, even with these measures in place, uncertainty about future climate conditions presents planning challenges. Some irrigators see value in "doubling down" on past investments in infrastructure by increasing reservoir capacity in anticipation of the next dry period. But, should future climate conditions exceed past patterns of variability, it is possible that enhanced reservoir capacity could still prove insufficient (see Chapter 3 by Wheaton et al. in this volume). Under this sort of scenario, making additional investments in infrastructure would be a costly mistake.

At the same time, some proponents of expanded irrigation in Saskatchewan take a more optimistic view, speculating that climate change could bring a warmer and longer growing season to the Canadian Prairies. Under irrigation, such conditions could facilitate production of higher-value crops such as corn, sugar beets, and soybeans, which are not particularly well suited to current climate conditions in Saskatchewan (SIPA 2008a, 2008b; Saskatchewan AgriVision 2004). This line of thinking underlines the idea that a changing climate could generate beneficial as well as adverse outcomes depending on the social and geographical context being affected—a notion that Hadarits et al. touch on in Chapter 13 of this volume.

What we are reasonably certain about is that climate change forecasts currently indicate that the Palliser Triangle region will experience more intense droughts in coming decades than have been experienced over the course of the twentieth century (Sauchyn 2010; St. Jacques et al. 2010; Sauchyn and Kulshreshtha 2008; Lemmen et al. 1997; see also Chapter 3 by Wheaton et al. in this volume). Notwithstanding forecasts based on anthropogenic global warming scenarios, paleoclimatic research suggests that the droughts experienced in the Palliser Triangle during the period of agricultural settlement (from approximately 1885 until today) were a virtual walk in the park compared to some of the severe, decades-long droughts of preceding centuries (Sauchyn 2010: 35–37; see also Chapter 2 by Sauchyn and Kerr in this volume).

Planning for the future of irrigation in southwestern Saskatchewan would clearly benefit from additional climate change research that reduces the level of uncertainty (RCAD 2012: 48). Nonetheless, some features of the planning problem seem reasonably certain. For example, most observers assume it is relatively safe to predict that severe multi-year droughts will occur over the course of coming decades, making irrigation both more necessary and more difficult (especially given current infrastructure limitations). It is also reasonable to predict that intense precipitation events, such as the Maple Creek flood of 2010, could reoccur over coming decades. Infrastructure needs to be designed and/or modified accordingly (see Chapter 3 by Wheaton et al. in this volume). Academic assessments of the vulnerability of communities to climate change indicate that resilience is a function of the levels of adaptive capacity available to a given community (Department for International Development 2009; IPCC 2001). The RCAD project found that agricultural communities in the Palliser Triangle had considerable access to certain forms of adaptive capital—forms of capital that tend to be far less accessible to people in less developed parts of the world. Notwithstanding these regional advantages, irrigators in the southwestern corner of Saskatchewan are experiencing an increase in vulnerability. Their coping capacity has been impacted along three principal dimensions. First, changing climate conditions have frustrated their ability to irrigate. Second, they have experienced a decades-long economic struggle, whereby increases in the income they receive for the products they produce have often lagged behind increases in input costs. Indeed, difficult economic conditions in agriculture have contributed to a significant decline in the number of farmers and viable communities in southwestern Saskatchewan (Diaz et al. 2003; Stabler and Olfert 2002). The Rural Municipality of Reno, where three of the PFRA irrigation projects are located, is the largest rural municipality in Saskatchewan, yet it has only 154 farms, down from over 300 in the 1970s (RCAD 2012: 52–53). Long-term trends have more recently been exacerbated by drought and the BSE crisis. These challenges have reduced the amount of capital available to producers for withstanding losses caused by climate hazards and for investing in resilience-enhancing infrastructure. Third, Canada's federal government has walked away from its longstanding commitment in support of irrigation in southwestern Saskatchewan, and the provincial government is apparently reluctant to assume responsibility for the functions abandoned by Ottawa.

The decline in government involvement in management and financial support is symptomatic of a wider process of offloading on the part of Canada's federal government. Under Prime Ministers Jean Chrétien, Paul Martin, and Stephen Harper, the role of the federal government in supporting agriculture on the Canadian Prairies has significantly declined (Warren 2013; Conway 2006; Diaz et al. 2003). Conway (2006), among others, contends that Canadian federal and provincial governments, including New Democratic Party governments in Saskatchewan, have increasingly become associated with neo-liberal economic maxims since the 1980s (see also Brown et al. 1999). A symptom of the neo-liberal turn of governments is the wide acceptance of balanced budget orthodoxy, low taxes, and minimal government, which has limited government's willingness to fund new initiatives and has encouraged cost cutting across a range of programs, including the decommissioning of the PFRA in 2013 and the elimination of the Canadian Wheat Board's marketing monopoly in 2012 (see Chapter 7 by Fletcher and Knuttila in this volume).

Oddly enough, in supposedly conservative Alberta, the provincial government has responded to declining federal government support, ensuring that the province's irrigation sector remains viable. Governments in Saskatchewan have typically been far less active in supporting irrigation agriculture (SIPA 2008a; Saskatchewan AgriVision 2004). Without stronger support from the province and/or the federal government, it is questionable whether irrigation projects in southwest Saskatchewan can survive, let alone prosper.

References

Alberta Agriculture and Rural Development. 2013. "Map of Irrigation Districts in Alberta." Alberta Agriculture and Rural Development website. http://www1.agric.gov.ab.ca/$department/deptdocs.nsf/all/irr12911. Accessed 21 November 2013.

Archer, J. 1980. *Saskatchewan: A History*. Saskatoon: Western Producer Prairie Books.

Brown, L., J.K. Roberts, and J. Warnock. 1999. *Saskatchewan Politics from Left to Right '44–'99*. Regina: Hinterland Press.

Brownsey, K. 2008. "Enough for Everyone: Policy Fragmentation and Water Institutions in Alberta." Pp. 136–55 in M. Sproule-Jones, C. Johns, and T. Heinmiller (eds.), *Canadian Water Politics: Conflicts and Institutions*. Montréal, QC, and Kingston, ON: McGill-Queens University Press.

Conway, J. 2006. *The West: The History of a Region in Confederation*. Toronto: James Lorimer and Company.

Department for International Development (UK). 2009. "Sustainable Livelihoods Framework." In *Livelihoods Connect: Sustainable Livelihoods Guidance Sheets from DFID*. London: UK Department for International Development.

Diaz, H., J. Jaffe, and R. Stirling. 2003. *Farm Communities at the Crossroads: Challenge and Resistance*. Regina: Canadian Plains Research Center Press.

Gray, J. 1967. *Men Against the Desert*. Saskatoon: Western Producer Prairie Books.

Herriot, T. 2000. *River in a Dry Land: A Prairie Passage*. Toronto: Stoddart.

Hill, D. 1996. *A History of Engineering in Classical and Medieval Times*. New York: Routledge.

IACC (Institutional Adaptation to Climate Change) Project. 2009. *Integration Report: The Case of the South Saskatchewan River Basin, Canada*. Regina: Canadian Plains Research Center Press.

IPCC (Intergovernmental Panel on Climate Change). 2001. *Climate Change 2001: Impacts, Adaptation, and Vulnerability. Contribution of Working Group II to the Third Assessment Report of the Intergovernmental Panel on Climate Change*. Cambridge: Cambridge University Press.

Kulshreshtha, S. 1991. *Economic Assessment of PFRA Irrigation Projects and Reservoirs in Southwest Saskatchewan*. Regina: Prairie Farm Rehabilitation Administration.

Lemmen, D.S., R.E. Vance, S.A. Wolfe, and W.M. Last. 1997. "Impacts of Future Climate Change on the Southern Canadian Prairies: A Paleoenvironmental Perspective." *Geoscience Canada* 24, no. 3: 121–33.

Monk, S. 2013. "How the Cookie Crumbles." Sheri Monk (blog), 25 July 2013. http://www.sherimonk.com. Accessed 26 July 2013.

PFRA (Prairie Farm Rehabilitation Administration). 1992. *A Summary of the Draft Report: An Initial Environmental Evaluation of the Battle Creek Reservoir Proposal.* Regina: PFRA.

RCAD (Rural Communities Adaptation to Drought Project). 2012. *RCAD Research Report August 2012.* Regina: Canadian Plains Research Center Press.

Saskatchewan AgriVision Corporation Inc. 2004. *Water Wealth: A Fifty Year Water Development Plan for Saskatchewan.* Saskatoon, SK: Saskatchewan AgriVision Corporation.

Sauchyn, D. 2010. "Prairie Climate Trends and Variability." In D. Sauchyn, H.P. Diaz, and S. Kulshreshtha (eds.), *The New Normal: The Canadian Prairies in a Changing Climate.* Regina: Canadian Plains Research Center Press.

Sauchyn, D., and S. Kulshreshtha. 2008. "Prairies." In D.S. Lemmen, F.J. Warren, J. Lacroix, and E. Bush (eds.), *From Impacts to Adaptation: Canada in a Changing Climate 2007.* Ottawa: Government of Canada.

SIPA (Saskatchewan Irrigation Projects Association). 2000. "Irrigation Pioneers: From Small Beginnings." *SIPA News* (November): 8–9.

———. 2008a. *A Time to Irrigate Volume I: The Economic Social and Environmental Benefits of Expanding Irrigation in the Lake Diefenbaker Region.* Outlook, SK: SIPA.

———. 2008b. A *Time to Irrigate Volume II: The Economic Social and Environmental Benefits of Expanding Irrigation in Saskatchewan.* Outlook, SK: SIPA.

Spry, I. 1995. *The Palliser Expedition: The Dramatic Story of Western Canadian Exploration 1857–1860.* Saskatoon: Fifth House.

SSFIG (Southwest Saskatchewan Flood Irrigation Groups). 2013. "Southwest Saskatchewan Irrigation Projects Divestiture Process Stalled." *SSFIG Newsletter* (August): 1.

St. Jacques, J.M., D. Sauchyn, and Y. Zhao. 2010. "Northern Rocky Mountain Stream Flow Records: Global Warming Trends, Human Impacts or Natural Variability." *Geophysical Research Letters* 37 (March 26): 1–5.

Stabler, J., and R. Olfert. 2002. *Saskatchewan's Communities in the 21st Century: From Places to Regions.* Regina: Canadian Plains Research Center Press.

Suderman, D. 1966. "Unpopular Oasis." *Family Herald* (July 14): 17–18.

Thraves, B., M.L. Lewry, J. Dale, and H. Schlichtmann (eds.). 2007. *Saskatchewan: Geographic Perspectives.* Regina: Canadian Plains Research Center Press.

Warren, J. 2013. "Rural Water Governance in the Saskatchewan Portion of the Palliser Triangle: An Assessment of the Applicability of the Predominant Paradigms." PhD dissertation, University of Regina.

Warren, J., and H. Diaz. 2012. *Defying Palliser: Stories of Resilience from the Driest Region of the Canadian Prairies.* Regina: Canadian Plains Research Center Press.

CHAPTER 7

GENDERING CHANGE: CANADIAN FARM WOMEN RESPOND TO DROUGHT

Amber J. Fletcher and Erin Knuttila

Climate change is one of the most profound environmental, political, and social issues of our era, particularly given its global scope and direct regional impacts. The threats and vulnerabilities associated with climate change and climate extremes, including drought, are not gender-neutral, and actors' everyday responses to these events can both challenge and reinforce existing gender roles and ideologies. As climate change scenarios become a reality, residents of the Canadian Prairies can expect more dramatic climate extremes—particularly severe, prolonged drought. There is a need for context-specific analyses of gender and drought, which can inform effective and gender-attentive strategies for preparedness and response.

In this chapter, we present a contextualized analysis of gender and drought in the Canadian Prairie province of Saskatchewan. Such contextualization is important because vulnerability and adaptation are shaped by much more than climatological factors. Social, political, and economic stressors interact to create unique situations of vulnerability to climate extremes. We begin our analysis by briefly situating farmers' vulnerability and adaptive strategies within the broader macro-level political-economic context of Prairie agriculture. Next, drawing on a recent qualitative study of farm women in Saskatchewan, we present a micro-level analysis that reveals the gendered dynamics of vulnerability and adaptation in everyday life. Drawing together both levels of analysis, we illustrate that particular political, economic, and social conditions have resulted in unique forms of vulnerability and adaptation on the Canadian Prairies. We suggest several policy implications to strengthen the adaptive capacity of Canadian farm families in the future.

Gendering Climate Extremes

Social factors can significantly shape how humans are affected by climate extremes (Adger 2003; see Chapter 1 by Wandel et al. in this volume) and how they respond to these extremes. Kelly and Adger (2000: 325–26), for example, noted the importance of examining the social dimensions of climate events:

> Climate impact studies have tended to focus on direct physical, chemical or biological effects, yet a full assessment of consequences for human well-being clearly requires evaluation of the manner in which society is likely to respond through the deployment of coping strategies and measures which promote recovery and, in the longer-term, adaptation.

The Intergovernmental Panel on Climate Change (IPCC) has stated that social inequality based on gender, race, age, socio-economic status, and ability can determine a person or system's vulnerability to climate extremes (Field et al. 2012). Disasters can exacerbate existing inequalities in society (Enarson et al. 2007), resulting in different levels of vulnerability and unequal access to resources before, during, and after the climate event. Women have historically played integral roles in food preparation, childcare, and healthcare—roles that are critical during disaster situations but that become more difficult to carry out during such events (Enarson and Chakrabarti 2009). Entrenched gender roles can therefore create different experiences of disaster for women than for men (Dankelman 2010; Enarson and Chakrabarti 2009).

At the same time, some feminist scholars have observed a tendency in the gender and climate change literature to portray "women" and "men" as homogeneous categories while ignoring differences caused by race, socio-economic class, geography, ability, and education (Enarson et al. 2007). Moosa and Tuana (2014) documented the growing importance of intersectional and contextual studies that examine how gender interacts with other forms of social difference, such as socio-economic class or rurality, to create different experiences of climate extremes even within the social categories of "women" and "men." Arora-Jonsson (2011) disputed sweeping and universal statements about the vulnerability of women, calling instead for more contextualized analyses of gender and climate change that address local gender roles and ideologies in specific locations. This includes situations where hegemonic masculinity may render men vulnerable to climate extremes.

Few academic studies have been conducted specifically on gender and drought on the Canadian Prairies. However, the experiences of farm women in the region provide an important window into the gendered dimensions of drought. Because they depend on the land for their livelihoods, farmers are directly and dramatically affected by drought and other climate extremes. Many family farms in the Prairies are structured by a gendered division of labour, in which men are more likely to be positioned as the "main farmer" while farm women's work is construed as "helping" (Fletcher 2013; Faye 2006). In addition to farm work, women are primarily responsible for childcare and other caregiving work (Jaffe and Blakley 1999), household tasks such as cooking and cleaning, and yard work (Fletcher 2013; Kubik and Moore 2005). Through their association with social reproduction tasks and their relative detachment from day-to-day farm decisions (Fletcher 2013; Reinsch 2009), farm women may experience drought disasters differently than men.

In her research on Manitoba farm women's experiences of the bovine spongiform encephalopathy (BSE) crisis, which occurred only one year after the drought of 2001–2, Reinsch (2009) found that farm women experienced high levels of stress as a result of the disaster. The women's stress was due, in part, to their lack of control over major farm decisions and coping strategies (Reinsch 2009). Historical sources, including farm women's own accounts of past droughts, also provide some insight into the gendered dimensions of drought in a historical framework. In an analysis of her great-grandmother's letters from the 1930s, Bye (2005) argued that farm women actively reinforced gendered roles and ideologies during a drought, which had the effect of reproducing gendered inequalities and, therefore, gendered vulnerability. Other authors have documented the importance of farm women's adaptive strategies during the Great Depression, including household resource management strategies, subsistence food production, and off-farm work (Gilbert and McLeman 2010; Schwieder and Fink 1988). In a recent article based on interviews with environmental migrants during the 1930s drought, Laforge and McLeman (2013) suggested that women may have experienced drought-related migration differently than men and that women may have experienced increased isolation due to gender roles that limited their social interaction.

The literature suggests that while women and men can be similarly exposed to the same climate conditions, contextual differences—including gendered divisions of labour—can produce different degrees of sensitivity and different forms of adaptation (Leichenko and O'Brien 2008; Milne 2005). Climate scenarios indicate that the Canadian Prairies will face significant water scarcity as the result of warmer and drier weather and that the region will be exposed to extreme droughts of long duration in the future (Sauchyn et al. 2010; Sauchyn and Kulshreshtha 2008). Programs to reduce vulnerability and encourage adaptation are necessary, but they are only helpful if they are attentive to local social and gender orders. Programs should not exacerbate existing forms of inequality in the community and should, when possible, challenge these inequalities. It is necessary to understand the gendered dimensions of extreme events to create culturally appropriate and gender-attentive approaches to future climate extremes.

Drought on the Canadian Prairies

The Canadian Prairies have the most variable and drought-prone climate in Canada (Sauchyn 2010; Bonsal and Regier 2007; see also Chapter 8 by Marchildon in this volume), yet the region is also one of Canada's key agricultural areas. The province of Saskatchewan, for example, contains 40% of Canada's farmland and exports more than half of the world's lentils, peas, and flaxseed (Government of Saskatchewan 2012a). The region also produces over 30% of the durum, canola seed, and mustard consumed worldwide (Government of Saskatchewan 2012a). The neighbouring Prairie province of Alberta is known for its cattle industry, producing 40% of Canadian beef cattle (Statistics Canada 2011a).

Over the past decade, this important agricultural region has experienced a series of droughts, most notably in the 1930s, 1960s, 1980s, and early 2000s, and most recently in 2009 (Warren and Diaz 2012; Marchildon et al. 2008; Bonsal and Regier 2007; see also Chapter 4 by Kulshreshtha et al. and Chapter 8 by Marchildon in this volume). Dendroclimatic records indicate that even worse droughts occurred before European settlement of the Prairies and thus before instrumental recording began (Sauchyn et al. 2003; see also Chapter 2 by Sauchyn and Kerr in this volume), which suggests that similar extreme droughts could potentially reoccur in the future.

The social and economic impacts of drought on agriculture can be dramatic. Saskatchewan farmers sustained crop production losses of $925 million in 2001 and $1.49 billion in 2002, and the province reported negative net farm income in 2002 (Wheaton et al. 2008). Farmers relied heavily on crop insurance to cope. Insurance payments in Saskatchewan jumped from $331 million in 2001 to $1.1 billion as the drought continued into 2002 (Wheaton et al. 2005).

Climate scientists predict more dramatic changes for the Prairie region in the future. More severe and protracted droughts are expected as overall temperatures continue to rise (Sushama et al. 2010; Bonsal and Regier 2007; see also Chapter 3 by Wheaton et al. in this volume), but the region will also experience more fluctuations and a greater range of extreme climate events as anthropogenic climate change interacts with natural cycles (Sauchyn 2010: 38; Sauchyn and Kulshreshtha 2008). These events will test residents' abilities to cope and adapt.[1]

There is a need for context-specific analyses that highlight the unique forms of vulnerability and adaptation at play in certain locations. Such localized understandings can facilitate policies that are attentive to the strengths and needs of actors in unique circumstances. In the following section, we provide a contextualized analysis of the gendered dimensions of vulnerability and adaptation on the Canadian Prairies. We present the results of a qualitative research project conducted with 30 Saskatchewan farm and ranch women between August and December 2011. The project involved 30 semi-structured interviews, most of which occurred at the participants' farms. For a detailed description of the methods and participant demographics, see Fletcher (2013).

The study's findings reveal the importance of the broader political, social, and economic context shaping farm women's lives, as well as the uniquely gendered dynamics affecting both vulnerability and adaptation. The following sections discuss both of these contexts, macro and micro. Based on these findings, we present recommendations for gender-attentive policies that should benefit Prairie farmers facing climate extremes in the future.

Managing Uncertainty: Vulnerability, Adaptation, and Gender on Saskatchewan Farms

Context: The Changing Face of Prairie Agriculture

Farm women's experiences of climate change must be understood within the broader political and economic context of Prairie agriculture. Vulnerability is not simply a product of climatic factors (see Chapter 1 by Wandel et al. in this volume). Factors such as market prices, input costs, policies, and population trends can increase or decrease farmers' access to much-needed resources, and this subsequently shapes their vulnerability and adaptive capacity in the face of climate extremes.

Farmers have been "squeezed" between high production costs and low commodity prices since the early days of agricultural settlement in the Prairies (McCrorie 1964; Fowke 1957). However, contemporary farmers face a new kind of cost-price squeeze. Large, vertically integrated agricultural corporations have become dominant forces in multiple links of the food chain, from the production of patented seed varieties to processing and export (Fletcher 2013; Kuyek 2007). At the same time, deregulatory policy changes and the elimination of farm support programs have increased both farmers' costs and their susceptibility to the vagaries of international market prices (Roppel et al. 2006).

Farmers have adapted to these macroeconomic changes through farm-based economies of scale and increased production. Farm size growth has reached new heights in recent years, as Prairie farms grow at the fastest rate in history (Figure 1). However, we cannot assume that small or lower-income farming operations are more or less vulnerable to climate

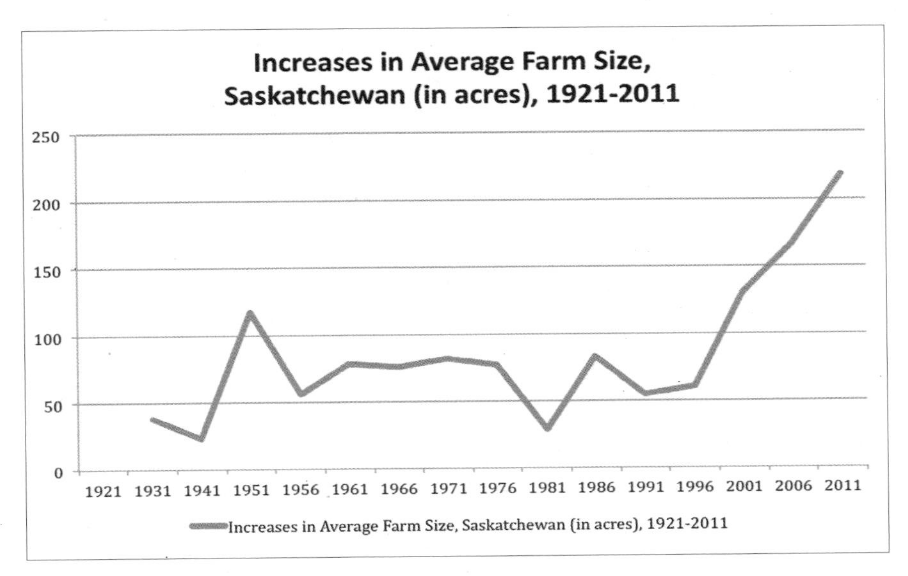

Figure 1. Increase in average farm size in Saskatchewan (in acres), 1921–2011 (Source: Statistics Canada 2011a)

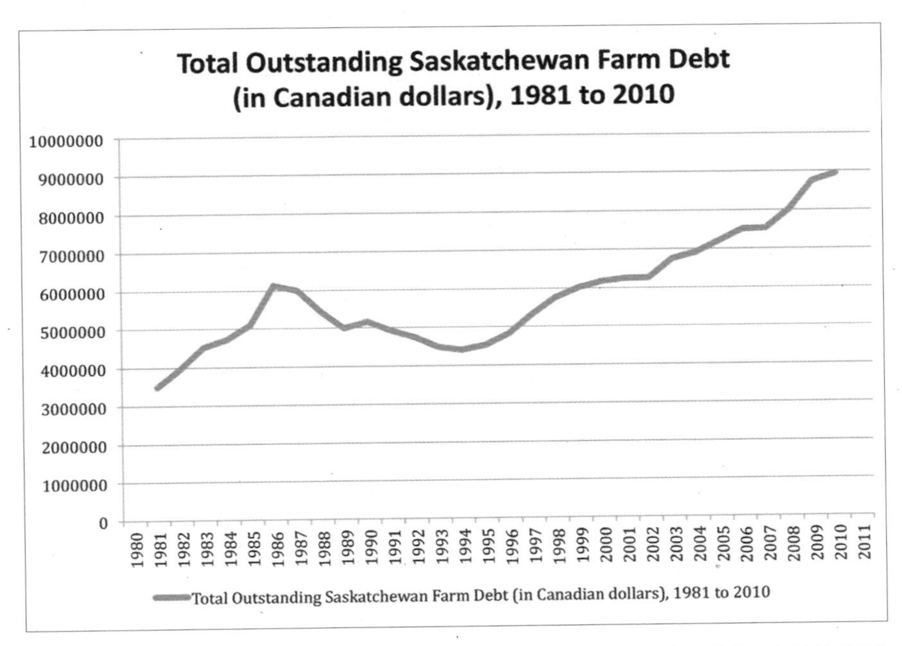

Figure 2. Total outstanding Saskatchewan farm debt (in Canadian dollars), 1981–2010 (Source: Statistics Canada 2011b)

extremes than large, "successful" farms. In fact, the debt levels associated with farm expansion, as well as the cost of expensive inputs intended to enhance productivity, can leave larger or highly industrialized farms even more financially precarious and vulnerable to climate disasters (Figure 2).

Farm women in the study identified debt and high input costs as having "make or break" power during extreme weather events. One participant said the following:

> If you've put it all in your land and you don't get a profit back, every year that you lose is a year that you don't get back. It takes you longer to regain what you've lost. So, I know those years of drought, yes, that's a farm crisis: when you have nothing to sell but your bills are still coming in. (Fletcher 2013: Interview 20)

The participant also added, "When it's a drought, of course your income's down but your expenses still stay the same: the price of fuel, the price of repairs, the price of everything" (Fletcher 2013: Interview 20).

Another participant described the challenges for farmers who had expanded their operations through increased debt: "We've got men in their early 20s that are in debt $2.5 million, but they're one of the biggest farmers in the area" (Fletcher 2013: Interview 7). For some, decreasing their debt was an adaptive strategy to prepare for crises: "So we're small but at least we own it, we don't owe any money, we're not in debt like a lot. Some, they go big and they're fine, but some go too big and they crash" (Fletcher 2013: Interview 17).

Many participants relied on insurance and government disaster programs in times of environmental crisis. Despite some concerns about the administration of these programs, such as the length of time to receive a payment, participants generally saw the programs as important and necessary. However, insurance is a viable coping mechanism only if it remains affordable for farmers. Many participants expressed concern that current insurance programs were not keeping pace with the rising cost of inputs. Others contemplated the future cost of insurance in a changing climate: "I think that crop insurance is possibly going to get more expensive. It may, it just may not, but I also think it will get more expensive just because of the increased variability in weather. It's an insurance program and if your weather gets weird it's going to get more expensive" (Fletcher 2013: Interview 8).

Drought also affects farmers' workloads, although the effects are different for cattle and crop producers. Dry years create more work for cattle producers, as they are forced to pump or haul water for the herd. Women are often involved with hauling water for the farm or household during a drought. One participant, who was the sole farmer on her operation, relied on her daughter for assistance with water collection: "When you have cattle, it causes more work because there was no water. We had to water them. [Daughter] was the water girl. Five-hundred gallons a day, every day" (Fletcher 2013: Interview 23). Drought tended to have the opposite effect on grain and oilseed producers, who often found their workload reduced. When crops did not grow, there was simply nothing to be done.

In the case of both cattle and crop producers, the most dramatic consequences of climate extremes were felt internally. Stress was the most commonly mentioned issue in discussions about drought and other climate extremes. As we discuss below, vulnerability and adaptation to these psychological effects takes gendered forms.

The Gendered Dimensions of Vulnerability and Adaptation to Drought

Existing research has documented the historical invisibility and marginalization of North American farm women's contributions to agriculture (e.g., Fletcher 2013; Faye 2006; Kubik and Moore 2005; Kubik 2005, 2004; Rosenfeld 1985; Sachs 1983; Ireland 1983; Koskie 1982). This invisibility persists despite the importance of women's work and despite their rising participation in activities often considered masculine, such as driving large machinery (Martz 2006). The lack of recognition is mostly due to the persistent notion that farming is a "man's job." Participants in the study were asked to name all job titles they identified with. Despite the fact that "farmer" was selected most often (n = 14), it was very common for participants to identify their male partners as the "main" or "primary" farmer while describing their own role as that of "helper," "employee," or "go-for."

With this "helper" identity comes a relative lack of control over the day-to day farm decisions, which are often made by men. As one participant stated, "I think farm women tend to be more supportive, rather than the decision makers. I think, as far as me personally, I'm the sounding board. I do lend some opinions that alter the end decision, but the end decision is generally [husband's]" (Fletcher 2013: Interview 15). Another participant described the phenomenon this way:

> The division of farm work is: he's in charge. He does everything with the farm work, except when he needs me to help him fix something, hold a part, or put it this way—he chooses what he has to do to make money on the farm and I do everything else that he doesn't want to do. (Fletcher 2013: Interview 4)

The farm household is also structured by rigid gender roles. The women in the study performed an average of 88% of all domestic, household, and caregiving work; this is 20% more than the national average for women according to Statistics Canada (Milan et al. 2011). The gendered division of labour is partly due to the concrete realities of farm work. Farming is not a "9 to 5" job, and farmers may work from 4 or 5 a.m. until midnight during busy seasons, such as harvest time. Many rural areas in Saskatchewan lack childcare services, and even if these exist, few service providers can accommodate farming schedules. These material realities combine with historically ingrained gender ideologies that position women as "natural" caregivers and men as providers for the family.

Farm women's roles are thus structured at the confluence of both ideological and material factors. These gendered roles result in different experiences of environmental crisis for farm women and farm men. Only three participants felt that the response to a climate event depends on individual personality and is not gendered; the remainder felt that gender roles make a difference. In contrast to Reinsch's (2009) findings, several participants described how men's closeness to the farm causes them to be more negatively affected by the psychological impacts of climate extremes. Men's typical position as the "main" farmer is a privileged position when times are good, but it can increase their personal vulnerability during a drought. It is they, the "main" farmers, who watch closely as crops wither or livestock suffer. As one farm woman explained, "I'm not in contact with it 24/7 like they [her husband and sons] are. It affects their appetite. It affects their outlook for the next day. They don't rest properly, you know. It's just, it's a battle" (Fletcher 2013: Interview 29).

Farm men's psychological distress is also caused by dominant ideals of masculinity and, in particular, a stoic and independent form of masculinity commonly found in Prairie agricultural communities. Participants described the gendered expectations placed on men to be "providers" and the vulnerability this can cause in times of crisis. One said, "I think when you're the man of the household, ultimately it's your responsibility no matter how much you're supported by your wife and how much she helps, ultimately . . . it sits on your shoulders" (Fletcher 2013: Interview 19).

It should not be assumed that these differences are somehow natural or inherent to men and women. Different forms of vulnerability are the product of entrenched gender ideologies and roles. This is clearly shown in the case of one farm woman who was the main farmer on her operation while her husband worked full-time off the farm. As the main farmer, she was more severely affected by the mental turmoil of a multi-year drought. Her words illustrate not only this mental turmoil but also the interaction of financial and climatological factors in shaping vulnerability:

> It's not the weather itself, it's what the weather does to the bottom line . . . I find it way, way too stressful . . . the financial part is what stresses me . . . I think there were two or three years in there, probably three years there, where I was on anti-depressants. (Fletcher 2013: Interview 23)

More commonly, environmental crises further entrench farm women's historical role as a caregiver for the farm family and community. Participants often used terms like "nurturer," "mediator," "buffer," and even "counsellor" to describe their role during a drought. One farm woman described this role:

> We try and keep everybody on the level and you know, don't try to irritate them. Try and keep it a peaceful atmosphere, like 'maybe tomorrow will be better,' 'next year will be better,' or . . . 'we can deal with this.' . . . You have to take the role of a matriarch kind of thing, you know? I don't know if that's a good word, but . . . try and be the buffer I guess. (Fletcher 2013: Interview 3)

This raises the question of who supports the supporters. To whom did farm women turn for support? For many, gendered ideologies made it easier for women, as opposed to men, to talk about their concerns with friends or family members. As one participant said, "I think farm men tend to keep more inside and I think farm women tend to network" (Fletcher 2013: Interview 15). These social networks, however, are rapidly disappearing. As agricultural production becomes more competitive and industrialized, many small and medium-size farms have disappeared, unable to compete in the current conditions. Many farm women reported the loss of neighbours and other social support networks.

Although women's relative disconnection from farm control could help buffer them somewhat from the psychological effects of climate extremes, it can also give them less agency over practical strategies for coping and adaptation on the farm. As the "main" farmers, men are more likely to make coping and adaptation decisions, such as which crops to plant or whether to spend more on new seeds that promise drought resistance. Although participants felt that gender roles made it easier for women to network and talk about their concerns with friends and family, the constant pressure to support others resulted in hidden stress for many women. As one woman described it, "The physical strain and the emotional strain that the wife carries is not something that can ever be measured, but it is something that she wears all the time" (Fletcher 2013: Interview 7).

Drought also exacerbated women's work responsibilities. Women often hauled water for the farm and household, a task they combined with their existing responsibilities. One participant described the difficulties of raising a small child during the drought of 2001 and 2002:

> I had to get a water tank and haul water from the city to fill up my well so I could bathe and do laundry. It was terrible. About once every three days I had to haul water from the city. One load of water would take me about two hours . . . at the time I had a child who was in diapers. Then you're trying to entertain her and I'd bring along games, tic-tac-toe. (Fletcher 2013: Interview 5)

Participants often tried to save money during difficult times by changing household practices. Strategies included growing more vegetables and preserving them through canning or freezing, mending clothes and other household items instead of buying new things, and accepting temporary work or self-employment to bring in extra income. According to one farm woman, "We put off farm-related purchases and applied heavy restrictions on personal entertainment, fuel, power, telephone, groceries. . . " (Fletcher 2013: Interview 30).

Some women had taken an off-farm job as a form of adaptation to the uncertainties of farming. However, few participants saw their off-farm income as directly supporting the farm operation. This income was usually seen as a way to pay household expenses when the farm could not sustain such expenses or as money to support lifestyle preferences and extra "wants." For many women, off-farm work was also a source of self-fulfillment and a place to pursue their own goals. However, working off-farm is rarely accompanied by a decrease in household or farm responsibilities. Despite the personal fulfillment it offers, farm women who find their work increasingly stretched in multiple directions can experience increased levels of stress.

These findings support Arora-Jonsson's (2011) argument about the importance of context; they respond to her recommendation for a nuanced analysis of gender and climate change. We cannot assume that vulnerability can be neatly mapped onto social categories of difference like gender. This study showed that gender is indeed a key dimension that shapes vulnerability and adaptation to extreme climate events; however, vulnerability should not be uncritically attached to women. In the case of Saskatchewan farmers, gender roles and ideologies made men more vulnerable to the psychological consequences of drought, challenging conventional discourses that feminize vulnerability. Women "pick up the pieces" (Fletcher 2013: Interview 1) as caregivers, farm workers, or off-farm wage earners. They play a critical role in coping and adaptation. At the same time, however, environmental crises tended to further entrench historical gender roles. Material and ideological factors position women as the "caregivers" and "nurturers" for the family during times of environmental crisis while giving them less agency over concrete adaptation strategies.

Policy Implications

Beyond strategies at the farm level, there is a need for government intervention to reduce vulnerability and facilitate adaptation over the long term (Marchildon et al. 2008). Indeed, Marchildon et al. (2008) showed that government disaster programs were a crucial coping mechanism used by farm families dealing with climate extremes. At the same time, such programs must be appropriately and accessibly designed in order to be useful. Programs must be attentive to both the macro- and micro-level political, economic, and social conditions affecting the community they serve. Disaster assistance and insurance coverage, for example, must keep up with the cost of production as farms grow larger and more industrialized.

Attention must also be paid to financial vulnerability caused by high levels of farm debt. The findings of the current study challenge the idea that small farming operations are necessarily more vulnerable than large farming operations; in fact, debt levels are a key determinant of vulnerability. Agricultural policies and programs must not uncritically promote farm growth—which is often premised on high debt levels—as a positive step for long-term farm sustainability.

Mental health and psychological stress emerged as key forms of vulnerability in the study. Rural residents often lack access to mental health support services due to geographical constraints and the urbanization of health services (Kubik and Moore 2005; Jaffe and Blakley 1999). Further, the stigma associated with use of these services can be a barrier in small and tightly knit communities (Fraser et al. 2005). Until 2012, Saskatchewan had a publicly funded, peer-based telephone helpline, the Farm Stress Line, which was well known among farmers as a source of mental and emotional support. In July 2012, the operation of the helpline was transferred to an urban-based community organization to save government expenditures of $100,000 per year (Government of Saskatchewan 2012b). Mental health services will become more important as climate extremes become more frequent and severe. These services must be provided with attention to appropriateness; that is, they should be provided by individuals with knowledge of, and experience in, agriculture. Further, support services should be designed with attention to gendered dimensions that create different experiences for farm men and women.

Government programs must also extend beyond just coping and disaster assistance to facilitate adaptive capacity over the long term. A dual focus on coping *and* adaptive capacity will help reduce future public expenditure in the event of an environmental disaster. Recent changes to federal infrastructure such as the erosion of the Prairie Farm Rehabilitation Administration, which was established as an institutional adaptation to the extreme droughts of the 1930s, suggest a decreased governmental emphasis on long-term adaptation.

Conclusion

A recent non-governmental organization handbook on climate adaptation stated that "one of the challenges of working at the local level on climate change adaptation is the lack of scaled-down information on impacts" (Dazé et al. 2009: 2). In this chapter, we have presented a scaled-down analysis of gendered impacts of drought in the Canadian Prairies. The analysis reveals the gender dynamics of vulnerability and adaptation—dynamics that are often invisible from a macro-level perspective and thus give the (mistaken) impression that climate change is a gender-neutral phenomenon that affects everyone equally. It is through a gendered lens that important social factors, such as psychological stress and the importance of social support networks, become visible.

At the same time, we emphasize the importance of situating micro-level understandings of climate vulnerability and adaptation within larger political and economic conditions. The current emphasis on industrialization and rapid farm expansion through debt has resulted in particular vulnerabilities for Prairie farm families; these vulnerabilities can be exacerbated by climate events and threaten the future sustainability of food production. Policies aimed at enhancing adaptive capacity must consider these broader economic challenges while simultaneously attending to differences within and between farm families. Such multi-scale, gendered analyses can inform more effective adaptation policies that are relevant and aligned with the realities of everyday life on the Canadian Prairies.

NOTE

1 A recent study (in which the authors are currently involved) found that unanticipated fluctuations between flood and drought, as well as dramatic departures from the "expected" extremes, can be particularly difficult for agricultural producers. For example, among producers in the drought-prone Palliser Triangle region of southern Saskatchewan and Alberta (see Chapter 8 by Marchildon in this volume), drought is generally expected and prepared for; therefore, the occurrence of an extremely wet year in 2010–11, which followed immediately after the drought of 2009, challenged producers' abilities to cope and adapt (VACEA Forthcoming).

References

Adger, N.W. 2003. "Social Aspects of Adaptive Capacity." Pp. 29–50 in J.B. Smith, R.J.T. Klein, and S. Huq (eds.), *Climate Change: Adaptive Capacity and Development*. London: Imperial College Press.

Arora-Jonsson, S. 2011. "Virtue and Vulnerability: Discourses on Women, Gender and Climate Change." *Global Environmental Change* 21: 744–51.

Bonsal, B., and M. Regier. 2007. "Historical Comparison of the 2001/2002 Drought in the Canadian Prairies." *Climate Research* 33: 229–42.

Bye, C.G. 2005. "'I Like to Hoe My Own Row': A Saskatchewan Farm Woman's Notions about Work and Womanhood during the Great Depression." *Frontiers: A Journal of Women Studies* 26, no. 3: 135–67. doi: 10.1353/fro.2006.0001.

Dankelman, I. (ed.). 2010. *Gender and Climate Change: An Introduction*. London: Earthscan Publishing.

Dazé, A., K. Ambrose, and C. Ehrhart. 2009. *Climate Vulnerability and Capacity Analysis Handbook*. CARE.

Enarson, E., and P.G.D. Chakrabarti (eds.). 2009. *Women, Gender and Disaster: Global Issues and Initiatives*. New Delhi: Sage Publications.

Enarson, E., A. Fothergill, and L. Peek. 2007. "Gender and Disaster: Foundations and Directions." Pp. 130–46 in H. Rodriguez, E.L. Quarantelli, and R. Dynes (eds.), *Handbook of Disaster Research*. New York: Springer.

Faye, L. 2006. "Redefining 'Farmer': Agrarian Feminist Theory and the Work of Saskatchewan Farm Women." MA thesis, Memorial University of Newfoundland.

Field, C.B., V. Barros, T.F. Stocker, D. Qin, D.J. Dokken, K.L. Ebi, M.D. Mastrandrea, K.J. Mach, G.K. Plattner, and S.K. Allen. 2012. *Managing the Risks of Extreme Events and Disasters to Advance Climate Change Adaptation*. Cambridge: Cambridge University Press.

Fletcher, A.J. 2013. "The View from Here: Agricultural Policy, Climate Change, and the Future of Farm Women in Saskatchewan." Ph.D. dissertation, University of Regina.

Fowke, V.C. 1957. *The National Policy and the Wheat Economy*. Toronto: University of Toronto Press.

Fraser, C.E., K.B. Smith, F. Judd, J.S. Humphreys, L.J. Fragar, and A. Henderson. 2005. "Farming and Mental Health Problems and Mental Illness." *International Journal of Social Psychiatry* 51: 340–49.

Gilbert, G., and R. McLeman. 2010. "Household Access to Capital and Its Effects on Drought Adaptation and Migration: A Case Study of Rural Alberta in the 1930s." *Population and Environment* 32, no. 1: 3–26. doi: 10.1007/s11111-010-0112-2.

Government of Saskatchewan. 2012a. "Agriculture in Saskatchewan." Government of Saskatchewan website. http://www.agriculture.gov.sk.ca/Default.aspx?DN=7b598e42-c53c-485d-b0dd-e15a36e2785b. Accessed 24 November 2013.

———. 2012b. "News Release: Mobile Crisis Services to Take on Farm Stress Calls." Government of Saskatchewan website. https://www.saskatchewan.ca/government/news-and-media/2012/june/27/mobile-crisis-services-to-take-on-farm-stress-calls. Accessed 10 December 2015.

Ireland, G. 1983. *The Farmer Takes a Wife*. Chesley, ON: Concerned Farm Women.

Jaffe, J.M., and B. Blakley. 1999. *Coping as a Rural Caregiver: The Impact of Health Care Reforms on Rural Women Informal Caregivers*. Winnipeg: Prairie Women's Health Centre of Excellence.

Kelly, P.M., and W.G. Adger. 2000. "Theory and Practice in Assessing Vulnerability to Climate Change and Facilitating Adaptation." *Climatic Change* 47: 325–52.

Koskie, S. 1982. *The Employment Practices of Farm Women*. Saskatoon, SK: National Farmers Union.

Kubik, W. 2004. "The Changing Roles of Farm Women and the Consequences for Their Health, Well Being, and Quality of Life." PhD dissertation, University of Regina.

———. 2005. "Farm Women: The Hidden Subsidy in Our Food." *Canadian Woman Studies* 24: 85–90.

Kubik, W., and R.J. Moore. 2005. "Health and Well-being of Farm Women: Contradictory Roles in the Contemporary Economy." *Journal of Agricultural Safety and Health* 11: 249–56.

Kuyek, D. 2007. "Sowing the Seeds of Corporate Agriculture: The Rise of Canada's Third Seed Regime." *Studies in Political Economy* 80: 31–54.

Laforge, J.M.L., and R. McLeman. 2013. "Social Capital and Drought-Migrant Integration in 1930s Saskatchewan." *The Canadian Geographer / Le Géographe Canadien* 57, no. 4: 488–505. doi: 10.1111/j.1541-0064.2013.12045.x.

Leichenko, R.M., and K.L. O'Brien. 2008. *Environmental Change and Globalization: Double Exposures*. New York: Oxford University Press.

Marchildon, G.P., S. Kulshreshtha, E. Wheaton, and D. Sauchyn. 2008. "Drought and Institutional Adaptation in the Great Plains of Alberta and Saskatchewan, 1914–1939." *Natural Hazards* 45: 391–411.

Martz, D.J. Forsdick. 2006. "Canadian Farm Women and Their Families: Restructuring, Work and Decision Making." PhD dissertation, University of Saskatchewan.

McCrorie, J.N. 1964. *In Union Is Strength*. Saskatoon: Centre for Community Studies, University of Saskatchewan.

Milan, A., L.-A. Keown, and C.R. Urquijo. 2011. *Families, Living Arrangements and Unpaid Work (Women in Canada: A Gender-Based Statistical Report)*. Ottawa: Statistics Canada.

Milne, W. 2005. "Changing Climate, Uncertain Future: Considering Rural Women in Climate Change Policies and Strategies." *Canadian Woman Studies* 24, no. 4: 49–54.

Moosa, C.S., and N. Tuana. 2014. "Mapping a Research Agenda Concerning Gender and Climate Change: A Review of the Literature." *Hypatia* 29, no. 3: 677–94. doi: 10.1111/hypa.12085.

Reinsch, S. 2009. "'A Part of Me Had Left': Learning from Women Farmers in Canada About Disaster Stress." Pp. 152–64 in E. Enarson and P.G.D. Chakrabarti (eds.), *Women, Gender and Disaster: Global Issues and Initiatives*. New Delhi: Sage Publications.

Roppel, C., A.A. Desmarais, and D. Martz. 2006. *Farm Women and Canadian Agricultural Policy*. Ottawa, ON: Status of Women Canada. http://www.aic.ca/gender/pdf/Farm_Women.pdf.

Rosenfeld, R. 1985. *Farm Women: Work, Farm, and Family in the United States*. Chapel Hill, NC: University of North Carolina Press.

Sachs, C.E. 1983. *The Invisible Farmers: Women in Agricultural Production*. Totowa, NJ: Rowman & Allanheld.

Sauchyn, D. 2010. "Prairie Climate Trends and Variability." Pp. 32–40 in D.J. Sauchyn, H.P. Diaz, and S. Kulshreshtha (eds.), *The New Normal: The Canadian Prairies in a Changing Climate*. Regina: Canadian Plains Research Center Press.

Sauchyn, D., H. Diaz, and S. Kulshreshtha (eds.). 2010. *The New Normal: The Canadian Prairies in a Changing Climate*. Regina: Canadian Plains Research Center Press.

Sauchyn, D., and S. Kulshreshtha. 2008. "Prairies." Pp. 276–328 in D.S. Lemmen, F.J. Warren, J. Lacroix, and E. Bush (eds.), *From Impacts to Adaptation: Canada in a Changing Climate 2007*. Ottawa: Government of Canada.

Sauchyn, D.J., J. Stroich, and A. Beriault. 2003. "A Paleoclimatic Context for the Drought of 1999–2001 in the Northern Great Plains of North America." *Geographical Journal* 169: 158–67.

Schwieder, D., and D. Fink. 1988. "Plains Women: Rural Life in the 1930s." *Great Plains Quarterly* 8, no. 2: 79–88.

Statistics Canada. 2011a. *2011 Census of Agriculture*. Ottawa: Government of Canada. http://www.statcan.gc.ca/daily-quotidien/120510/dq120510a-eng.htm. Accessed 28 September 2013.

———. 2011b. *Farm Debt Outstanding: Agriculture Economic Statistics*. Ottawa: Minister of Industry. Cat. No. 21-014-X, vol. 10, no. 2. http://www.statcan.gc.ca/pub/21-014-x/21-014-x2011002-eng.pdf. Accessed 28 September 2013.

Sushama, L., N. Khaliq, and R. Laprise. 2010. "Dry Spell Characteristics over Canada in a Changing Climate as Simulated by the Canadian RCM." *Global and Planetary Change* 74: 1–14.

Vulnerability and Adaptation to Climate Extremes in the Americas (VACEA) Project Integrated Report. Forthcoming.

Warren, J.W., and H.P. Diaz. 2012. *Defying Palliser: Stories of Resilience from the Driest Region of the Canadian Prairies*. Regina: Canadian Plains Research Center Press.

Wheaton, E., S. Kulshreshtha, V. Wittrock, and G. Koshida. 2008. "Dry Times: Hard Lessons from the Canadian Drought of 2001 and 2002." *The Canadian Geographer/Le Géographe Canadien* 52: 241–62.

Wheaton, E., V. Wittrock, S. Kulshreshtha, G. Koshida, C. Grant, A. Chipanshi, and B. Bonsal. 2005. *Lessons Learned from the Canadian Drought Years of 2001 and 2002: Synthesis Report*. 11602-46E03. Saskatoon: Saskatchewan Research Council.

PART 4

GOVERNANCE SYSTEMS FOR PRAIRIE DROUGHT AND WATER MANAGEMENT

CHAPTER 8

DROUGHT AND PUBLIC POLICY IN THE PALLISER TRIANGLE: THE HISTORICAL PERSPECTIVE

Gregory P. Marchildon

Introduction

The Canadian Prairies have had a distinct climate since the last Ice Age, characterized by extreme seasonal temperatures with short, hot summers alternating with long, cold winters, and by a semi-arid climate with cyclical bouts of severe, multi-year droughts (Davison 2001). Following the region's settlement and use for agricultural production, the Great Depression of the 1930s generated the extreme conditions that made this region well known to North Americans. Collectively remembered as an ecological and human disaster, the prolonged drought of the Dirty Thirties triggered responses by governments at the federal, provincial, and local levels that attempted to address the physical damage and mitigate the human suffering caused by the most prolonged drought in the region in the twentieth century (McLeman et al. 2013; Jones 2002). This chapter reviews the most important of these policy interventions to extract some lessons for the future of the region, a future likely to involve prolonged droughts

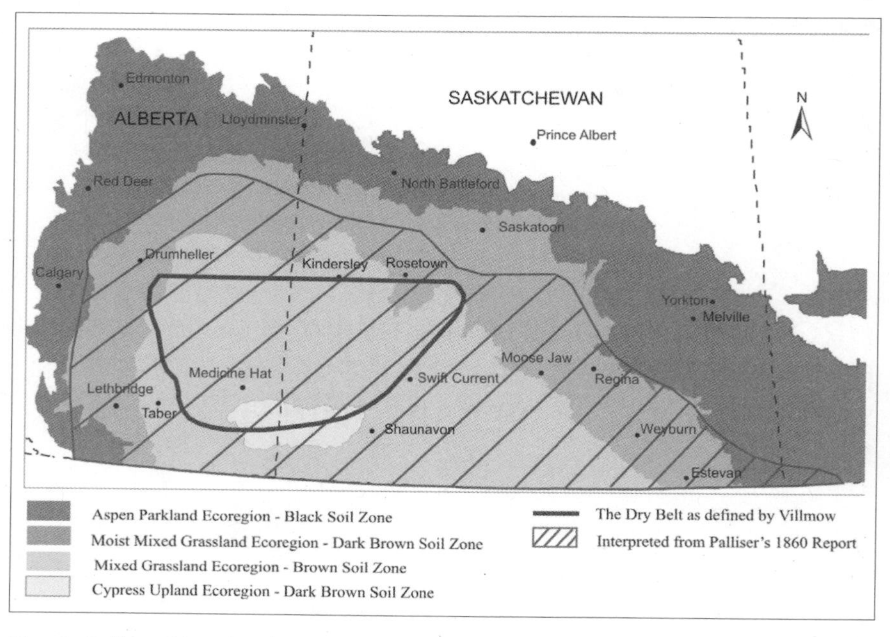

Map 1. Palliser Triangle with Prairie ecoregions and soil zones

due to human-induced climate change, especially in the drier sub-region known as the Palliser Triangle.

After the arrival of Europeans, and after the international boundary between Canada and the United States was set, subsequent explorers and surveyors notionally subdivided the Canadian portion of the North American Plains into sub-regions. The southernmost sub-region was named the Palliser Triangle (Map 1) after the leader of the British North American Exploring Expedition of 1857–60, Captain John Palliser (Spry 1963). One of this area's longest droughts in the entire nineteenth century occurred during Palliser's expedition on behalf of the British government, leading him to declare the southern Canadian Prairies unsuitable for agriculture. In the twentieth century, the dry inner core of the Palliser Triangle was labelled the Dry Belt by climatologists, a term subsequently used by historians to describe the same area (Marchildon et al. 2009).

A History of Drought in the Palliser Triangle

Given the extreme climate and water scarcity that marks the Canadian Prairies, it is not surprising that vulnerability has been an integral part of the human experience in the Palliser Triangle. This vulnerability also helps to explain the sparse population pattern of the Canadian Prairies in general, and the Palliser Triangle in particular, relative to other southern regions of Canada. Similar to today, low population density was a feature of the Canadian Prairies during its pre-history. Indigenous agriculture ranged from extremely limited to non-existent in the southwestern portion of the Canadian Prairies, even during the relatively warm centuries preceding the dry and cold period of the Little Ice Age, more formally known as the Pacific Climate Episode (AD 1250–1550). However, the grasslands did support the enormous herds of bison that were the mainstay of Indigenous communities. Based on extended clan networks speaking a common language, these communities migrated by necessity, moving their buffalo-skin shelters and minimal belongings to follow the bison herds (Dawson 2003; Thomas 1976).

While hunting and gathering was not as water-intensive as farming, water was still required in this dry environment, and there is some evidence that the Indigenous inhabitants of the Palliser Triangle "developed a water management strategy that buffered them from the effects of even long-term drought" (Daschuk 2009: 17). In a semi-arid environment, this meant protecting non-river water sources, such as beaver ponds by restricting beaver hunting. Bison herds would not move from river valleys to their usual summer ranges in the open prairie during the worst droughts, so protecting river-based water sources was an absolute necessity. During prolonged droughts when river tributaries ran dry, Indigenous populations and bison sought forced refuge along the main river channels and beside bodies of water dammed by beavers. It is interesting to note that the Indigenous restriction on hunting beaver lasted long after the arrival of Europeans, despite the economic incentives for Plains tribal groups to engage in large-scale beaver trapping during the fur trade (Daschuk 2009).

The first European occupation of the Palliser Triangle was based on open-range cattle ranching. By the 1870s, the western bison herds were nearing extinction because of the demand for bison hides and bison meat, including pemmican and luxury items such as tongues, which was met

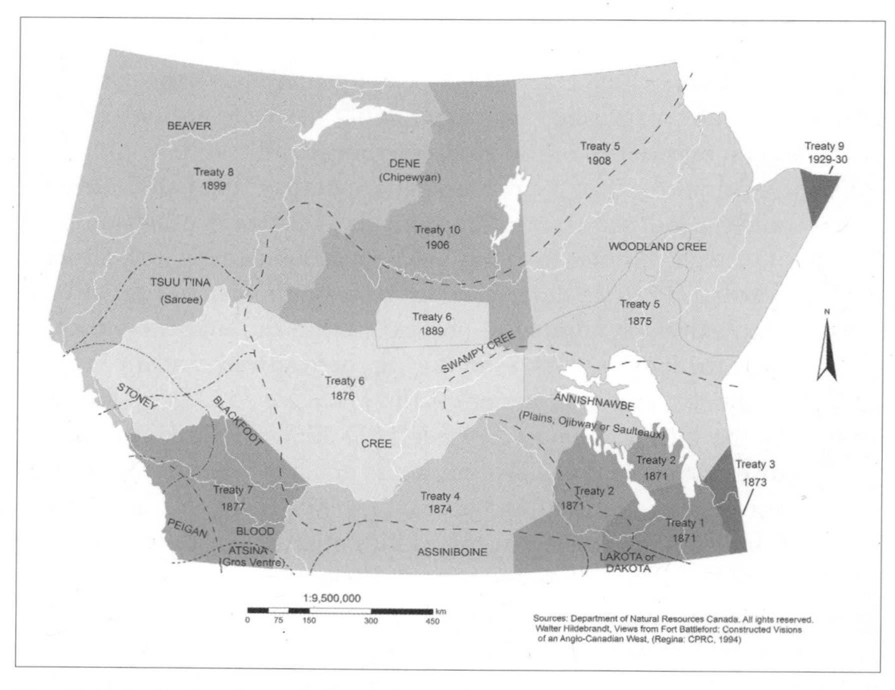

Map 2. Plains Indians boundaries, ca. 1850, showing Treaty areas. (Source: Marchildon 2009c: 5)

by faster-loading and increasingly accurate rifles, resulting in the collapse of the herds. As a consequence, the Indigenous occupants of Palliser Triangle—predominantly the Plains Cree and the Blackfoot Confederacy, made up of the Siksika, Peigan, and Kainai (Blood) Nations—faced widespread famine. In exchange for food and medical supplies from the newly established Government of Canada, these First Nations signed Treaty 6 (1876) and Treaty 7 (1877), relinquishing possession of most of their traditional bison-hunting territories in exchange for much smaller parcels of reserve land (Daschuk 2013; Marchildon 2009a; Map 2).

When these treaties were signed, the US Plains were already experiencing a ranching boom that would spill over the border into the southwestern portion of the Canadian Prairies (Olefson 2000; Breen 1983). Eager to establish a cattle industry, the Government of Canada passed an order in council to permit 21-year leases of land up to 100,000 acres (approximately 40,500 hectares) for the highly subsidized price of one cent per acre. The original leases prohibited homestead farm settlement to facilitate open (unfenced) ranges. To encourage the northern migration of cattle, the Canadian government also permitted ranchers to import cattle duty-free for two years from the United States. These policies favouring open-range ranching ensured that it expanded rapidly in the last two decades of the nineteenth century (Wandel and Marchildon 2010).

The cattle boom ended abruptly in the first decade of the twentieth century. Three major factors seem to have each played a role in bringing this era to an end. First, the introduction of new refrigeration technologies allowed for major import markets, such as Great Britain, to receive less expensive chilled beef from Argentina. Second, an extreme weather event known as the "Killer Winter of 1906–7" decimated the cattle herds in the short-grass prairie of the Palliser Triangle, killing up to 65% of cattle in the Dry Belt. Third, the Canadian government reversed its open-range subsidized lease policy and instead supported and subsidized fenced-off homestead settlement (Evans 1983).

Although cattle ranching remained viable in the western long-grass prairie of the foothills that received higher precipitation, most of the drier short-grass lands of the Palliser Triangle were opened to farm settlement after the Killer Winter of 1906–7. Under the Dominion Lands Act, settlers were given 160 acres (65 hectares) of land under the condition that they cultivate that parcel and establish a permanent homestead on it within three years. In 1909, the Canadian government officially opened the Dry Belt to homesteaders. In conjunction with local real-estate speculators and the Canadian Pacific Railway, the Government of Canada unleashed a major publicity campaign to attract settlers, despite the fact that the Dry Belt received less average rainfall than all other parts of the Palliser Triangle (Marchildon 2007).

A growing British market for imported wheat, coupled with a high world price, encouraged farmers in the Palliser Triangle to cultivate wheat to the exclusion of almost all other grains. The wheat boom brought in both settlers and "suitcase" farmers—individuals from other locales who only worked the land to make a quick profit. The growing population in the region was reinforced by a doubling in the world price of wheat during the First World War. In addition, the region received higher than average rainfall, with even the Dry Belt experiencing bumper crops in 1915 and

1916. However, this boom was the beginning of the end in the Dry Belt in particular, as a prolonged drought took hold in the years that followed (Marchildon 2007; Gorman 1988).

From 1917 until the unusually wet year of 1927, Dry Belt wheat farmers would suffer repeated crop failures due to a lack of rainfall. Drought became an almost permanent feature of the area, recurring year after year. Maps based on a gridded database of mean monthly temperature and total precipitation derived from the Canadian Climate Archive for the Prairie provinces indicate that the Alberta side of the Dry Belt was even more drought-stricken than the Saskatchewan side. These maps also reveal that the extent to which the region was affected by the droughts after 1928 was far larger than the Dry Belt. Indeed, the drought of the Dirty Thirties blanketed the Palliser Triangle and slightly beyond (Marchildon et al. 2008), affecting a far larger population and segment of the Canadian economy. Known within Canada as the "breadbasket of the world," the Palliser Triangle saw wheat yields plummet and residents migrate to British Columbia, Manitoba, and the forest fringe of the Canadian Prairies (McLeman and Ploeger 2012; McLeman et al. 2010).

The droughts resulted in widespread bankruptcy and poverty for farm families. Many left the devastation in the Palliser Triangle to begin new lives in other parts of Canada. As tax revenues plummeted, local governments were unable to meet their obligations to finance schools, maintain roads, and provide relief for the thousands of destitute farm families (Marchildon and Black 2006; Jones 2002).

The government of Alberta intervened long before that of Saskatchewan because the initial impact of the drought had been greater on its side of the Dry Belt, although some of the policies adopted would be the same in both provinces. The first step was to force banks and other financial institutions to negotiate settlements on farm debt. The next step was to defray the cost of relocating farm families and support local governments in their efforts to provide relief assistance to the families remaining on the land. However, the Alberta government would go further than its provincial neighbour by actively promoting changes in land tenure and, where necessary, replacing some local governments with a provincially appointed administration in the Dry Belt.

The environmental shock caused by the prolonged droughts was considerably exacerbated by the collapse in commodity and stock prices in the Great Depression. In Alberta, per capita income fell by 61%, while in Saskatchewan, where the wheat economy remained dominant throughout the 1930s, per capita income fell by an astounding 72% between 1929 and 1932 (Marchildon 2005). To be sure, there was also a collapse in industrial production affecting central Canada, but the decline in per capita income in Ontario and Quebec (44%) was far less. Having only a small area included in the Palliser Triangle, Manitoba suffered less than Saskatchewan or Alberta: per capita income dropped 49% in the same period, less a result of drought than the decline of business suffered by grain companies and traders headquartered in Winnipeg.

This decline was exacerbated by a collapsing global market in wheat, a market on which Prairie wheat producers depended for the sale of almost all their grain. Beginning in 1928, falling agricultural prices contributed to the stock market crash one year later and would become a major feature of the 1930s (Marchildon 2013). The precipitous decrease in wheat and other grain prices, combined with institutional weaknesses in the banking sectors of numerous advanced industrial countries, initiated a deflationary spiral, which drove a redistribution of income and displaced populations en masse from agricultural regions of countries to non-agricultural regions. Of the wealthier nations in the world, this movement was most pronounced in Canada and the United States, in no small part because of the impact of prolonged drought in the Great Plains of both countries (Madsen 2001).

With the provinces of Alberta and Saskatchewan teetering on the edge of bankruptcy, the federal government intervened, first through large-scale transfers to the provinces for relief payments to thousands of farm families (Marchildon and Black 2006). Eventually, well after similar initiatives in the United States, the federal government created a regional organization to spearhead land and water reclamation initiatives throughout the Palliser Triangle (McLeman et al. 2013).

The remainder of this chapter focuses on two case studies of policy responses to the drought crisis described above. The first summarizes the Alberta government's response to the earlier drought in the Dry Belt and the actions that ultimately led to the establishment of the Special Areas Administration. The second case study focuses on the Government of Canada's response to the more expansive drought of the 1930s and the creation of the Prairie Farm Rehabilitation Administration (PFRA) to reclaim and conserve both soil and land resources in the Palliser Triangle.

The Special Areas Administration

The Special Areas of Alberta refer to a large—currently 5.2 million acres (2.1 million hectares)—and sparsely populated region on the Alberta side of the Dry Belt. Since the late 1930s, the Special Areas has been governed and managed by a provincially appointed administrative board rather than democratically elected local governments. Although nothing on the order of the droughts of the 1920s and 1930s has recurred, the residents have shown limited desire to eliminate the Special Areas Board and revive the old rural municipality system, in large part because of a continuing fear of drought (Marchildon 2007).

Even by the early 1920s, mounting evidence already suggested that the farm settlement of the Dry Belt had been a mistake. Not only was there less precipitation on average than in the rest of the Palliser Triangle, but the Dry Belt seemed even more prone to sustained episodes of drought than the rest of the Palliser Triangle. In 1921, the United Farmers of Alberta (UFA) formed the provincial government, elected in part to address the drought catastrophe in the Dry Belt. According to historian David Jones (2002), the Dry Belt was likely the greatest problem that faced the UFA government in the 1920s and would remain one of its most intractable problems until its defeat in 1935.

Initially, the UFA encouraged the renegotiation of bank loans made to farmers by empowering a government commissioner to negotiate the settlement of debts. By 1922, most farmers had endured the misfortune of five successive years of drought, which in turn had exacted a toll on local businesses, municipalities, and school districts. The purpose of negotiating settlements between debtors and creditors was to save the farms, businesses, school districts, and local governments in the Dry Belt.

However, even with debt rescheduling, only a minority of farms and businesses remained viable, so the UFA government then offered free transportation to destitute farm families who were willing to leave the Dry Belt. Sharing one-third of this cost with the federal government and railway companies, the provincial government provided each family with up to two railway cars to transport its machinery, farm supplies, livestock,

Table 1. Vacant or abandoned farms in the dry belt, 1926

	Population	Vacant or abandoned farms (number)	Vacant or abandoned farms (acres)
Alberta Census Divisions 3 and 5	39,365	5,124	1,287,594
Saskatchewan Census Division 8	44,667	916	212,091

Source: Derived from Tables 1,3,4 and 6 in Jones (2002), pp. 254–57.

and furniture. By 1926, almost 2,000 farm families had taken advantage of the assistance to move north of Calgary or further west to the irrigated districts near Lethbridge (Marchildon 2007).

That same year, the provincial and federal governments established a commission to study the Red Deer and Saskatchewan Rivers, from the town of Tilley in the west to the Saskatchewan border in the east. Covering 1.5 million acres, the Tilley East area (subsequently known as Special Area No. 1) had lost 80% of its peak population by 1926, the result of continual crop failures. Farms were abandoned at such a rate that the viability of the few remaining farms was further threatened by blowing topsoil from the untended fields encircling them. As indicated in Table 1, deserted farms were far more prevalent on the Alberta side of the Dry Belt (roughly contained in Alberta Census Divisions 3 and 5) than on the Saskatchewan side.

The federal-provincial commission recommended that a single board manage all land and water resources throughout the Tilley East Area so that the government could repossess abandoned land for non-payment of taxes. This practice would then allow the government to lease the better land at subsidized rates to the smattering of viable farmers and ranchers left in the area and reseed the worst land, converting it to community pastures to be used by mixed farmers and ranchers for minimal cost. However, implementing the commission's recommendation was difficult because all public (Crown) land was owned by the Government of Canada and thus not available for allocation by the provincial government.

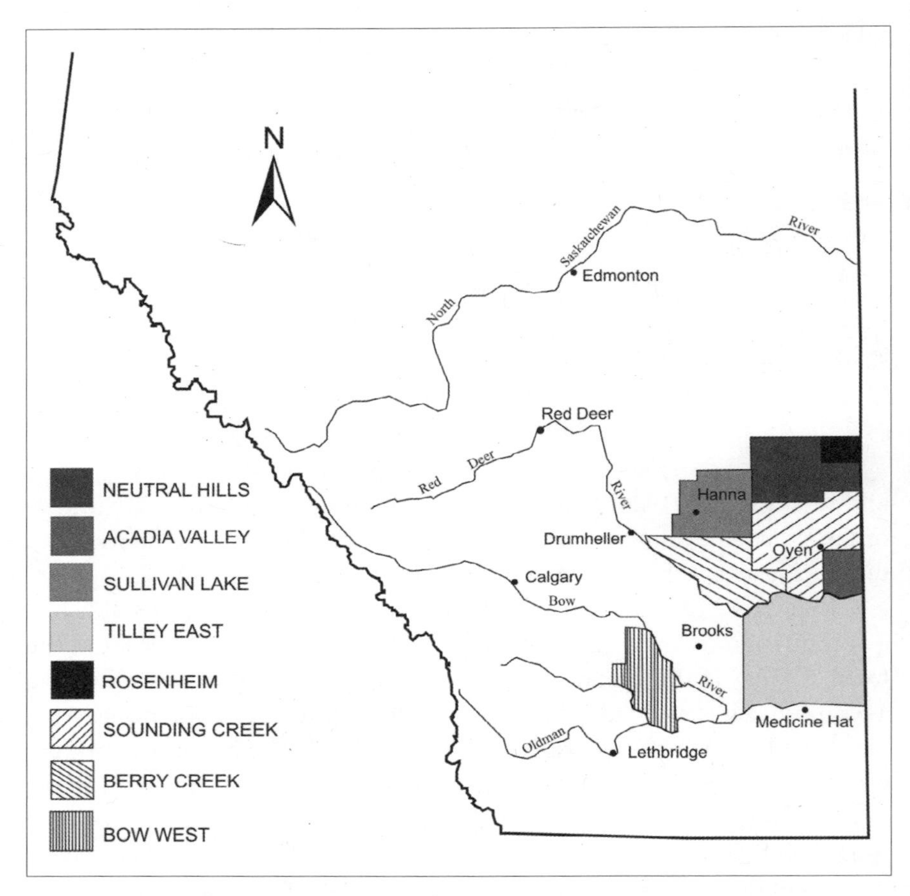

Map 3. Special Areas, ca. 1942.

It was only through a constitutional change—the Natural Resource Transfer Agreement of 1930—that it became possible for the provincial government to create the Tilley East Area Board and assign it the power to own and reallocate lands. With its new powers, the Tilley East Area Board leased and sold land to enlarge the most viable ranch or mixed ranch-farm operations, and actively discouraged farmers who were attempting to continue a wheat monoculture. The board also converted abandoned farms into community pastures. The experiment proved so successful that the provincial government created a similar body in the Berry Creek Area, northwest of Tilley East. In addition, the school districts were also dissolved, and schools were placed under the administrative control of the Berry Creek Special Area Board. This was followed by the establishment of the Neutral Hills, Sounding Creek, and Sullivan Lake Special Areas in 1935. In the next two years, the provincial government also set up the Acadia Valley, Rosenheim, and Bow West Special Areas (Map 3).

In 1938, during one of the worst drought years of the 1930s, all of these areas were consolidated under a single Special Areas Board. Although appointed by the provincial government in Edmonton, the board and its members were headquartered in the Dry Belt community of Hanna. The provincial government dissolved the 34 separate municipalities and improvement districts, effectively eliminating local government and putting all legal and governmental control in the hands of the new board. The rationale behind the change was to ensure that the Special Areas Board had all the necessary tools at its disposal to manage land and water resources, as well as roads, schools, and other physical and social infrastructure, for almost one-third of the province's agricultural land base. The three-member board was conferred a remarkably broad mandate to manage the Special Areas in the "manner it deemed most efficient for the remaining residents" of the Alberta Dry Belt (Marchildon 2007: 263; Gorman 1988).

The provincial government's chief policy objective was to reduce the drought vulnerability of the Dry Belt by thinning out both population (Table 2) and infrastructure, and transforming land tenure from small and unsustainable wheat farms to larger ranches and ranch-farms (Marchildon 2007; Jones 1978). Private ownership was increasingly supplanted by public ownership, under the managerial control of the Special Areas Board. Ranchers and mixed farmers obtained access to the land through inexpensive Crown leases and community pastures. In its first year of operation, the Special Areas Board leased grazing lands for 2.5 cents per acre and rented crop lands for a one-sixth share of the annual crop. Both rates were well below prevailing market values in the rest of the province (Marchildon 2007).

In 1936, farms in the Alberta Dry Belt were already 1.7 times the size of the average Alberta farm. However, with the intervention of the Special Areas Board, these Dry Belt farms would grow to 3.6 times the size of the average Alberta farm by 1956, even though the absolute size of the average farm or farm-ranch had also grown considerably over this period

Table 2. Rural and urban populations in the Special Areas, census years 1916–76

	Rural	Urban	Total
1916	21,715	2,449	24,164
1921	26,031	3,658	29,689
1926	19,344	3,529	22,873
1931	20,320	3,754	24,074
1936	14,976	3,038	18,005
1941	11,794	3,325	15,119
1946	9,542	3,504	13,046
1951	8,430	4,076	12,506
1956	8,723	4,657	13,380
1961	8,799	5,256	14,055
1966	7,974	5,354	13,328
1971	7,050	5,250	12,300
1976	5,854	5,128	11,036

Source: Martin (1977), p. 49.

(Marchildon 2007; Gorman 1988). Thus, the policy objective of improving the viability of farm-ranch operations by increasing their size was attained.

Despite the fact that the policy came at the price of residents not having democratically elected rural governments, residents in the Special Areas have consistently rejected a return to local rural governments. Although there have been no sustained multi-year droughts since the 1930s, enough residents continue to fear the possibility of prolonged drought to support this institutional arrangement, one that is unique in the Canadian Prairies. Despite at least two major reviews by the provincial government, one in 1953 and another in 1960, residents rejected a return to more local democratic control (Marchildon 2007).

The Prairie Farm Rehabilitation Administration

In contrast to the Alberta government, the federal government failed to establish any institutional mechanisms to address recurrent drought in the Palliser Triangle until the mid-1930s. Prior to this, the federal government directed its resources to help the provinces fund relief for the Triangle's rural residents. In July 1931, Prime Minister R.B. Bennett described the drought ravaging the Triangle as perhaps "the greatest national calamity that has ever overtaken this country" (Marchildon and Anderson 2008: 79). Relief was essential to provide the basic foodstuffs and clothing, as well as seed and other essential farm supplies, to ensure that farm families had sufficient nutrition and were also able to feed their remaining livestock and plant another crop. However, most municipalities in the Palliser Triangle lacked sufficient revenues to fund relief. This situation forced the provincial governments to intervene with relief paid for out of provincial revenues, but they too were unable to sustain the relief efforts without assistance from the federal government.

It was impossible to predict how long the droughts—or the Great Depression—would persist, so the federal government transferred money to the provinces for relief payments on a year-to-year basis. Saskatchewan was the province that received the most relief funding, because of the greater number of wheat farmers in the Palliser Triangle. In the 1931–32 season, some 305,000 Saskatchewan residents, nearly one-third of the population of the province, received relief (Marchildon and Black 2006).

The Dirty Thirties became synonymous with the Palliser Triangle because of the tendency of lighter soil types in the Triangle to blow and drift (McLeman and Ploeger 2012). Governments and agricultural experts had been encouraging dryland farmers to allow a portion of their land to go fallow each year to amass moisture for the following year's crop. However, this practice would prove disastrous on the light lands in the Palliser Triangle. The frequent cultivation required to clear the surface of moisture-robbing vegetation pulverized the soil to a powder, making it highly susceptible to wind erosion during a prolonged drought. These lighter soils, combined with high winds, resulted in dust storms that blackened the prairie skies (Wheaton 1992).

One of the main purposes of rural relief was to encourage farmers to "stay on the land" rather than drift into the cities seeking what turned out to be non-existent employment, a situation that could lead to civil unrest. However, even with relief, farm families were still abandoning their farms in the areas of the Palliser Triangle that had been rendered a desert by the drought and topsoil erosion. Although the Alberta government had concluded that wheat farming alone was no longer tenable in the Palliser Triangle, a contrary view was held by decision makers in Saskatchewan and Ottawa, who felt that with a few exceptions, most of the Palliser Triangle could be reclaimed and once again made productive for grain farming. As such, the exodus of thousands of farm families from southern Saskatchewan to the southern edge of the boreal forest was a source of disquiet to both governments (Marchildon 2009b).

In 1934, in response to pressure from political leaders in Saskatchewan and Manitoba, farm groups, the agricultural press, and segments of the general public, the federal government began working on a concerted effort to reclaim the Palliser Triangle. Early the next year, the Prairie Farm Rehabilitation Act was passed in Parliament to allocate money to the federal Department of Agriculture to plant grass in blown-out areas, build small earthen dams to conserve water, and establish demonstration farms in some of the most drought-stricken parts of the Palliser Triangle. Although the Bennett government was defeated mere months after the Act came into force, these initiatives were actually augmented over the next few years. In 1937, the PFRA was established as a separate agency of the federal government with its head office in Regina—at the time the largest city in the Palliser Triangle (Gray 1967).

As part of this expansion, the PFRA was mandated to take possession of drought-stricken land offered up by the provinces for the purpose of creating community pastures. The Saskatchewan government supported the scheme from its inception, but the Manitoba government would not agree to transfer heavily eroded lands in the southwest part of the province to the PFRA for community pastures until 1939. Alberta refused, permanently, to support the PFRA's community pasture program, in part because of its own extensive administration of community pastures through the Special Areas Board. However, the Alberta government eventually co-operated with the federal government to allow the PFRA to develop large-scale irrigation and dam projects. These projects captured the water flowing from the eastern slopes of the Rocky Mountains to the Canadian Prairies. The earliest irrigation projects were in the Lethbridge area but were soon extended to the 30,000-acre Rolling Hills project near Brooks (Balkwill 2002).

By the end of the Great Depression and the extensive droughts of the 1930s, the PFRA had facilitated the construction of thousands of dugouts—artificial farm ponds—and earthen dams for watering livestock. Dozens of PFRA community pastures were providing inexpensive access to grass for mixed farmers and ranchers in southern Saskatchewan and southwestern Manitoba. In addition, the PFRA had conducted a comprehensive soil survey of 90% of the Palliser Triangle. With its 200 agrologists, engineers, hydrologists, soil scientists, field husbandmen, and other highly trained staff, the PFRA would become a fixture in the southern Canadian Prairies for the remaining decades of the twentieth century. By 2010 the PFRA had ceased to exist as a separate branch within the federal Department of Agriculture and Agri-Food, and its community pasture program had been dismantled by the federal government.

Conclusion

The two case studies reflect the extent to which governments, both provincial and federal, were capable of intervening to facilitate more effective adaptation to the extreme drought conditions, first in the Dry Belt in the 1920s and then in the whole of the Palliser Triangle in the 1930s. Both the Special Areas Board and the PFRA altered existing institutional arrangements to reduce individual and community vulnerability in the most vulnerable part of the Canadian Prairies.

In both cases, governments initially intervened with programs and policies that were more incremental in nature. Only later, after it was clear that the drought was not a temporary phenomenon, did the provincial and federal governments intervene to facilitate more radical changes to the institutional environment.

Where governments did not feel they needed to act, they did not do so, as illustrated in Saskatchewan's portion of the Dry Belt during the 1920s. In any case, no government acted proactively in advance of the drought crisis. Once established, however, the organizations created out of the crisis continued to operate with considerable public support for decades afterward, despite the fact that multi-year droughts on the scale of the 1920s and 1930s did not reoccur in the Palliser Triangle. While the Special

Areas continue to operate in Alberta, the same is not true for the PFRA, only recently dismantled by the federal government. One can only surmise that the policy assumption underlying this decision is that the severe and prolonged drought conditions of the 1930s will never again return to the Palliser Triangle, a questionable assumption at best given the cyclical nature of prolonged periods of drought in the region and future climate change effects, which are likely to exacerbate these extreme climate conditions (McLeman et al. 2013).

References

Balkwill, D.M. 2002. "The Prairie Farm Rehabilitation Administration and the Community Pasture Program, 1937–1947." Master's thesis, University of Saskatchewan.

Breen, D.H. 1983. *The Canadian Prairie West and the Ranching Frontier, 1874–1924*. Toronto: University of Toronto Press.

Daschuk, J. 2009. "A Dry Oasis: The Canadian Plains in Late Prehistory." *Prairie Forum* 34, no. 1: 1–29.

———. 2013. *Clearing the Plains: Disease, Politics of Starvation and the Loss of Aboriginal Life*. Regina: University of Regina Press.

Davison, G.T. 2001. "An Interdisciplinary Approach to the Role of Climate in the History of the Prairies." Pp. 101–18 in R. Wardhaugh (ed.), *Toward Defining the Prairies: Region, Culture, and History*. Winnipeg: University of Manitoba Press.

Dawson, B. 2003. "The Roots of Agriculture: A Historiographical Review of First Nations Agriculture and Government Indian Policy." *Prairie Forum* 26, no. 1: 99–115.

Evans, S.M. 1983. "The End of the Open Range Era in Western Canada." *Prairie Forum* 8, no. 1: 71–87.

Gorman, J. 1988. *A Land Reclaimed: A Story of the Special Areas of Alberta*. Hanna, AB: Special Areas Board.

Gray, J.H. 1967. *Men Against the Desert*. Saskatoon: Western Producer Prairie Books.

Jones, D.C. 1978. "Schools and Social Disintegration in the Alberta Dry Belt of the Twenties." *Prairie Forum* 32, no. 2: 1–20.

———. 2002. *Empire of Dust: Settling and Abandoning the Prairie Dry Belt*. Calgary: University of Calgary Press.

Madsen, J.B. 2001. "Agricultural Crises and the International Transmission of the Great Depression." *Journal of Economic History* 6, no. 2: 327–65.

Marchildon, G.P. 2005. "The Great Divide." Chapter 4 in Gregory P. Marchildon (ed.), *The Heavy Hand of History: Interpreting Saskatchewan's Past*. Regina: Canadian Plains Research Center Press.

———. 2007. "Institutional Adaptation to Drought and the Special Areas of Alberta." *Prairie Forum* 32, no. 2: 251–72.

———. 2009a. "Introduction." Pp. 1–9 in G.P. Marchildon (ed.), *Immigration and Settlement, 1870–1939*. Regina: Canadian Plains Research Center Press.

———. 2009b. "The Prairie Farm Rehabilitation Administration: Climate Crisis and Federal-Provincial Relations during the Great Depression." *Canadian Historical Review* 90, no. 2: 284–94.

———. 2009c. *Immigration and Settlement, 1870-1939*. Regina: Canadian Plains Research Center Press.

———. 2013. "War, Revolution and the Great Depression in the Global Wheat Trade, 1917–39." Pp. 142–62 in L. Coppolaro and F. McKenzie (eds.), *A Global History of Trade and Conflict since 1500*. London: Palgrave Macmillan.

Marchildon, G.P., and C. Anderson. 2008. "Robert Weir: Forgotten Farmer-Minister in R.B. Bennett's Depression-era Cabinet." *Prairie Forum* 33, no. 1: 65–98.

Marchildon, G.P., and D. Black. 2006. "Henry Black, the Conservative Party and the Politics of Relief." *Saskatchewan History* 58, no. 1: 4–17.

Marchildon, G.P., S. Kulshreshtha, E. Wheaton, and D.J. Sauchyn. 2008. "Drought and Institutional Adaptation in the Great Plains of Alberta and Saskatchewan, 1914–1939." *Natural Hazards* 45, no. 3: 399–403.

Marchildon, G.P., J. Pittman, and D.J. Sauchyn. 2009. "The Dry Belt and Changing Aridity in the Palliser Triangle, 1895–2000." *Prairie Forum* 34, no. 1: 32–33.

Martin, L.S. 1977. "The Special Areas of Alberta: Origin and Development." Unpublished report prepared for G.E. Taylor, MLA, Government of Alberta.

McLeman, R.A., J. Dupre, L. Berrang Ford, J. Ford, K. Gajewski, and G. Marchildon. 2013. "What We Learned from the Dust Bowl: Lessons in Science, Policy, and Adaptation." *Population and Environment* 35: 417–40. doi: 10.1007/s11111-013-0190-z.

McLeman, R., S. Herold, Z. Reljic, M. Sawada, and D. McKenny. 2010. "GIS-based Modeling of Drought and Historical Population Change on the Canadian Prairies." *Journal of Historical Geography* 36, no. 1: 43–56.

McLeman, R., and S.K. Ploeger. 2012. "Soil and Its Influence on Rural Drought Migration: Insights from Depression-era Southwestern Saskatchewan, Canada." *Population and Environment* 33: 43–56.

Olefson, W.M. 2000. *Cowboys, Gentlemen and Cattle Thieves: Ranching on the Western Frontier*. Montréal, QC, and Kingston, ON: McGill-Queen's University Press.

Spry, I.M. 1963. *The Palliser Expedition: An Account of John Palliser's British North American Expedition*. Toronto: Macmillan Canada.

Thomas, L.H. 1976. "A History of Agriculture on the Prairies to 1914." *Prairie Forum* 1, no. 1: 31–46.

Wandel, J., and G.P. Marchildon. 2010. "Institutional Fit and Interplay in a Dryland Agricultural Socio-agricultural System in Alberta, Canada." Pp. 179–95 in D. Armitage and R. Plummer (eds.), *Adaptive Capacity and Environmental Governance*. New York and Berlin: Springer.

Wheaton, E. 1992. "Prairie Dust Storms—A Neglected Hazard." *Natural Hazards* 5, no. 1: 53–63.

CHAPTER 9

THE GOVERNANCE OF DROUGHTS

Margot Hurlbert

Introduction

An important determinant of a community's ability to adapt to future climate change impacts and current climate variability is its institutional setting and the degree to which this setting facilitates or hinders the community's adaptive capacity (Willems and Baumert 2003; see also IPCC 2001: 891, 897 and Chapter 10 by Hurlbert on water governance in this volume). Institutions contribute to managing a community's assets and, in the case of drought, the assets relating to rural agricultural producers' livelihoods: land, soil, crops, and income. Institutions also contribute to the community members' relationships with natural resources—for example, the provision of drinking water, property rights to land, or access to community pastures. Both formal institutions (e.g., government, non-profit organizations, and civil society organizations) and informal institutions (e.g., social norms, values, and contexts) contribute to the relationships of people to each other and natural resources.

This chapter focuses on government policy in relation to drought—one facet of the institutional context of adaptive capacity and the governance setting. Governance encompasses laws, regulations, and institutions, as well as governmental policies and actions, national activities, and networks of influence, including international market forces, the private sector, and civil society (Demetropoulou et al. 2010: 341). In this chapter, I describe government policies and programs that assist, or enhance, the adaptive capacity of rural agricultural producers in preparing for and responding to drought in Saskatchewan and Alberta, and then analyze their potential effectiveness at doing so. These policies and programs are divided into three categories in this chapter. The first category includes policies and programs that have been developed to assist agricultural producers in building adaptive capacity to withstand drought. An example is a program facilitating the building of dugouts or water pipelines. The second category includes policies and programs that assist agricultural producers in times of drought, for example, an income-stabilization program. The third category includes climate change and adaptation; this would include regulations reducing greenhouse gas emissions.

Policies Assisting Adaptation to Drought

Drought response and adaptation have been a constant reality for the people of the Canadian Prairie provinces, and for all levels of government, since the beginning of the settlement period. The region has one of the most variable natural climates (ranging from extreme heat to extreme cold) and variable hydrological resources. Droughts and floods are frequent, and the frequency and intensity of droughts are anticipated to increase in the future (Sauchyn and Kulshreshtha 2008; see also Chapter 3 by Wheaton et al. in this volume). Policies and programs that respond to this increased risk of drought will become increasingly important. These policies and programs can be divided into two groups: those that assist rural agricultural producers in adapting to more intense water shortages of longer durations and those that help producers respond to a drought after it has been declared as such.

The federal government's strategy to support farm programs entitled Growing Forward was reintroduced in December 2012 as Growing Forward 2. This second iteration continued to offer a suite of business risk-management programs aimed at helping farmers manage risks from income declines resulting from drought, flood, low prices, and increased input costs. These programs include the following:

- AgriInvest: This program helps cover small margin declines. It is a self-managed producer-government savings account whereby producers can set aside up to 1% of their allowable net commodity sales, and the federal government will match it (up to $15,000 per year). Funds can be withdrawn at any time.

- AgriStability: This program assists producers in cases of large margin declines in farm income, which may have resulted from low prices and rising input costs. If a producer's margin (allowable revenue less allowable expenses) drops below their average margin from previous years (a historical reference margin) by more than 30%, governments will provide a share of the lost income.

- AgriInsurance: This program protects against production losses related to specific crops or commodities caused by drought, flood, hail, disease, or other natural hazards. Delivered by provincial agriculture departments, this crop insurance program provides for cost sharing of premiums between the producer, the province, and the federal government. Producers receive a payment when their production is below their guaranteed insured level of protection. To address flooding, unseeded acreage benefits were expanded in 2012. Livestock price insurance coverage is being explored.

- AgriRecovery: This program helps farm businesses return to operation following disaster situations. It provides a framework for federal and provincial governments to work together and cost share (on a 60/40 basis) funding on a case-by-case basis in response to natural disasters (e.g., extreme weather, disease, pests). This program provides coverage when assistance is needed beyond that available from other existing programs.

Three new programs under the Growing Forward 2 strategy were created in 2013:

- AgriInnovation: This program is designed to accelerate the pace of innovation by supporting research and development activities and facilitating the adoption, demonstration, and commercialization of innovative products, technologies, processes, practices, and services. Two lines of support exist. An industry-led research and development stream provides non-repayable support for agri-science projects (individual research projects that can be local, regional, or national in scope) or projects that are in the agri-science cluster (aimed at mobilizing and coordinating a critical mass of scientific expertise in industry, academia, and government, which is national in scope). The second line of support provides loans to facilitate the demonstration, commercialization, and adoption of innovative agri-based products, technologies, processes, or services.

- AgriMarketing: This program invests in projects to enhance the agriculture sector's access to international markets or assist in developing assurance systems and standards to give Canadian products a competitive advantage internationally.

- AgriCompetitiveness: This program provides directed investments to help the agricultural sector adapt to rapidly changing and emerging global and domestic opportunities and issues, and respond to market trends.

When the Growing Forward strategy was reintroduced in 2012, it was reported that just over $10 billion had been expended through federal and provincial contributions and payments since 2007, and it was announced that over the ensuing five years (2013–17), $3 billion would be invested in the programs (Government of Canada 2012). Two of the business risk-management programs, AgriStability and AgriInvest, had benefits reduced in the 2012 iteration of the strategy.

Agricultural programming is an area of the Canadian federal system where both levels of senior government—federal and provincial—play roles in program financing and delivery. Over the course of the 1990s, government funding for programs such as AgriInsurance and AgriStability tended to reflect a 60/40 split between the federal and provincial governments, respectively, although for AgriInsurance a portion of the provincial share included in-kind contributions related to program delivery. The federal-provincial AgriInsurance program requires producers to pay premiums accounting for up to one-third of program costs. AgriStability does not require a cash contribution from farmers.

Field research undertaken prior to the 2012 reintroduction of the strategy identified considerable dissatisfaction among Prairie farmers with the AgriStability program (RCAD 2012; Warren and Diaz 2012). A common complaint was the onerous application process. Many farmers required the services of an accountant to complete the required forms, and the cost for these services runs from $1,000 to $3,000 per application. Another area of concern involved the five-year averaging system, which saw the likelihood of payments to producers reduced in conjunction with extended periods of weak commodity prices coupled with rising input costs. After paying to submit an application, a farmer had no assurance that a support payment would be forthcoming. Producers were also frustrated by the lack of agricultural knowledge on the part of program administrators located in large urban centres such as Winnipeg. Recently some provinces, including Saskatchewan and Alberta, have worked to improve the quality of program delivery for AgriStability by taking over program management. While more localized administration may reduce some of producers' frustrations, it is unlikely that the reductions in overall program support associated with the 2012 strategy will be welcomed.

The federal-provincial AgriInsurance system has received mixed reviews from producers in the drought-prone regions of the Prairies, although complaints have historically been more common in Saskatchewan than in Alberta (RCAD 2012; Warren and Diaz 2012). Frustration in Saskatchewan stemmed from the effects of severe drought in the late 1980s and 2001–2 on finances for the program. Following a succession of years when payouts overtook the value of farmer premiums and government contribution levels, the Saskatchewan program fell into deficit. In response, premiums were raised to levels that farmers found exorbitant, and payout levels were reduced during the 1990s and early 2000s. The Saskatchewan Party government, elected in 2007, addressed farmer concerns by injecting the cash required to make premiums and payouts more attractive. Since 2007, farmer participation in AgriInsurance in Saskatchewan has increased significantly. In Alberta, the provincial government has apparently been more consistently amenable to providing financial resources to maintain attractive premium rates in the wake of major drought events. The programs in both drought-prone provinces (Alberta and Saskatchewan) have benefited from the fact that, with a few localized exceptions, there has not been a severe region-wide drought on the Prairies since 2002.

Agriculture and Agri-Food Canada provides information on drought through the Drought Watch website (Agriculture and Agri-Food Canada, n.d.). Timely information on weather and climate relevant to the agriculture sector in Canada is posted, including historical weather and climate conditions; impacts of these conditions on the sector; short-term forecasting products; and information on mitigating and adapting to the impacts of weather and climate.

In 1935, the federal government established rural water programs to address drought, following the devastating multi-year droughts in the 1920s and 1930s. From 1935 to 1940, the Rural Water Development Program existed to provide funding to help develop secure on-farm water supplies in the Prairie provinces. Group and community projects were added after 1980. From 1980 to 2004, the program expended an estimated total of $154 million. The Prairie Farm Rehabilitation Administration (PFRA), an entity created by federal statute, managed the program from its inception (Government of Canada 2002; see also Chapter 8 by Marchildon in this volume). The National Water Supply Expansion Program (2002–9) expended approximately $102 million across Canada, with roughly $68 million on the Prairies (Wittrock and Koshida 2005: 9). These programs were most often shared with the provinces.

The Saskatchewan Farm and Ranch Water Infrastructure Program (FRWIP) continued this type of programming from 2008 onward. The FRWIP supports the development of secure water sources in Saskatchewan to expand the livestock industry, encourage rural economic activity, and mitigate the effects of future drought. Projects such as community wells, large and small diameter wells, shallow or deep buried pipelines, and dugouts are eligible for funding. Project costs are shared between the proponent (i.e., producer or municipality) and the federal and provincial governments (Government of Saskatchewan 2011, 2012). This program was designed specifically to deal with hydro-climate extremes (i.e., drought) by providing producers and rural communities with increased access to water resources through infrastructure developments.

The Canada-Saskatchewan and the Canada-Alberta Farm Stewardship Programs (FSPs) assist agricultural producers in adapting to water shortages. Specifically, these programs assist agricultural producers in responding to environmental risk and water supply threats, thereby potentially reducing producers' vulnerability to climate and environmental change by increasing their adaptive capacity. The FSPs are designed specifically with the stated goal of helping producers address on-farm environmental risk (not directly responding to climate change). The programs provide eligible producers with financial assistance to implement beneficial management practices (BMPs) to help maintain or improve the quality of soil, water, air, or biodiversity resources. These BMPs are intended to ensure the long-term health and sustainability of ecological resources used for agricultural production, positively impact long-term economic and environmental viability of agricultural production, and minimize negative impacts and risks to the environment. Federal and provincial funds are available to assist in implementing BMPs. Although they are not specifically designed to improve adaptive capacity for climate variability, there are a number of complementary benefits associated with BMPs (e.g., reduced soil erosion, improved pasture management) that augment producer capacity to deal with variations in climate.

Drought Response Policies

The *Agriculture Drought Risk Management Plan for Alberta–2010* plans for and responds to drought and weather extremes through strategies aimed at three situations: 1) normal or near normal conditions, 2) exceptional/notable conditions, and 3) extreme conditions. Drought is defined as "an extended period of below-normal precipitation resulting in decreased soil and subsoil moisture levels and diminished surface water supplies affecting crop growth, livestock water or irrigation water" (Alberta Agriculture and Rural Development 2010). This management plan integrates policies allowing adaptation and response to drought and establishes a drought advisory group, which provides advice and oversees the plan.

In Saskatchewan, an intergovernmental drought monitoring committee led by the Saskatchewan Department of Agriculture includes representatives of the Water Security Agency, Crop Insurance Corporation, and Ministry of Environment. This committee provides advice and meets weekly regarding agricultural drought. The committee has drafted drought plans, but they have never been finalized. The last documented plan was the 2002 draft "Drought Risk Management Plan for Saskatchewan," which was designed to help government agencies develop a coordinated response to prepare for, mitigate, and respond to drought (Agriculture and Agri-Food Canada 2002).

Cities and urban municipalities have adapted to water shortages for many years. The City of Regina developed contingency plans in 1988, including water conservation programs and expansion of water treatment and delivery capacity (Cecil et al. 2005). Many urban municipalities have found voluntary alternate watering guidelines very effective (Warren and Diaz 2012).

Watershed groups have commenced planning for drought and excessive moisture. Plans have been developed for the North Saskatchewan River watershed (Rowan et al. 2011) and the Upper Souris River watershed (East et al. 2012); these plans were facilitated by the provincial Water Security Agency and Natural Resources Canada. For the North Saskatchewan plan, representatives mapped their watershed by identifying key characteristics (e.g., where poor drainage, good drainage, and wells existed), reviewed potential future climate scenarios, and then identified vulnerabilities and adaptations to these future scenarios. This adaptation planning exercise was then organized by actions for producers, municipalities, and for policy and programs. For the Upper Souris Watershed Plan, representatives identified components of the plan that were key action items related to preparing for drought and excessive moisture, and began implementing them through three activities: 1) an Ecological Change Workshop was held to document past changes in adaptive capacity using participatory mapping; 2) cattle producers participated in a drought planning workshop; and 3) a survey established a baseline for assessing watershed understanding in the community. So far, these drought planning exercises have only occurred in a handful of situations. No strategy currently exists for conducting planning exercises, integrating planning among watersheds, and coordinating planning with other interested groups (e.g., civil society organizations). Although these exercises are an important beginning for drought planning, much is left to be done.

The provincial drought response committees offer timely, responsive problem solving in a drought situation. The institutional context for various government ministries is established so decisions can be made quickly. However, in Saskatchewan, priority should be given to finalizing a drought plan for the entire province to allow for coordination of not only the government ministries but also civil society organizations, non-governmental organizations, municipalities, producer associations, and businesses.

Climate Change and Adaptation Policies

As outlined in Chapter 10 the Prairie provinces have had specific policies surrounding climate change and adaptation for the past several years. Saskatchewan's previous New Democratic Party Government issued an *Energy and Climate Change Plan* in 2007—a cross-governmental vision in response to climate change and the development of a province-wide climate change adaptation strategy, which included working with research organizations and supporting critical local research on climate change and adaptation (Government of Saskatchewan 2007). These goals have been reiterated in the *25 Year Saskatchewan Water Security Plan* (Water Security Agency 2012). Several watershed groups have developed drought plans, as outlined above. Currently, climate legislation relating to mitigation remains on the legislative agenda, but it is yet to be proclaimed.

In Alberta, legislation has existed since the Climate Change and Emissions Management Act (2003), a precursor for *Alberta's 2008 Climate Change Strategy* (Government of Alberta 2008). In addition to establishing a carbon offset market and providing consumer rebates in relation to energy efficient products, two programs were introduced, a greenhouse gas reporting program and a greenhouse gas reduction program. These programs relate to the establishment of a greenhouse gas limit and in 2015 a carbon tax was announced (Bakx 2015). In 2003, the Alberta government also created a *Water for Life* strategy focusing on issues of quantity, quality, and conservation of water—all important issues in preparation for and during drought (Government of Alberta 2003). The strategy initiated three important activities: 1) planning for future management of water via the provincial Climate Change Adaptation Strategy, 2) developing land-use frameworks, and 3) watershed planning through local watershed groups.

Manitoba legislation acknowledges climate change considerations and adopts the precautionary principle and sustainable resource management practices. Recently, the Government of Manitoba announced that the International Institute for Sustainable Development would assist the province in updating its climate and green economy plan to address public concerns about reducing emissions and preparing for climate impacts. The initiative will engage representatives of key sectors, including agriculture, transport, industry, academic, civil society, and others (Pelletier 2013). Sector-wide adaptation as outlined in Manitoba's strategy makes provisions for increasing reliance on energy efficiency and minimizing reliance on fossil fuels (Government of Manitoba 2015).

Alberta and Manitoba are the only two Prairie provinces with policies in place to mitigate climate change. Alberta has passed legislation requiring large emitters to reduce their emissions by 12% using an average of 2003 as a baseline. These requirements apply to emitters making up 70% of Alberta's emissions. Manitoba's legislation requires a reduction of 6% of Manitoba's total 1990 emissions. These requirements are to be achieved in numerous ways, including embracing more renewable sources of energy and developing technology in things such as geothermal and other energy sources and developing hydrogen technologies for transportation.

Canada embraces many measures in these areas as well, but it has no legislated reduction targets for greenhouse gases. The most recent communication filed by Canada in 2010 with the secretariat for the United Nations Framework Convention on Climate Change states that Canada expects to be 802 Mt above its Kyoto Protocol target of 2,792 Mt during the 2008 to 2012 period (Government of Canada 2010). In December 2011, Canada withdrew from the Kyoto protocol. The Conservative government blamed the previous Liberal government for having made an error by committing to the protocol. Prime Minister Stephen Harper has set a target of reducing annual emissions to 17% below 2005 levels by 2020. This threshold is much lower than the Kyoto Protocol target to cut emissions to below 1990 levels (CBC 2011; De Souza 2012). Publicly Stephen Harper rejected carbon pricing or a carbon tax (supporting regulating each sector instead). However, in Privy Council documents obtained under access to information, Canada stated its support for the development of new market-based mechanisms expanding the scale and scope of carbon markets (De Souza 2013). The new government of Justin Trudeau has spent much time in climate change discussions with other world leaders and the premiers. It would be safe to conclude that we shall see a change in the federal government policy.

Discussion

It is expected that the impacts of climate change in the future will be increased variability of climate with longer durations of drought and extreme moisture (see Chapter 3). This review of policies and programs relevant to climate change and related problems of mitigation, adaptation, drought, and disaster shows that short-term drought strategies are planned at the federal and provincial levels. Farm income stabilization policies do offer a level of protection in the event of both drought and flood. The economic impacts are clearly planned for with a suite of agricultural producer programs available. Research in southern Alberta and Saskatchewan confirms that available protection assists producers for a time frame of only a few years. Given that future droughts are expected to be of longer duration, these policies are not likely to protect producers. If these policies are not redesigned to respond to longer, more severe droughts, it is probable that many producers will not be able to continue farming. Further, long-term drought strategies are missing.

The absence of policy responding to long-term drought appears to be due in part to uncertainty surrounding *when* such an event might occur, which may reflect disagreement on the certainty of climate change science. Alternatively, difficulty in preparing and implementing strategy and policy to respond to long-term drought could relate to values and norms. Government has competing priorities in terms of its attention and its budget, which must be addressed through bargaining. Given these two competing characterizations of the policy problem, it would appear that work needs to be done to overcome both issues. Thus, attention should be given to increasing dialogue and focus on climate change science, specifically in relation to the needs and requirements of policy makers, and bargaining within the policy system for increasing focus, attention, and priority on climate change and its impacts.

Government attention and funding need to address adaptive measures. These measures might include additional water storage, irrigation infrastructure, and programs to incentivize water conservation. Prioritizing these initiatives needs to be done through public engagement and dialogue, wherein conflicts resulting from different values and norms surrounding these decisions can be resolved. Currently, programs that encourage adaptive measures (e.g., FRWIP and FSPs) are "sold" on the basis that they enhance efficiencies and improve profitability of farm operations. These programs are not directly marketed to the public and producers as assisting in adaptation to climate change. This allows the policy problem with which these programs are attempting to assist to be structured as improving farm profitability rather than adapting to climate change. Incorporating the climate change problem into these policies would enhance them by encouraging producers to incorporate climate change science into planning for a longer term, thus improving their adaptive capacity.

A challenge surrounding drought policy is the fact it is "creeping" in time (over several weeks, months, or even years) and space (occurring often in a dispersed manner within various rural municipalities). This creeping characteristic accentuates the policy problem of drought. The goals of government are somewhat uncertain as governments are hesitant to allocate today's resources to what could be tomorrow's (or the next government's) problems.

Although provincial governments have an apparatus of intergovernmental committees ready to respond in the event of a drought, the federal government is absent in the field of this policy problem in relation to long-term proactive planning. Although droughts were once listed as four of the five top disasters in Canada (Public Safety Canada 2007), droughts no longer appear in the listing, and other than several droughts in the 1990s, total costs are not estimated for droughts. The federal government's lack of policy on drought is notable and cause for consternation. Responding to droughts without formalized institutional relationships and policy is problematic. Although the federal programs associated with Growing Forward offer individual producers some income protection, research has shown this to be inadequate for droughts lasting longer than two years.

The federal response to climate change, climate mitigation, and adaptation to climate change is even more problematic. Canada's performance in relation to the Kyoto Protocol is dismal. Canada's plans for greenhouse gas reduction are confusing. A void in policy responding to climate change problems exists.

Many municipal governments and individual agricultural producers have plans in place for adaptation to climate change. Plans for disaster response to floods, plans for conservation of water in the event of dry years, and plans to deal with drainage access issues have always been part of the Prairie landscape; ensuring that these strategies meet the future anticipated climate is the challenge. Policies exist to encourage best farm practices (e.g., FSPs), many of which allow producers to adapt to climate change by building infrastructure such as dugouts and pipelines (e.g., FRWIP). Although these individual initiatives are important, more concerted planning needs to occur at community and regional levels for responding to flood and drought. This planning would alleviate the pressure placed on individual adaptive initiatives.

Often, policy that responds to flood does not consider drought, and vice versa. For instance, when infrastructure is built and considerations of flood are paramount, communities and government may construct dams or weirs to retain water and protect communities. When infrastructure is built and considerations of drought are paramount, communities and government may construct water storage facilities. Often water storage infrastructure constructed for one of these events is not appropriate for the other. For example, when irrigators in southern Alberta were confronted with significant flooding, their irrigation infrastructure, constructed for water retention in times of drought, was not effective in times of flood (Hurlbert et al. 2015). Predictions of increased variability and more rapid swings between drought and flood should result in a holistic approach to water planning and policy aimed explicitly at responding to both flood and drought and this new condition of extreme variability.

The governments have not holistically responded to our changing climatic future with proactive policy changes. Nevertheless, Canadian climate change policy exhibits some strengths. These strengths relate to long-standing programs, such as crop insurance programs, the FRWIP, and FSPs. However, a comprehensive policy consideration of future climate change has not yet occurred. From this brief overview, it is apparent that policy response is fragmented and considered only in relation to the structured policy problems of impacts (droughts and floods). Reduction of greenhouse gases in the future, or mitigation, is not even being considered as one long-term adaptation to future climate change. To date, Canada is far from achieving its Kyoto commitments and has in fact given up and removed itself from the Kyoto Protocol. Sparse lip service is paid by the federal government to mitigation of climate change, with mixed messages about tools and strategies. To effectively respond to future climate change, a comprehensive strategy is required that uses the policy framing approach identified herein (see Hisschemöller and Hoppe 1996; Hisschemöller and Gupta 1999; Hoppe 2011). Continuing in a fragmented manner as has been done in the past clearly will not work in the future.

Conclusion

Producers in the Prairie provinces have a long history of adapting to droughts. Future climate change is expected to result in increasing climate variability, including increasing duration and intensity of droughts and floods. One of the key determinants of rural agricultural producers' ability to adapt to drought is the capacity of institutions interacting with these producers to assist with adaptation. Government policies and programs relating to drought are key determinants of whether producers will be able to adapt to future climate change.

This chapter reviewed the institutional governance setting, specifically in relation to drought and flood policies and programs, that impacts a producer's ability to adapt to climate change. This institutional setting is informed by government policies and programs appropriate to water shortages or drought that draw from agricultural policy, water governance, and disaster response. These policies and programs are many and varied when one considers the totality of programs relating to climate change and climate change adaptation, as well as the policy problems of building resilience through drought and flood infrastructure, anticipating future floods and droughts, and responding to present-day droughts. This chapter assessed the successes and challenges that exist in this institutional framework in relation to helping producers adapt to one impact of climate change—drought.

Although policies and programs for responding to present-day droughts and floods have existed for some time, these initiatives have not been reinvigorated to respond to droughts lasting more than two years, as is anticipated with future climate change. Many policies and programs do assist with adaptations, but they are not currently structured around responding to this larger issue. Framing these programs and policies in relation to future climate change may assist in their implementation, allowing producers to plan for a longer term. Local watershed planning is a perfect forum for pursuing discussions of anticipated future climate change and appropriate community and watershed adaptations.

The federal government's lack of attention to drought and climate change mitigation and adaptation is cause for concern. Leadership is required at the national level to comprehensively tackle future climate change, especially in the areas of climate mitigation and greenhouse gas reductions. Provinces, municipalities, and local watershed groups have led the way with comprehensive, sectoral initiatives. These important policies and programs need to be expanded with federal government support. As well, the federal government needs to enter into the policy and program space in relation to climate change adaptation and mitigation, not only in its national coordinating and planning role but also in relation to all sectors under federal jurisdiction, including international and interprovincial trade, energy, and waters.

References

Agriculture and Agri-Food Canada. n.d. "Drought Watch." Agriculture and Agri-Food Canada website. http://www.agr.gc.ca/eng/?id=1326402878459.

Agriculture and Agri-Food Canada, 2002. "Drought Risk Management Plan for Saskatchewan, 2002." In Canada-Saskatchewan MOU on Water Committee Minutes, June 24, 2002. Regina: Agriculture and Agri-Food Canada, Prairie Farm Rehabilitation Administration.

Alberta Agriculture and Rural Development. 2010. *Agriculture Drought Risk Management Plan for Alberta*. http://www1.agric.gov.ab.ca/$department/deptdocs.nsf/all/ppe3883.

Bakx, K. 2015. Alberta carbon tax could bring whopper of a bill to cities and towns. CBCnews. CBC website, November 2015. http://www.cbc.ca/news/canada/calgary/alberta-climate-change-calgary-edmonton-medicine-hat-1.3333666

CBC (Canadian Broadcasting Corporation). 2011. "Canada Pulls Out of Kyoto Protocol." CBC website, December 2011. http://www.cbc.ca/news/politics/canada-pulls-out-of-kyoto-protocol-1.999072

Cecil, B., H. Diaz, D. Gauthier, and D. Sauchyn. 2005. *Social Dimensions of the Impact of Climate Change on Water Supply and Use in the City of Regina*. Report prepared by the Social Dimensions of Climate Change Working Group for the Canadian Plains Research Center. Regina: University of Regina..

Demetropoulou, L., N. Nikolaidis, V. Papadoulakis, K. Tsakiris, T. Koussouris, N. Kalogerakis, K. Koukaras, A. Chatzinikolaou, and K. Theodoropoulos. 2010. "Water Framework Directive Implementation in Greece: Introducing Participation in Water Governance – The Case of the Evrotas River Basin Management Plan." *Environmental Policy and Governance* 20: 336–49.

De Souza, M. 2012. "Canada Officially Pulls Out of Kyoto Agreement." *The Leader-Post*, 15 December 2012. Regina, Saskatchewan.

De Souza, M. 2013. "Federal Report Supports Carbon Pricing. At Odds with PM's Public Position." *The Leader-Post*, 20 August 2013. Regina, Saskatchewan.

East, V., J. Pitman, J. Kylie, K. McRae, and E. Soulodre. 2012. *Upper Souris River Watershed Drought and Excessive Moisture Preparedness Plan*. Moose Jaw, SK: Saskatchewan Watershed Authority.

Government of Alberta. 2003. *Water for Life – Alberta's Strategy for Sustainability.* Edmonton: Queen's Printer.

———. 2008. *Alberta's 2008 Climate Change Strategy.* Edmonton: Queen's Printer.

Government of Canada. 2002. "Backgrounder on Drought Measures. Farm Financial Assistance Programs." Ottawa: Government of Canada.

———. 2010. *Fifth National Communication on Climate Change*. Submitted to the UNFCCC Secretariat on 12 February 2010. Ottawa: Government of Canada. http://unfccc.int/resource/docs/natc/can_nc5.pdf.

———. 2012. "Backgrounder on AgriInnovation, AgriMarketing and AgriCompetitiveness Programs." Ottawa: Government of Canada.

Government of Manitoba. 2015. Adapting to Climate Change: Preparing for the Future. https://www.gov.mb.ca/asset_library/en/beyond_kyoto/adapting_to_climate_change.pdf Accessed November 24, 2015.

Government of Saskatchewan. 2007. *Energy and Climate Change Plan 2007.* Regina: Premier's Office.

———. 2011. "Farm and Ranch Water Infrastructure." Government of Saskatchewan website. http://www.agriculture.gov.sk.ca/FRWIP_2009. Accessed 29 September 2011.

———. 2012. Website Archive—"2008 Farm and Ranch FRWIP." Government of Saskatchewan website. http://www.gov.sk.ca/news?newsId=a06a158f-551b-4c10-9f01-ba20e122bc42.

Hisschemöller, M., and J. Gupta. 1999. "Problem-Solving through International Environmental Agreements: The Issue of Regime Effectiveness." *International Political Science Review* 20, no. 2: 151–74.

Hisschemöller, M., and R. Hoppe. 1996. "Coping with Intractable Controversies: The Case for Problem Structuring in Policy Design and Analysis." *Knowledge and Policy* 8, no. 4: 40–61.

Hurlbert, M., Hague, S., Diaz, H. 2015. Vulnerability to Climate Extremes in the Americas Project (VACEA) Governance and Policy Assessment (Unit IC) (167 pp.) Available at: www.parc.ca/vacea/assets/PDF/reports/institutional%20 governance%20report%20final.pdf Accessed November 24, 2015.

Hoppe, R. 2011. *The Governance of Problems, Puzzling, Powering and Participation.* Briston: The Policy Press, University of Bristol.

IPCC (Intergovernmental Panel on Climate Change). 2001. *Climate Change 2001: Impacts, Adaptation, and Vulnerability. Technical Summary.* A Report of Working Group II of the Intergovernmental Panel on Climate Change 2001. Geneva: IPCC, World Meteorological Organization, and United Nations Environment Programme.

Pelletier, N. 2013. "IISD to Lead Engagement with Manitobans in Climate Change and Economic Development." Correspondence posted to the listserver (iisdnews@ lists.isd.ca), 30 July, 2013.

Public Safety Canada. 2007. "Canadian Disaster Database." Public Safety Canada website. http://www.publicsafety.gc.ca/cnt/rsrcs/cndn-dsstr-dtbs/index-eng. aspx. Accessed 10 September 2007.

RCAD (Rural Community Adaptation to Drought Project). 2012. *Rural Communities Adaptation to Drought: Research Report.* Regina: Canadian Plains Research Center.

Rowan, K., J. Pittman, V. Wittrock, K. Finn, and J. Kindrachuk. 2011. *North Saskatchewan River Watershed Drought and Excessive Moisture Preparedness Plan.* Moose Jaw, SK: Saskatchewan Watershed Authority.

Sauchyn, D., and S. Kulshreshtha. 2008. "Prairies." In D.S. Lemmen, F.J. Warren, J. Lacroix, and E. Bush (eds.), *From Impacts to Adaptation: Canada in a Changing Climate 2007.* Ottawa: Government of Canada. http://www. adaptation.nrcan.gc.ca.assess.2007/index_e.php. Accessed 24 July 2013.

Warren, J., and H. Diaz. 2012. *Defying Palliser: Stories of Resilience from the Driest Region of the Canadian Prairies.* Regina: Canadian Plains Research Center Press.

Water Security Agency. 2012. *25 Year Saskatchewan Water Security Plan.* Moose Jaw, SK: Water Security Agency.

Willems, S., and K. Baumert. 2003. *Institutional Capacity and Climate Actions.* Paris: Organisation for Economic Co-operation and Development.

Wittrock, V., and G. Koshida. 2005. *Canadian Droughts of 2001 and 2002: Government Response and Safety Net Programs–Agriculture Sector.* SRC Publication No.: 1101-2E03. Saskatoon: Saskatchewan Research Council.

CHAPTER 10

WATER GOVERNANCE IN THE PRAIRIE PROVINCES

Margot Hurlbert

Introduction: Water Governance and Adaptive Capacity

Water resources, water infrastructure, and livelihoods that depend on water (e.g., agriculture, forestry, and recreation) are expected to be significantly impacted by climate change in many regions of the world. An important determinant of a community's ability to adapt to future climate change impacts and current climate variability is its institutional setting and the degree to which this setting facilitates or hinders the community's adaptive capacity (Willems and Baumert 2003). As the Intergovernmental Panel on Climate Change (IPCC) argues, nations with "well developed institutional systems are considered to have greater adaptive capacity," and accordingly, developed countries have a better "institutional capacity to help deal with risks associated with future climate change" (2001: 896 and 897). Institutions contribute to the management of a community's assets, the community members' interrelationships, and in turn their relationships with natural resources. Both formal institutions (e.g., government, non-profit organizations, and civil society organizations) and informal institutions (e.g., social norms, values, and contexts) contribute to the relationships of people to each other and natural resources.

The institutional context of adaptive capacity can be studied through an investigation of the institutions involved in governance. Governance encompasses laws, regulations, and organizations, as well as governmental policies and actions, domestic activities, and networks of influence, including international market forces, the private sector, and civil society (Demetropoulou et al. 2010: 341). It entails the interactions among structures, processes, rules, and traditions that determine how people in societies make decisions and share power, exercise responsibility, and ensure accountability (Cundhill and Fabricius 2010: 14; Raik and Decker 2007; Lebel et al. 2006). Thus, governance involves institutions through which citizens and groups articulate their interests, exercise their legal rights, meet their legal obligations, and mediate their differences (Kiparsky et al. 2012; Armitage et al. 2009). A rich literature has developed regarding adaptive governance, adaptive water governance, and specifically how the wider institutional context of governance can facilitate adaptation and improve adaptive capacity of communities. Adaptive capacity is especially important in responding to drought events. The governance framework surrounding drought (constituted by such things as water allocation laws, programs and policies facilitating drought preparation, and income stabilization in the event of drought) plays an important role.

A large body of literature is available on the adaptive governance of water and the subsumed institution of water law. Water law establishes the formal framework of rules within which people and organizations operate in relation to water, and it constitutes a foundation of water governance. Water governance refers to the range of political, social, economic, and administrative systems that develop, manage, and distribute water resources (GWP 2009: 14). It involves public and civil society organizations and comprises norms, programs, regulations, and laws relevant to the management of water resources (Hall 2005; see also Conference Board of Canada 2007; UNDP 2007; de Loe and Kreutzwiser 2007).

This chapter reviews adaptive institutional design principles applicable to water governance, the structure of water governance in the Canadian Prairie provinces, and the legal tools and instruments most germane to water and the occurrence of drought. These legal instruments are then analyzed in relation to the principles of adaptive governance.

Adaptive Institutional Design Principles

How do we recognize a system of water governance as adaptive? Within the adaptive capacity literature, several dimensions have been identified as important characteristics called institutional design principles, or features of governance systems that define an institutional system as adaptive. These dimensions include such things as "availability of information," "openness for experimentation," "flexibility," "learning," and others. The discussion in some cases is generic and applies to institutions in general (Gupta et al. 2010; Olsson et al. 2006; Folke et al. 2005; Gunderson and Holling 2002;) and in other cases applies to specific institutional regimes, such as water governance (Huntjens et al. 2012; Hill 2012; Cook et al. 2011; Young 2010; Mollenkamp and Kastens 2009; Huitema et al. 2009). The literature refers to a proper understanding of the complexities of the phenomenon of climate change, which include the requirements imposed by boundaries, levels, sectors, and diverse stakeholders, as well as the uncertainties surrounding, and long-term time frame of, climate change (Gupta et al. 2010; Frohlich and Knieling 2013; Cook et al. 2011). Table 1 outlines these various dimensions.

Adaptive governance entails a more flexible, participatory, experimental, collaborative, and learning-based design and approach to policy making and governance to increase adaptive capacity of institutions and sustainability of natural resources (Pahl-Wostl 2010; Pahl-Wostl et al. 2007a, 2007b, 2007c; Kallis et al. 2006; Tompkins and Adger 2004; Walters and Holling 1990; Lee and Lawrence 1986; Walters 1986). Adaptive governance shifts focus from rule-based, fixed organizations to a view of institutions as dynamic, flexible, pluralistic, and adaptive in order to cope with present and future uncertain climatic conditions and the limits of predictability (IISD 2006: 5; Carpenter and Gunderson 2001; Levin 1999). Adaptive governance then becomes a means to achieve adaptive capacity (Cook et al. 2011). Assessing whether a governance regime is adaptive entails a consideration of its institutional structure and its most important constituent parts (or instruments). For instance, crop insurance is an instrument that helps producers stabilize income in times of drought.

Table 1. Institutional design principles of adaptive governance

	Institutional design principle of adaptive governance	Related principles / sub-principles	Explanation	Literature
1	Responsiveness		The ability of governance networks, organizations, and actors to respond appropriately and in a timely manner to climate variability, hazards, and extreme events in a manner that accounts for ecosystem dynamics	Hatfield-Dodds et al. 2007; Kjaer 2004; Dietz et al. 2003
		Robust and flexible process	Institutions and policy processes that continue to work satisfactorily when confronted with social and physical challenges but at the same time are capable of changing	Huntjens et al. 2012: 73; Mollenkamp and Kastens 2009
2	Variety of problem frames		Openness to multiple frames of reference, opinions, and problem definitions offering diverse and sometimes competing solutions and options to assess a problem as well as resolve conflict	Gupta et al. 2010
		Multi-level – redundancy	A variety of problem frames inherently involves participation of a variety of different actors, levels of government, and sectors in the governance process and collective choice arrangements, without redundant overlapping costly systems.	Huntjens et al. 2012
		Polycentric governance	Different centres of management and control should exist (as opposed to hierarchical systems).	Ostrom 2010
3	Learning and institutional memory		Past experiences must be remembered, learned from, and routines improved.	Huntjens et al. 2012; Gupta et al. 2010; Armitage 2005; Olsson et al. 2004; Pretty 2003; Dietz et al. 2003; Pretty and Ward 2001
		Participation	Participation by non-state actors	Folke et al. 2005
		Collective choice arrangements	Enhance participation of those involved in making decisions about the system in how to adapt	Huntjens et al. 2012

		Monitor and evaluate	Institutional evaluation processes must monitor and evaluate policy experiences.	Huntjens et al. 2012
4	Trust		Institutional patterns must exist to promote mutual respect and trust so that participants continue to be involved in the process of governance.	
		Open to uncertainty / open to experimentation	Policy experiments allow feedback loops so policy can be changed quickly in response to changed conditions.	Mollenkamp and Kastens 2009
		Constructive conflict resolution	Timely response to problems, careful sequencing, transparency	Huntjens et al. 2012
5	Capacity building	Information Leadership Resources	The informational, human, and social capital must exist within the governance regime to respond appropriately to climate variability, hazards, and extreme events. Leadership must exist to act as a catalyst to change. Appropriate resources (financial, political, human) must be available for this change.	Gupta et al. 2010; Olsson et al. 2004
		Information	Rigorous up-to-date information; sufficient and reliable	Mollenkamp and Kastens 2009
6	Equity	Legitimate Accountable Fair	The governance regime must be perceived as legitimate and accountable, as well as fair in its process and impact, so that there is an equal and fair (re-)distribution of risks, benefits, and costs.	Gupta et al. 2010; Huntjens et al. 2012; Ostrom 2011
7	Political support		Responding to climate change is a long-term policy challenge that requires solid political support for plans longer than election cycles.	Molenkamp and Kastens 2009
8	Clearly defined boundaries		Clarity over who has rights; who has responsibility, capacities, access to resources, and information in times of climate events	Huntjens et al. 2012

Source: Hurlbert and Diaz 2013.

Similarly, water infrastructure programs assist producers in building water-retention facilities and shallow pipelines, which also increase adaptive capacity in times of drought.

Institutional Structure of Water Governance

Water governance in the Prairie provinces involves many actors, including the government (all levels) and civil society organizations. Water in Canada is essentially the mandate of the provinces; however, there are shared jurisdictional roles with the federal government (e.g., transboundary flow, environmental protection) and some delegated function to local municipal governments (e.g., drinking water, land use, environmental protection). Nineteen federal government agencies are involved in water governance issues across Canada (Hurlbert et al. 2009). Environment Canada prescribes national drinking-water standards, monitors interprovincial streamflows, and facilitates the work of the Prairie Provinces Water Board (an agency overseeing the agreement apportioning flows between Alberta, Saskatchewan, and Manitoba). The International Joint Commission administers the Canada-US Boundary Waters Treaty. The number of federal agencies involved in water governance on the Prairies will be reduced as the Agri-Environmental Services Branch (formerly the Prairie Farm Rehabilitation Administration) winds down through government layoffs and program terminations. This institution assisted rural adaptation and water infrastructure development and management in the Canadian Prairies and its dismissal clearly will affect the adaptive capacity of agricultural producers (see Chapters 5, 6, and 8 in this volume).

At the provincial level, each province has an entity responsible for water: in Saskatchewan it is the Water Security Agency, in Alberta it is the ministry of Alberta Environment and Parks, and in Manitoba it is the Manitoba Water Stewardship Division. However, other departments and government organizations also play a role in water. In Saskatchewan and Manitoba, government branches responsible for the environment and health also play a lesser role in relation to water. In Alberta, a 24-member, non-profit Alberta Water Council oversees the province's water strategy and facilitates water disputes between sectors. In Saskatchewan, 19 members of the Saskatchewan Watershed Advisory Committee advise on water issues. All provinces have a host of watershed associations (some constituted pursuant to legislation, others non-profit) or conservation districts (Manitoba) involved in source water protection planning. Table 2 lists these institutions.

These organizations manage day-to-day decisions pertaining to water, including water allocation and decisions impacting water quality. Considerable similarity exists between the provincial organizations (as outlined in Table 2); however, the structure of water law used in each province differs. Table 3 summarizes the major features and differences between the legal institutional structures of water law in the three Prairie provinces.

This table is organized around the "principle" of water management for each province, which has been categorized by the author. Alberta states that the purpose of its water legislation is to support and promote the conservation and management of water balanced with the need to manage and conserve water resources to sustain a healthy environment, and the need for Alberta's economic growth and prosperity (Water Act, R.S.A., c. W-3). The *25 Year Saskatchewan Water Security Plan* states its vision of water as "supporting economic growth, quality of life and environmental well-being" (Water Security Agency 2012: 3). Water is considered a finite resource requiring a long-term perspective managed adaptively through collaborative processes. Although this plan mentions the interests of future generations, the legislation envisions management for economical and efficient use, distribution, and conservation of the water without mention of these future interests (the Saskatchewan Water Security Agency Act). The Manitoba Water Protection Act specifically states in its preamble that an abundant high-quality water supply is essential to sustain life now and in the future and is a "fundamental right of citizens"; the Water Resources Conservation Act states in its preamble that water is to be administered based on the precautionary principle and sustainable water resource management practices and that legislated priority is given to domestic, municipal purposes over agricultural, industrial, irrigation, and other purposes. Because of these principles, the Manitoba legislation has been termed as treating water as "public property." These principles of water governance structure determine the nature of the instruments created by legislation and policy surrounding water covered in the next section.

Table 2. Key provincial and federal government agencies with water mandates

Provinces	
Alberta Environment and Parks / Saskatchewan Water Security Agency / Manitoba Water Stewardship	Responsible for water allocations; licensing; oversight of municipal treatment of drinking water and wastewater; watershed management in partnership with watershed groups; and planning, monitoring, and protection of water quantity and quality in surface water and groundwater systems in the environment. In the 1930s, the federal government transferred responsibility of water resources to the provinces via Natural Resource Transfer Agreements.
Health Ministries	Responsible for protecting public health (e.g., drinking water, wastewater management); acts as decentralized authority to regional health authorities.
Agriculture	Irrigation, drought management; encourages adoption of agricultural best management practices to protect water supplies from agricultural contamination; assistance for on-farm agricultural and domestic water supplies.
Alberta Emergency Management Agency / Saskatchewan Ministry of Government Relations	Coordinates, collaborates, and co-operates with all organizations in prevention, preparedness, and response to disasters
Alberta Drought Management Committee / Saskatchewan Drought and Excess Moisture Committee	Committees monitor, plan for, and provide alerts for drought conditions; committees focus on reporting, monitoring, and response actions
Government of Canada	
Environment Canada	Surveys and monitors water quality and quantity; regulates trans-boundary flow; enforces and protects the aquatic environment; and conducts water and climate research. Environment Canada and provincial ministers of the environment set the Canadian Environmental Quality Guidelines. Guidelines pertinent to water include limits established for the protection of aquatic ecosystems, municipal uses of water (community supplies), recreational uses of water, and agricultural uses of water (Canadian Council of Ministers of the Environment). Environment Canada leads the Prairie Provinces Water Board.

Health Canada	Sets Guidelines for Canadian Drinking Water in partnership with provinces; sets health-based standards for materials in contact with drinking water; assists First Nations with drinking water safety on their lands; provides drinking water guidance to other departments, governments, and citizens; regulates the manufacture and sale of pesticides in the Pest Control Products Act, co-leads the Canadian Environmental Protection Act with Environment Canada
Agriculture and Agri-Food Canada	Encourages adoption of agricultural best management practices to protect water from agricultural contamination; PFRA responsible for applied research and rural water management (water supply/quality, irrigation, climate, drought adaptations)
Natural Resources Canada	Conducts groundwater mapping and monitoring; conducts water and climate research; responsible for climate programs and activities with Environment Canada (e.g., lead for Canada's now defunct Climate Change Secretariat)
Fisheries and Oceans	Responsible for protecting, managing, and controlling inland and marine fisheries; conserving, protecting, and restoring fish populations and fish habitat; preventing and responding to pollution; and regulating navigation
Public Safety Canada	Responsible for disaster planning, recovery, and response
Coordinating water management institutions	
Prairie Provinces Water Board	Federal-provincial board to manage inter-jurisdictional water issues in the Prairie provinces (Alberta, Saskatchewan, and Manitoba); board includes representatives from Environment Canada, Agriculture Canada, Alberta Environment and Parks, Saskatchewan Water Security Agency, and Manitoba Water Stewardship; board addresses inter-provincial water issues (allocations, flows, water quality)
Watershed advisory councils and boards / conservation districts	A variety of watershed councils, conservation districts, and groups exist in each province. Their focus is on water management by landscape boundary (defined as a watershed for surface water and an aquifer for groundwater). Watershed groups involve all water users, local government, and provincial and federal governments, each working to identify and address water management issues unique to each watershed.
Irrigation districts	Irrigation districts in the South Saskatchewan River basin (SSRB) manage water for irrigated agriculture for scale field crops. Because these are large water users, the districts work with provincial agencies to manage water in the SSRB. Irrigation in the SSRB accounts for 90% of consumptive water use in the SSRB.

Table 3. Institutional legal water structures of the Prairie provinces

Province / principle under which water is managed	Alberta	Saskatchewan	Manitoba
Principle	Most beneficial use	Common property	Public property, future generations, and precautionary principle included
Allocation of water rights	Statutory legislated model	Licensed interests allocated by the Water Security Agency on conditions considered appropriate	Statutory legislated priorities
Priorities	First-in-time, first-in-right principles	No statutory priority scheme	Order of priority: domestic, municipal, agricultural, industrial, irrigation, and then other purposes
Water market	Transfers of water independent of land allowed	None	None

Water Instruments

Within the context of laws, regulations, and policy, specific policy instruments are used to influence behaviour and effect a certain response (Anderson 2010: 242). Although many types of instruments exist (Gupta et al. 2013: 45; Baldwin et al. 2011; McManus 2009), this chapter focuses on market or economic interests—the property interest of water. Instruments can be classified into four categories: regulatory, economic and market-based, suasive, and management (Gupta et al. 2013: 45). Although this classification is not ideal because there is much overlap and potential for errors in deciding on a classification, examples of these instruments in the case of water (and drought specifically) are listed in Table 4.

Table 4. Classification and description of instruments

Instrument	Description	Example
Regulatory	Adopted by the state authority; binding; determining what is permitted and what is illegal, including sanctions for non-compliance; without a market component (McManus 2009; Baldwin et al. 2011)	Holdback for minimum river flow requirements on water transfers Water licences with terms and conditions
Economic / market-based	Encourage behaviour through market signals rather than explicit directives (Stavins 2003)	Tradable water rights Water tariffs
Suasive	Measures that internalize environmental awareness and responsibility into individual decision making through persuasion (OECD 1994) Public and private information, research, and public awareness	Public participation in watershed planning Drinking water quality reports and alerts Drought prediction and alerts
Management	Includes mostly self-management by private actors but could be hybrid management processes	Local watershed governance Source water protection plans Irrigation association constitutions

Source: Adapted from Gupta 2013: 45.

There are three major instrumental contexts relating to the bundle of property rights associated with water; these contexts concern whether water is privately owned (as a saleable interest as in Alberta or Chile [Bauer 1998]), is public property (freely available to all), or is common property (owned by the water users). In the Prairie provinces, because the Crown owns all water and because water rights are allocated by licence, this property ownership distinction is not applicable; however, the property distinction is illustrative, as parallels can be seen in the characteristics of bundles of water rights received by way of water licence. Based on the three models of property rights (see Table 3), the three instrument models are as follows:

- Government agency management, generally associated with water regarded as public property: Government defers its authority for managing water to an agency, which assumes authority for directing who receives water rights in accordance with bureaucratic policies and procedures. In Canada, water is owned by the State (or Crown), and interests are allocated by licence. Often a first-in-time, first-in-right priority scheme applies (Hurlbert 2008). This model is implemented through water licences with terms and conditions, or regulatory instruments.

- User-based management, generally associated with water regarded as common property: Water users, or those with licence or rights to water, join together and coordinate their actions in managing water resources. Decision making is collective among users. Irrigation associations are an example of this type of ownership; another example is co-managed water resources (Plummer 2009). This model is an example of the use of management instruments to manage water (i.e., water is managed by private actors).

- Market, generally associated with water owned as private property: Water is allocated and reallocated through private transactions. Users can trade water rights through short-term or long-term agreements or temporary or permanent transfers, reallocating rights in response to prices (Bruns and Meinzen-Dick 1995). This model is an example of an economic or market-based instrument.

Sometimes these instruments are used in combination. Alberta has led the provinces in developing a water market where transfer of water rights is allowed in accordance with an approved water management plan or by Cabinet order in the absence of such a plan. These transfers are possible only within six districts. However, water continues to be owned by the Crown; a licence is granted to property owners in respect of a parcel of land and then transferred with the land. It is possible to transfer a water interest. For example, Alberta's water management plan for the South

Saskatchewan River basin allows the director to consider applications to transfer water allocations within the basin (Alberta Environment and Parks 2015). This market-based management model used by Alberta is not a true laissez-faire market with vendors and purchasers conducting transactions purely based on market rules; a certain amount of oversight is retained in the review of these transactions, and, as such, the predictability of a market model is reduced somewhat (Hurlbert 2009a). This market model aligns with the principle of most beneficial use (outlined in the structure of water governance above). In Manitoba and Saskatchewan, the government agency management model is used, with the government allocating licences and determining priorities. All three provinces have employed a degree of user-based water management with the development of source water protection plans by local watershed committees. The persuasiveness of these plans and the permanence of this activity have yet to be determined.

Analysis

The provincial structures of water governance, with a specific focus on the property rights of water, are analyzed in this section in relation to the institutional design principles of adaptive governance. This analysis is carried out based on the characteristics of the provincial water governance structures described in Tables 3 and 4. This section discusses the economic or market instruments used in Alberta, but this description is perhaps overgeneralized. The Alberta water governance structure predominantly uses regulatory or government agency management instruments, but also makes considerable use of water management instruments (e.g., source water protection planning by irrigation associations and local watershed groups). Although Saskatchewan does not have tradable water interests, it uses government agency management instruments, but also makes considerable use of management instruments (e.g., source water protection planning by local watershed groups and irrigation associations).

Manitoba's system has been characterized as using a government agency regulatory instrument and user-based management approach because it embraces both source water protection planning and principles of future sustainability. This assessment is summarized in Table 5.

Table 5. Assessment of institutional principles in each province

			Alberta (economic)	Saskatchewan (government)	Manitoba (government / user)
	Institutional design principle of adaptive governance	*Related principles/ sub-principles*			
1	Responsiveness	Robust and flexible process	Market instruments provide poor response to social conditions. Remainder of Alberta's water governance structure is similar to that of Saskatchewan and Manitoba.	Government agency has legislative ability to respond in timely fashion.	Government agency has legislative ability to respond in timely fashion. User groups are context specific.
2	Variety of problem frames	Multi-level – redundancy Polycentric governance	Concerns with federal government withdrawal from water governance	Concerns with federal government withdrawal from water governance	Concerns with federal government withdrawal from water governance
3	Learning and institutional memory	Participation Collective choice arrangement	Not applicable because market instruments operate in real time.		

4	Trust	Open to uncertainty Constructive conflict resolution	Trust – beyond the scope of the paper. Poor access to justice—court remedies after a drought are slow and expensive	Ability to provide timely response in legislation	
5	Capacity building	Information Leadership Resources	Further research required	Further research required	Further research required
6	Equity	Legitimate Accountable Fair	Further research required	Further research required	Further research required
7	Political support		Present	Present	Present
8	Clearly defined boundaries		Interjurisdictional issues unclear in face of increasing drought		

Responsiveness

A tradable water interest, or market instrument, responds to the terms and conditions created within the market and the regulation of that market. The Alberta market model was developed specifically to more efficiently allocate and price water. The statutory provisions allowing transfer are touted by some researchers, and the Alberta government, as advancing the goals of efficient allocation of water interests and conservation in encouraging the transfer of surplus interests. This process is also described as creating a non-regulatory method of reducing wasteful use by creating an incentive to save water and transfer its marginal value for compensation (Percy 2004). Many would argue this market instrument does not capture the community value of water, nor does it facilitate political and ethical considerations in allocation decisions. The risk of the market instrument is that impacts on third parties not directly involved in a market transaction are neglected, and third parties have difficulty enforcing their interests in a court of law. These characteristics make the market instrument in relation to water property rights not as responsive as a system whereby governments and all users can hear and determine water issues. It should be kept in mind that only a small fraction of Alberta's water governance structure entails tradable water interests.

However, studies of water governance that have focused on how the institutional context of the regulatory tools of government have managed water structures in Alberta and Saskatchewan have concluded that challenges in relation to responsiveness exist. One study concluded that improvements are needed to increase the efficacy and effectiveness of organizations and processes of water governance, as much fragmentation impedes setting clear policy objectives and implementing, assessing, monitoring, and evaluating policy (Hurlbert et al. 2009: 123; see also Bakker 2007). Further, there is limited institutional coordination and integration, which is a result of management rigidity (Hurlbert and Diaz, 2013). To improve responsiveness, a robust channel of communication between local communities and water governance organizations is needed (Hurlbert et al. 2009: 124).

An abundance of academic literature concludes that management instruments effected by local watershed governance and participatory resource co-governance (such as that practised by irrigation associations) are more responsive (e.g., Hickey and Mohan 2004; Brooks 2002). More research is needed to determine conclusively which structures respond in more timely and appropriate manners. It would appear that a market instrument might allow timely response to certain economic interests, whereas water user conflicts in relation to scarcity of water in times of drought might be best addressed in a timelier manner by regulatory government agency tools or user-based management tools.

Variety of problem frames

The multitude of government agencies involved in water management results in a variety of problem frames in relation to water issues. In the Canadian constitution, water is not treated as a single topic assigned to one level of government (federal versus provincial). The provincial government has powers that relate to water, including property (generally including water in its definition).[1] The federal government also has certain powers in relation to water, albeit historically somewhat more limited than the provinces.[2] Limits would include powers in relation to water allocation to facilitate navigation and in relation to water quality and quantity to maintain and preserve fish populations and their habitat. The federal government takes control of water once it crosses an interprovincial or international boundary, in accordance with the federal head of power relating to interprovincial works and undertakings (Kennett 1991). Often overlap exists and both levels of government share jurisdiction in relation to certain aspects.

Although the multitude of water organizations existing at each level of government would appear to give rise to the possibility of a variety of problem frames, this is not the case in practice. When the federal government developed a Federal Water Policy in 1987, it was not fully supported with the necessary resources and never fully implemented (Hurlbert and Cokal, 2009). Although there have been numerous calls for a renewed Canadian water strategy (e.g., Barlow 2011), a comprehensive strategy has not been formulated and does not appear on the federal government's agenda. As a result, water is increasingly governed provincially. In addition, the federal government has withdrawn from many water governance activities it had historically been involved in, such as irrigation infrastructure (see Chapter 6 by Warren on irrigation in this volume) and community pastures. This withdrawal has negatively impacted the variety of problem frames in relation to water as well as the polycentric nature of Canadian water governance.

In the event of future water shortages, the lack of a federal water mandate could also have significant implications if interprovincial conflicts arise. The current Master Agreement on Apportionment between Canada, Alberta, Saskatchewan, and Manitoba contains a strict formula for sharing water.[3] In the event of severe water shortage, the lack of drinking water for Saskatchewan residents will be inconsequential, as the formula is the only mechanism of allocation. This strict formula was developed partly as a response to disagreement between Saskatchewan and Alberta on what developments should occur and to a mandate change several decades ago. This historical impasse should not be forgotten as water shortages loom on the horizon. Research confirms that having response mechanisms in place is important in addressing issues and potential conflicts (Adger 2003).

The addition of a tradable water interest adds an important economic tool for capturing surplus water and creating financial incentives to conserve and realize efficiencies in relation to water allocations. More research is required to ascertain if these market instruments solve these problems in relation to fully allocated watersheds. A tradable market water instrument allows only the considerations built into the legislated regulatory fabric of the market to be reflected in the problem frame. Many issues could arise if and when shortages of water are so severe—as is in the case of extreme drought—that the traded water interests cannot be met within Alberta while honouring the historic water agreement between Canada, Alberta, Saskatchewan, and Manitoba.

Learning and institutional memory

The market instrument—the tradable water interest—responds to current conditions at the time it is used. As such, any learning and institutional memory would relate to the actors participating in the market. At present, trades of these interests are sparse, and details such as this require further research. Studies have been conducted on the institutional context of water governance in relation to learning and institutional memory in the Prairie provinces—the regulatory instrument or government agency–based water management (Diaz et al. 2009)—and some of the findings detailed below arise from this work.

The Prairie provinces have been managing the water resource since its transfer to them by the federal government in the 1930s. One of the biggest challenges facing all three provinces is the aging workforce and the retirement of key personnel who have the institutional memory of managing this resource. It will become increasingly important to develop strategies to document this knowledge, transfer it through mentorship to the emerging younger workforce, and maintain access to the retiring workforce through novel retention arrangements.

Alberta has a long history of water policy, strategy, and planning through its Water for Life initiative. Manitoba's history relates to its drainage and conservation district management. Saskatchewan's first water strategy was issued in 2012, but one Crown corporation has been tasked with water management in Saskatchewan for decades. The relatively recent use of the management instrument—local watershed-source protection planning—should facilitate the transfer of knowledge of water governance between these local watershed groups and the water users (i.e., the public and other stakeholders) interacting with these groups.

This process will provide an additional strategy to transfer knowledge and retain past learnings to address the issue of pending civil-servant retirements.

The federal government's lack of involvement in water and water strategy since 1987 leaves an important gap in jurisdictional strategy, which potentially hinders long-term learning and institutional memory. The Prairie Provinces Water Board's mandate relates to implementing a historic water sharing arrangement. Particularly given anticipated future drought, the absence of a long-term national plan limits the possibility of a flexible institutional governance environment able to identify social needs and problems in relation to impending climate change, balance competing interests, and execute and implement solutions. As a result, drought or extreme climate events will be addressed in a reactive manner, instead of using a flexible, proactive policy response, which would stimulate learning.

Trust

The market instrument—the tradable water interest—creates a market for the transfer of water interests. If market rules are clear and transparent, those able to access the market will in all likelihood have a degree of trust in the market. However, the broader institutional structure of water governance in this context is arguably different. Those without access to water interests would in all likelihood not experience the same trust. Further research defining and exploring trust and the perspectives of participants in water governance is needed. Previous studies have expressed some scepticism as to how trustful participants might be of market and government contexts concerning water governance, specifically in relation to the resolution of conflicts over water (e.g., Hurlbert and Diaz 2013).

The increasing spectre of water shortages is expected to amplify potential conflicts among current water rights holders. The current institutional context appears not well situated to respond to these conflicts. The Saskatchewan and Manitoba system appears situated within a government review and reconciliation framework; Alberta's within a court and litigation-based framework. Albeit the former may be more conducive to timely resolution of conflict with less expense, both systems are in need of improved access to justice. Failure to provide this access may erode trust and ultimately legitimacy.

Capacity Building

It is difficult to postulate how the market instrument—the tradable water interest—might impact information, leadership, and resource capacity. Research methods teasing out insights in this regard also raise many challenges in relation to both choice of method and implications of results. However, the following case study uncovered in previous research studies is informative.

The Institutional Adaptation to Climate Change Project (http://www.parc.ca/mcri) uncovered a case study wherein Alberta's water transfer provisions—or market instruments—were used in the 2001 drought, but this case also illustrated the usefulness of user associations—or management instruments. Usually during years of water shortages, regional staff of Alberta Environment had to advise junior licensees (or last-in-time licensees) that they needed to shut down their pumps and were being cut off. In the St. Mary's River in 2001, there was a severe water shortage, which was going to allow only six or seven licences to operate. Stop orders would have had to be issued on 500 to 600 licences, which could have dried up the river. Sharing provisions that were put into the Water Act between 1993 and 1996 allowed two licences to share water back and forth (if physically possible), as long as no other licensee complained that it hurt their right.

Irrigation districts sent out letters to their licensees and held meetings to discuss water shortages. A smaller percentage of water allocation for each licence was agreed on (approximately 60%). However, because irrigators and other users of water could not meet their agricultural or business needs with this smaller allocation of water, novel arrangements were made. Farmers transferred their allocation to other farmers in exchange for agreed-upon consideration, which allowed at least one farmer to irrigate and grow a crop that year. Approximately 70 licensees did not agree to the sharing arrangement and received stop orders (Hurlbert 2009b, 2009c; see also Chapter 11 by Corkal et al. in this volume).

This case study illustrates an important finding. Although institutional contexts are often portrayed as mutually exclusive totalities (as illustrated above in the characterization of the three Prairie provinces' water governance structures), the reality is that the Prairie provinces use a combination of institutional contexts and thus a combination of instruments that embrace these concepts. How these instruments are employed and accessed, and therefore how they operate in conjunction with one another, warrants further consideration and study.

Equity

As with many of the other indicia of adaptive governance, it is difficult to assess the equity in relation to water governance instruments without appropriate primary social science research. Perceptions of participants in the institutional water governance context on legitimacy and accountability would be particularly germane. However, failing this, the case of Chile, where tradable water property interests are the sole water instrument in relation to water property interests, sheds some light on the use of one sole instrument. In Chile, a Water Code established a market for water rights, where water rights are treated as any commodity, so they could be sold, rented, and transferred to other people. The government has a very limited role in administering water transactions and water conflicts, since they are defined as issues to be resolved between private individuals. Given that water resources are fully allocated in some areas, many local communities and small, medium, or poor farmers may be without water rights and without the means to purchase them (Reyes et al. 2009; Bauer 1998: 67).

The adoption of a neo-liberal Water Code—where water is considered a privately owned commodity—has been an imposition of a top-down system that has not only limited the capacity of governance to establish adaptive water strategies at the regional level but also has imposed a process of competition in a context characterized by an unequal distribution of power (Galaz 2003), resulting in an adaptive capacity to water scarcities that is concentrated in a small number of large producers with the ability to more easily obtain access to water rights. This situation has resulted in inequitable water governance structure in times of drought.

Political Support

The selection of water instruments predominantly used by the Prairie provinces would appear to have little relationship to a province's support for climate change action. Although the market-based beneficial-use water governance structure of Alberta places considerable onus on individuals to make informed decisions in relation to risks such as climate change, the Alberta government has had a climate change strategy for some time. The Climate Change and Emissions Management Act (2003) was a precursor to Alberta's Climate Change Strategy (2008) and focused on risks and vulnerabilities to water. In addition to establishing a carbon offset market and providing consumer rebates in relation to energy efficient products, two programs were also introduced, a greenhouse gas reporting program and a greenhouse gas reduction program. These relate to the establishment of a greenhouse gas limit. In 2003, Alberta also created its Water for Life strategy focusing on issues of quantity, quality, and conservation of water, which has continuously been reviewed and revised (AWC n.d., 2009, 2007, 2005) The strategy initiated three important activities: 1) planning for future management of water via the provincial Climate Change Adaptation Strategy; 2) developing land-use frameworks; and 3) watershed planning through local watershed groups. All of these activities are important for adaptation to climate change.

In Saskatchewan, a previous New Democrat Party government issued an *Energy and Climate Change Plan*, which was a cross-governmental vision in response to climate change and the development of a province-wide climate change adaptation strategy that included working with research organizations and supporting critical local research on climate change and adaptation (Government of Saskatchewan 2007). Currently, climate legislation relating to mitigation remains on the legislative agenda but is yet to be proclaimed. However, the *25 Year Saskatchewan Water*

Security Plan (Water Security Agency 2012) states that work with research partners on climate change impacts will continue to identify possibilities for adaptation.

In Manitoba, climate change considerations are acknowledged within legislation and the climate change strategy document *Adapting to Climate Change: Preparing for the Future* (2015). At the legislative level, climate change is acknowledged in the Water Resources Act. In its preamble, the Act states the following:

> In light of the fact that future domestic needs and the potential effects of climate change are unknown, such a scheme should be based on the precautionary principle and on sustainable water resource management practices. (n.p.)

In its climate change plans, the Government of Manitoba discusses actions implemented to date and future directions. Actions-to-date relating to climate change adaptation include developing integrated watershed management plans, revising flood protection plans, expanding Manitoba's hydrometric network, introducing incentive-based programs, and developing research relating to land-use planning (Government of Manitoba 2008: 47). The document addresses sector-based climate change adaptation. For example, within the agricultural sector, "climate friendly" best management practices are recommended, such as "improved handling, treatment, storage and application of manure to reduce CH_4 and N_2O emissions" (Government of Manitoba 2008: 3). Within the energy sector, the Manitoba government emphasizes minimizing reliance on fossil fuels and maximizing energy efficiency through programming (Government of Manitoba 2008: 4). The role of municipalities in promoting adaptive practices is also discussed through the idea of "climate friendly planning."

A challenge in the Prairie provinces' water governance structures in recent years relates to the long-term and comprehensive consideration of climate change adaptation within the water governance agenda (Hurlbert et al. 2009). Although some inroads have been made by each province, it would appear that a considerable opportunity exists for expanding policy in this area.

Clearly defined boundaries

User-based management instruments can result in sandbox politics and can fail to provide clearly defined boundaries with respect to water interests, resolution of uncertain water relations, and water strategies into the future. This is due to the participatory, iterative nature of user-based management processes. However, use of this form of instrument of governance, in combination with other approaches, such as that of government agency management and perhaps a well-constructed and limited market instrument, can be highly beneficial. The key is establishing clear conditions of market instruments, well-conceived government management back stops, and functions within the water governance structure that facilitate success of user-based management instruments. Currently, employment of user-based management instruments in relation to source water protection planning and day-to-day management of irrigation districts has proven highly successful.

Market-based instruments in relation to water governance must be as clearly defined and transparent to the public as the mechanisms within the legislative, regulatory, and policy foundation establishing them. In Alberta, use of this instrument and fulfillment of these institutional principles require further research. Because its use has been relatively infrequent, the urgency of this research is reduced.

Bakker and Cook (2011) have concluded that there is an urgent need to establish clear roles for all of the various actors involved in water governance and coordinate their activities to avoid increasing balkanization of water management. As provincial strategies such as Alberta and Saskatchewan's become increasingly known and embraced by the public, it is anticipated that the need for involvement will be met. In recent years, the provinces have embarked on important initiatives to identify and coordinate actors involved in governance; however, further attention is warranted. Some uncertainty exists in relation to jurisdictional matters (such as First Nations' interests) and interprovincial issues, which may arise in the face of increasing water shortages and the federal government's withdrawal from involvement in water governance. Establishing and supporting local watershed groups are important components of comprehensive water planning and management, as in the geographical space of the watershed all actors, all levels of government, and all issues come together. This geographical space is the site of integrated watershed management. Although these groups embody the user-based management principle, they hold an important place within the water governance structure by helping to make boundaries clear, real, and understood by local people.

Conclusion

This analysis has illustrated some of the considerations pertinent to regulatory instruments (government-allocated water licence interests), user management instruments (local watershed groups and irrigation associations), and market instruments (transferable water interests) in the context of climate change and expected increasing variability in climate, specifically drought. An institutional context of water governance structure whereby multiple water instruments operate has been used because water property interests in Canada are best described as a bundle of entitlements effected through a combination of management, regulatory, and market instruments.

On their face, market instruments appear to respond poorly to all peoples' interests, reflect only economic problem frames, and exclude individuals who are without tradable interests. As a result, market instruments scored lower in relation to the institutional design principles of adaptive governance (trust, capacity building, and equity). However, positive examples of adaptation emerge when analyzing the use of market instruments in combination with regulatory instruments and management instruments. These cases, of course, are illustrative only; more research using different methods is required to provide additional evidence.

Assessments of water governance structures in the Prairie provinces have concluded that more effort is required to define institutional boundaries, communicate roles of water organizations, and coordinate among water organizations. The federal government's absence from the water policy field is worrisome given the prospects of increasing climate variability and drought in the future. As the impacts of future climate change add strain on water resources and the incidents of drought increase, more work will be required on comprehensive sectoral adaptation to leverage and optimize the initial work that has been done to date. Using an institutional framework and the institutional principles of adaptive governance in this preparation would help reduce vulnerabilities of individuals and communities.

NOTES

1 These headings include publicly owned lands, mines, minerals and royalties, property and civil rights, local works and undertakings, and natural resources. which include the right to make laws in relation to the development, conservation, and management of non-renewable natural resources and forestry resources in the province. It is through the first heading "lands" that the provincial jurisdiction to water primarily resides. In traditional Canadian common law, water rights transferred with the land with which it was associated. "Land" is defined as "every species of ground, soil or earth whatsoever, as meadows, pastures, woods, moors, waters, marshes, furs and heath" Jowitt (1959: 1053).

2 These include federal lands (national parks, Indian reserves), trade and commerce, navigation and shipping, seacoast and inland fisheries, works for the general advantage of Canada, entering into treaties, and matters not specifically assigned to the provinces (s. 91 of the Constitution Act, 1982). The federal government is responsible for ensuring the safety of drinking water within areas of federal jurisdiction, such as national parks and Indian reserves, and water quality in respect of inter-jurisdictional waters (Canada Water Act, R.S.C. 1985, c. C-11).

3 Water is shared such that 50% of flows must be passed to Saskatchewan, which in turn must pass the same proportion to Manitoba (Prairie Provinces Water Board, *The 1969 Master Agreement on Apportionment and Bylaws, Rules and Procedures*).

References

Adger, W. 2003. "Social Aspects of Adaptive Capacity." Pp. 29–49 in J. Smith, R. Klein, and S. Huq (eds.), *Climate Change, Adaptive Capacity and Development*. London: Imperial College Press.

Alberta Environment and Parks. 2015. South Saskatchewan River Basin Approved Water Management Plan. Available at: ep.alberta.ca/water/programs-and-services/river-management-frameworks/south-saskatchewan-river-basin-approved-water-management-plan/default.aspx. Accessed November 24, 2015.

Anderson, J.E. 2010. *Public Policymaking: An introduction*. Hampshire, UK: Cengage Learning.

Armitage, D. 2005. "Adaptive Capacity and Community-based Natural Resource Management." *Environmental Management* 35, no. 6: 703–15.

Armitage, D.R., R. Plummer, F. Berkes, R.I. Arthur, A.T. Charles, I.J. Davidson-Hunt, A.P. Diduck, N.C. Doubleday, D.S. Johnson, M. Marschke, P. McConney, E.W. Pinkerton, and E.K. Wollenburg. 2009. "Adaptive Co-management for Social-ecological Complexity." *Frontiers in Ecology and the Environment* 7, no. 2: 95–102.

AWC (Alberta Water Council). n.d. About Us. Alberta Water Council website. http://www.albertawatercouncil.ca/AboutUs/tabid/54/Default.aspx.

———. 2005. *Review of Implementation Progress of Water for Life, 2003–2004.* Edmonton, AB: AWC.

———. 2007. *Review of Implementation Progress of Water for Life, 2005–2006.* Edmonton, AB: AWC.

———. 2009. *Review of Implementation Progress of Water for Life, 2006–2008.* Edmonton, AB: AWC.

Bakker, K. (ed.). 2007. *Eau Canada: The Future of Canada's Water.* Vancouver: IBC Press.

Bakker, K., and C. Cook. 2011. "Water Governance in Canada: Innovation and Fragmentation." *Water Resources Development* 27, no. 2: 275–89.

Baldwin, R., M. Cave, and M. Lodge. 2011. *Understanding Regulation: Theory, Strategy and Practice.* 2nd ed. Oxford: Oxford University Press.

Barlow, M. 2011. "What You Don't Know About a Deal You Haven't Heard Of." *The Globe ad Mail.* January 6, 2011.

Bauer, C.J. 1998. *Against the Current. Privatization, Water Markets, and the State in Chile.* Boston: Kluwer Academic Publishers.

Brooks, D.B. 2002. *Water, Local-Level Management.* Ottawa: International Development Research Centre.

Bruns, B.R., and R. Meinzen-Dick. 1995. "Frameworks for Water Rights: An Overview of Institutional Options." Chapter 1 in B.R. Bruns, C. Ringler, and R. Meinzen-Dick (eds.), *Water Rights Reform: Lessons for Institutional Design.* Washington, DC: International Food Policy Research Institute.

Carpenter, S.R., and L.H. Gunderson. 2001. "Coping with Collapse: Ecological and Social Dynamics in Ecosystem Management." *BioScience* 51: 451–57.

Conference Board of Canada. 2007. *Navigating the Shoals, Assessing Water Governance and Management in Canada.* Toronto: Conference Board of Canada. http://www.conferenceboard.ca/documents.asp?rnext=1993. Accessed 23 December 2007.

Cook, J., S. Freeman, E. Levine, and M. Hill. 2011. *Shifting Course: Climate Adaptation for Water Management Institutions.* Washington: World Wildlife Fund. http://www.adaptiveinstitutions.org.

Cundill, G., and C. Fabricius. 2010. "Monitoring the Governance Dimension of Natural Resource Co-management." *Ecology and Society* 15, no. 1: Article 15.

de Loe, R., and R. Kreutzwiser. 2007. "Challenging the Status Quo: The Evolution of Water Governance in Canada." Pp. 85–103 in K. Bakker (ed.), *Eau Canada: The Future of Canada's Water.* Vancouver: University of British Columbia Press.

Demetropoulou, L., N. Nikolaidis, V. Papadoulakis, K. Tsakiris, T. Koussouris, N. Kalogerakis, and K. Koukaras. 2010. "Water Framework Directive Implementation in Greece: Introducing Participation in Water Governance –

The Case of the Evrotas River Basin Management Plan." *Environmental Policy and Governance* 20: 336–49.

Diaz, H., M. Hurlbert, J. Warren, and D. Corkal. 2009. *Saskatchewan Water Governance Assessment Final Report*. Regina, SK: Prairie Adaptation Research Collaborative. http://www.parc.ca/mcri/gov01.php.

Dietz, R., E. Ostrom, and P.C. Stern. 2003. "The Struggle to Govern the Commons." *Science* 302, no. 5652: 1907–12.

Folke, C., T. Hahn, P. Olsson, and J. Norberg. 2005. "Adaptive Governance of Social-Ecological Systems." *Annual Review of Environmental Resources* 30: 411–73.

Frohlich, J., and J. Knieling. 2013. "Conceptualizing Climate Change." Pp. 9–26 in J. Knieling and W. Leal Filho (eds.), *Climate Change Governance*. Berlin Heidelberg: Springer-Verlag.

Galaz, V. 2003. *Privatizing the Commons: Natural Resources, Equity and the Chilean Water Market*. Santiago: FLASCO.

Government of Manitoba. 2008. *Climate Change Strategy*. Winnipeg, MB: Government of Manitoba.

———. 2015. Adapting to Climate Change: - Preparing for the Future. https://www.gov.mb.ca/asset_library/en/beyond_kyoto/adapting_to_climate_change.pdf Accessed November 24, 2015.

Government of Saskatchewan. 2007. *Energy and Climate Change Plan 2007*. Regina: Premier's Office.

Gunderson, L.H., and C.S. Holling (eds.). 2002. *Panarchy: Understanding Transformations in Human and Natural Systems*. Washington: Island Press.

Gupta J., C. Termeer, J. Klostermann, S. Meijerink, M. van den Brink, P. Jong, S. Nooteboom, and E. Bergsma. 2010. "The Adaptive Capacity Wheel: A Method to Assess the Inherent Characteristics of Institutions to Enable the Adaptive Capacity of Society." *Environment Science and Policy* 13: 459–71.

Gupta, J., N. ven der Grijp, and O. Kuik (eds.). 2013. *Climate Change, Forests and REDD: Lessons for Institutional Design*. Abington: Routledge Publishers.

GWP (Global Water Partnership). 2009. *Integrated Water Resources Management*. Stockholm: GWP. http://www.gwp.org/Global/ToolBox/Publications/Background%20papers/04%20Integrated%20Water%20Resources%20Management%20(2000)%20English.pdf. Accessed 8 June 2010.

Hall, A.W. 2005. "Water: Water and Governance." In G. Ayre and R. Callway (eds.), *Governance for Sustainable Development: A Foundation for the Future*: London: Earthscan.

Hatfield-Dodds, S., R. Nelson, and D.C. Cook. 2007. "Adaptive Governance: An Introduction and Implications for Public Policy." Presentation at the Australia New Zealand Society for Ecological Economics Conference, Noosa, AU, 4–5 July 2007.

Hickey, S., and G. Mohan. 2004. "Towards Participation as Transformation: Critical Themes and Challenges." Pp. 3–24 in S. Hickey and G. Mohan (eds.), *Participation: From Tyranny to Transformation? Exploring New Approaches to Participation in Development*. London: Zed Books.

Hill, M. 2012. "Characterizing Adaptive Capacity in Water Governance Arrangements in the Context of Extreme Events." Pp. 339–65 in W. Leal Filho (ed.), *Climate Change and the Sustainable Use of Eater Resources*. Berlin Heidelberg: Springer-Verlag.

Huitema, D., E. Mostert, W. Egas, S. Moellenkamp, C. Pahl-Wostl, and R. Yulcin. 2009. "Adaptive Water Governance: Assessing the Institutional Prescriptions of Adaptive (Co)Management from a Governance Perspective and Defining a Research Agenda." *Ecology and Society* 14, no. 1: 26–45.

Huntjens, P., L. Leibel, C. Pahl-Wostl, J. Camkind, R. Schulzee, and N. Kranzf. 2012. "Institutional Design Propositions for the Governance of Adaptation to Climate Change in the Water Sector." *Global Environmental Change* 22: 67–81.

Hurlbert, M. 2008. "Canadian Legal Framework of Water and Governance in the Prairie Provinces, Critical Analysis of Adaptation to Climate Change." World Water Congress, Montpellier, France, September 2008.

———. 2009a. "Comparative Water Governance in the Four Western Provinces." *Prairie Forum* 34, no. 1: 181–207.

———. 2009b. "Integrating Climate Change Adaptation into the Law." *Retfaerd Argang* 32, no. 3: 23–39.

———. 2009c. "The Adaptation of Water Law to Climate Change." *International Journal of Climate Change Strategies and Management* 1, no. 3: 230–40.

Hurlbert, M., D.R. Corkal, and H. Diaz. 2009. "Government and Civil Society: Adaptive Water Management in the South Saskatchewan River Basin." *Prairie Forum* 34, no. 1: 181–210.

Hurlbert, M., and H. Diaz. 2012. "Governance in Chile and Canada – A Comparison of Adaptive Characteristics." The Governance of Adaptation Conference, Amsterdam, NL, 22–23 March 2012.

Hurlbert, M., and H. Diaz. 2013. "Water Governance in Chile and Canada: A Comparison of Adaptive Characteristics." *Ecology and Society* 18, no. 4: 61–76.

IISD (International Institute of Sustainable Development). 2006. *Designing Policies in a World of Uncertainty, Change and Surprise: Adaptive Policy-Making for Agriculture and Water Resources in the Face of Climate Change*. Winnipeg, MB, and New Delhi, IN: International Development Research Centre and The Energy and Resources Institute.

IPCC (Intergovernmental Panel on Climate Change). 2001. *Climate Change 2001: Impacts, Adaptation, and Vulnerability. Technical Summary*. A Report of Working Group II of the Intergovernmental Panel on Climate Change. Geneva: IPCC, World Meteorological Organization, and United Nations Environment Programme.

Jowitt, E. 1959. *The Dictionary of English Law*. London: Sweet & Maxwell.

Kallis, G., N. Videira, P. Antunes, G. Pereira, C.L. Spash, H. Coccossis, S.C. Quintana, L. del Moral, D. Hatzilacou, G. Lobo, A. Mexa, P. Paneque, B.P. Mateos, and R. Santos. 2006. "Participatory Methods for Water Resources Planning." *Environment and Planning C: Government and Policy* 24, no. 2: 215–34.

Kennett, S.A. 1991. *Managing Interjurisdictional Waters in Canada: A Constitutional Analysis*. Pp. 23–28. Calgary: Canadian Institute of Resources Law.
Kiparsky, M., A. Milman, and S. Vicuna. 2012. "Climate and Water: Knowledge of Impacts to Actions on Adaptation." *Annual Review of Environmental Resources* 37: 163–94.
Kjaer, A.M. 2004. *Governance*. Cambridge, UK: Polity Press.
Lebel, L., J.M. Anderies, C. Cambell, S. Folke, S. Hatfield-Dodds, T.P. Hughes, and J. Wilson. 2006. "Governance and the Capacity to Manage Resilience in Regional Social-ecological Systems." *Ecology and Society* 11, no. 1: Article 19.
Lee, K.N., and J. Lawrence. 1986. "Adaptive Management: Learning from the Columbia River Basin Fish and Wildlife Program." *Environmental Law* 16: 431–60.
Levin, S.A. 1999. "Towards a Science of Ecological Management." *Conservation Ecology* 3, no. 2: Article 6. http://www.consecol.org/vol3/iss2/art6/.
McManus, P. 2009. *Environmental Regulation*. Australia: Elsevier.
Mollenkamp, S., and B. Kastens. 2009. "Institutional Adaptation to Climate Change: Current Status and Future Strategies in the Elba Basin, Germany." In F. Luwig, P. Kabat, H. van Schaik, and M. van der Valk (eds.), *Climate Change Adaptation in the Water Sector*. London: Earthscan.
OECD (Organisation for Economic Co-operation and Development). 1994. *Reducing Environmental Pollution: Looking Back, Thinking Ahead*. Paris: OECD.
Olsson, P., L.H. Gunderson, S.R. Carpenter, P. Ryan, L. Lebel, C. Folke, and C.S. Holling. 2006. "Shooting the Rapids: Navigating Transitions to Adaptive Governance of Socio-Ecological Systems." *Ecology and Society* 11, no. 1: 1–18.
Olsson, P., C. Folke, and F. Berkes. 2004. "Adaptive Co-management for Building Resilience in Social-ecological Systems." *Environmental Management* 34, no. 1: 75–90.
Ostrom, E. 2010. "Beyond Markets and States: Polycentric Governance of Complex Economic Systems." *American Economic Review* 100, no. 3: 641–72.
Ostrom, E. 2011. "Background on the Institutional Analysis and Development Framework." *Policy Studies Journal*: 39, no. 1: 7–27.
Pahl-Wostl, C. 2010. "Water Governance in Times of Change." *Environmental Science & Policy* 13: 567–70.
Pahl-Wostl, C., P. Kabat, and J. Möltgen (eds.). 2007a. *Adaptive and Integrated Water Management. Coping with Complexity and Uncertainty*. Berlin: Springer-Verlag.
Pahl-Wostl, C., M. Craps, A. Dewulf, E. Mostert, D. Tabara, and T. Raillieu. 2007b. "Social Learning and Water Resources Management." *Ecology and Society* 12, no. 2: Article 5. http://www.ecologyandsociety.org/vol12/iss2/art5/.
Pahl-Wostl, C., J. Sendzimir, P. Jeffrey, J. Aerts, G. Berkamp, and K. Cross. 2007c. "Managing Change toward Adaptive Water Management through Social Learning." *Ecology and Society* 12, no. 2: 30–48.
Percy, D. 2004. "The Limits of Western Canadian Water Allocation Law." *Journal of Environmental Law and Practice* 14: 313–327.

Plummer, R. 2009. "The Adaptive Co-management Process: An Initial Synthesis of Representative Models and Influential Variables." *Ecology and Society* 14, no. 2: Article 24. http://www.ecologyandsociety.org/vol14/iss2/art24/.

Pretty, J., and H. Ward. 2001. "Social Capital and the Environment." *World Development* 29, no. 2: 209–27.

Pretty, J. 2003. "Social Capital and the Collective Management of Resources." *Science* 302: 1912–14.

Raik, D.B., and D.J. Decker. 2007. "A Multisector Framework for Assessing Community-based Forest Management: Lessons from Madagascar." *Ecology and Society* 12, no. 1: Article 14. http://www.ecologyandsociety.org/vol12/iss1/art14/.

Reyes, B., S. Salas, E. Schwartz, and E. Espinoza. 2009. *Chile Water Governance Assessment Final Report*. http://www.parc.ca/mcri/pdfs/papers/gov03.pdf.

Stavins, R.N. 2003. "Experience with Market-based Environmental Policy Instruments." Pp. 355–435 in K.-G. Mäler and J. Vincent (eds.), *Handbook of Environmental Economics*. Amsterdam: Elsevier.

Tompkins, E.L., and W.N. Adger. 2004. "Does Adaptive Management of Natural Resources Enhance Resilience to Climate Change?" *Ecology and Society* 9, no. 2: Article 10.

UNDP (United Nations Development Programme). 2007. *Energy and Environment – Water Governance*. New York: UNDP. http://www.undp.org/water/about_us.html. Accessed 25 September 2007.

Walters, C.J. 1986. *Adaptive Management of Renewable Resources*. New York, NY: McGraw Hill.

Walters, C.J., and C.S. Holling. 1990. "Large-scale Management Experiments and Learning by Doing." *Ecology* 71: 2060–68.

Water Security Agency. 2012. *25 Year Saskatchewan Water Security Plan*. Regina: Queen's Printer. https://www.wsask.ca/Global/About%20WSA/25%20Year%20Water%20Security%20Plan/WSA_25YearReportweb.pdf.

Willems, S., and K. Baumert. 2003. *Institutional Capacity and Climate Actions*. Paris: Organization for Economic Co-operation and Development.

Young, O. 2010. "Institutional Dynamics: Resilience, Vulnerability, and Adaptation in Environmental and Resource Regimes." *Global Environmental Change* 20: 378–85.

PART 5

STRATEGIC PLANNING AND DROUGHT

CHAPTER 11

VALUES ANALYSIS AS A DECISION SUPPORT TOOL TO MANAGE VULNERABILITY AND ADAPTATION TO DROUGHT

Darrell R. Corkal, Bruce Morito, and Alejandro Rojas

Introduction

At a basic level, the term "vulnerability" refers to a relatively weak capacity to adapt to potential harms to humans or nature. Individuals, human communities, and nature itself are vulnerable to both natural stressors (e.g., droughts, floods, extreme weather) and anthropogenic stressors (e.g., pollution from human activities, infrastructure failures, economic downturns). This chapter assumes that vulnerability is, to a great extent, a socially constructed concept that expresses people's orientation toward the harms that can befall them or the environment in which they live. This "vulnerability concept" is constructed in accordance with the values people hold, care about, and want or feel compelled to protect (Adger 2006). Values that individuals hold will influence group values (like-minded stakeholders, communities, institutions, government agencies) and vice versa. Established institutional values tend to be the most widely accepted values of a society or culture at a given location and in a historic place in time. Institutions are also guided by values associated with organizational culture, structure, mandates, and legal instruments established by the society in which they operate.

This chapter is based primarily on data collected from diverse stakeholders (water users, the agricultural sector, rural communities, and all orders of government). The research focused on stakeholders who experienced the 2001–2 drought, one of the most severe droughts to have affected western Canada in decades and which was particularly severe for those in the South Saskatchewan River basin (SSRB), Canada. Agricultural production dropped by about $3 billion, mostly in the Prairie region (Wheaton et al. 2008). This national drought caused a $5.8 billion drop in Canada's gross domestic product (GDP), 41,000 job losses, and a $3.6 billion drop in Canadian agricultural GDP (Wheaton et al. 2005). Drought and climate-induced water stress are recurrent, natural characteristics of the Prairies and affect the social, economic, and environmental fabric of the SSRB (Sauchyn et al. 2010; Marchildon 2009a, 2009b; Banks and Cochrane 2005 Gray 1967).

The task in this chapter is to present examples of value orientations of stakeholders and governance institutions with reference to stakeholder vulnerability to drought. Some of the normative concerns in this context are identified and a values analysis methodology is provided to help identify stakeholders' different values. A conceptual decision support tool is presented as a method to help stakeholders better understand and resolve conflicts, and develop better adaptive responses to drought risks.

Values and value commitments (or value systems) underlie all intentional, deliberate, and planned thought and action. They belong to a mostly implicit system of knowledge, beliefs, and common understandings that contribute to social, cultural, and institutional structures. In turn, the organizational culture and structure shapes the practices of people and the institutions they represent. Values are key factors that contribute to the expression of meaning, thought, and human action. Values, therefore, contribute to the manner in which people legitimize decision making and establish governance systems and policies.

Values analysis[1] helps us understand the underlying concerns and motivations of individuals, groups, communities, industry, institutions, and the wide spectrum of decision-making bodies. Values analysis is consistent with recent interest in including "stakeholder analysis" along with "shareholder interests" in resource management. In part, this relates to the concept of "a social licence to operate" of a particular sector such as agriculture; that is, the sector's activities must be acceptable to society, or else it may conflict with competing interests. Values analysis recognizes economic and environmental factors (Morito 2005), as well as social, ethnographic, and institutional factors (Morito 2008; Morito and Thachuk 2008). In our study context, Patiño and Gauthier (2009) provide an excellent overview of the complexity of SSRB stakeholders. They emphasize the importance of understanding who the stakeholders are and how they relate to each other. They suggest public engagement and participatory mapping to help integrate and foster dialogue and co-operation between diverse stakeholders and decision makers.

Hence, a values analysis helps us understand the "reasons" and motivations for decisions, policy, action, and conflict. It provides a framework for conducting a deeper analysis of conflict and may help guide approaches toward successful conflict resolution. It will help identify whether conflicts are relatively superficial and involve negotiable values or whether they are more deeply entrenched and may first require establishing common ground to allow stakeholders to better understand their differences before conflict resolution can begin. The initial phase of a values analysis is to create a context of mutual understanding to determine whether there is a basis for a common ground. Values analysis can help identify "institutional personalities," which may provide insight into understanding how individuals, groups, and institutions (including government agencies and industry) will act and interact. A values analysis can also help guide policies and initiatives and help determine whether they are achieving their intended results and are on an appropriate track.

Values Analysis Methodology: A Case Study with IACC Stakeholder Communities

This chapter is based on the field results of the Institutional Adaptations to Climate Change (IACC) research project. The project was conducted from 2004 to 2009 in Canada and Chile to improve our understanding of climate stress vulnerabilities and to consider how institutional adaptations may be useful in strengthening the resilience of rural communities and the agricultural sector. The study region in Canada was the SSRB, spanning the provinces of Alberta and Saskatchewan. Water was chosen as the focal point for our values analysis, as stakeholders confront impacts from extreme climate. The interconnectedness between water and stakeholder/institutional adaptations is more fully described in Rojas and Richer (2005).

We conducted numerous semi-structured interviews and focus groups involving diverse stakeholders with vested interests in water: individuals, rural communities, farmers, farm groups, agricultural industry, First Nations, non-government groups such as watershed organizations and environmental agencies, and all orders of government (local, provincial, and federal agencies). The interviews were audio-recorded, transcribed, and subsequently coded using NVivo software to allow analysis and interpretation. Interview respondents provided their own perspectives on risk, vulnerability, resilience, governance, and adaptation. We used an interview guide to elicit information and perspectives on drought, water management, water conflict, gaps in adaptive capacity, and opportunities to strengthen resilience.

This values-analysis case study focused on Canadian communities in Saskatchewan (Outlook, Stewart Valley, and Cabri) and Alberta (Hanna, Taber, and Blood Tribe First Nation), and on institutions responsible for water governance in the two provinces. Comments on groups and institutions are based mostly on aggregated results, although interesting contrasts between Alberta and Saskatchewan warrant the identification of specific institutions and in some cases specific individuals (anonymous, but cited by a code) or specific stakeholder groups. We provide only a summary of the research results and highlight those features that aid the value-analytic aspects of the research. It should be noted that the interviews and focus groups were structured to be as open as possible to allow stakeholders to indicate what they did in response to drought and what they thought was important for them to mention, and to allow other participants' responses to stimulate discourse of related concerns.

Main Observations of the Values Analysis from Respondent Interviews

Common and recurring themes of values occurred, and the data were organized in four categories, as developed by Rojas (2000), one of the

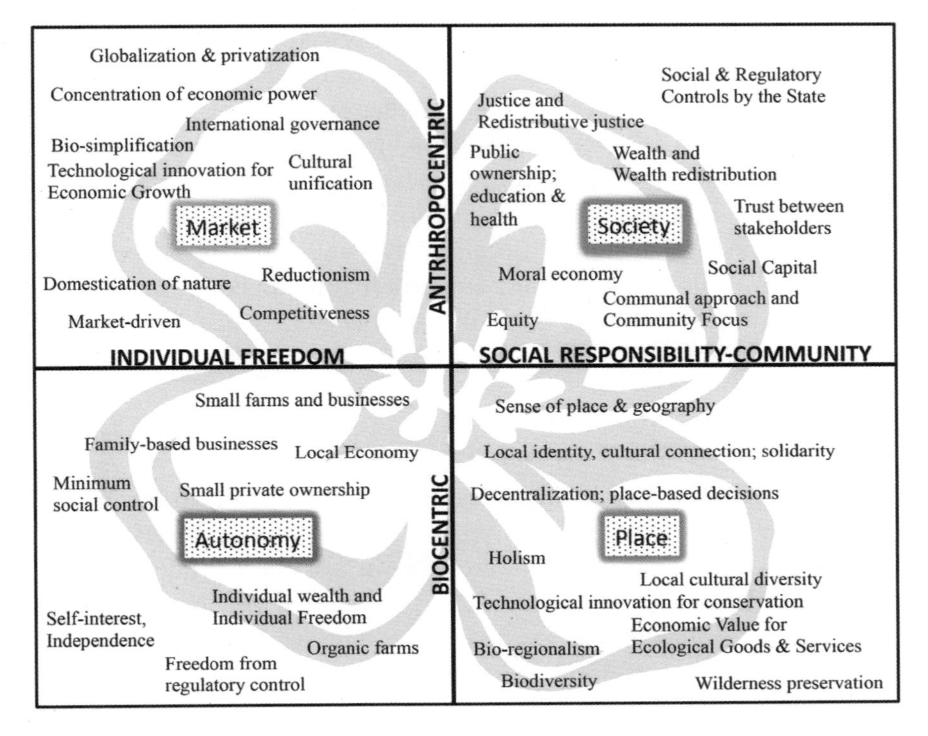

Figure 1. Flower of Values: values characteristics within four quadrants (Source: modified from Rojas 2000; Rojas, Magzul, et al. 2009; and Rojas, Reyes, et al. 2009)

co-authors of this chapter. The "Flower of Values" graphic (Figure 1) is presented as a key visual aid to explore the relationships between human values and the way water problems were defined by different stakeholders. Understanding the ethical basis of the way different stakeholders react to water risks is critical knowledge for those vested with the responsibility of making informed decisions for adaptive planning and action.

The "Flower of Values" graphic is a conceptual framework that combines cultural and ethical paradigms. Four paradigms are used: anthropocentrism, biocentrism, individual freedom, and social responsibility/community. The diagram helps situate the various value profiles and their relative commonalities and contrasts. These typical societal values characteristics are sorted into four quadrants: market, society, autonomy, and place. For example, in the upper-right quadrant (anthropogenic and social responsibility, labelled as "Society"), value is placed on public ownership, social regulation, and wealth redistribution. In the bottom-left quadrant (individual freedom and biocentric, labelled as "Autonomy"), value is placed on freedom of choice and small-scale operations sensitive to sustainability. In the upper-left quadrant (market competitiveness, globalization, labelled as "Market"), free market principles are most highly valued. In the bottom-right quadrant (local identity and geography, labelled as "Place"), local culture and ecology are most highly valued. In general, the most strongly held values are those depicted the greatest distance from the center of the graphic.

Market and Economic Values

The interviews were conducted relatively close in time to the 2001–2 drought, so its severe economic impacts were top-of-mind with stakeholders. The ecological impacts were more or less successfully managed largely because of two key factors: the drought lasted only two years, and many historical adaptations (e.g., low-tillage crop seeding, irrigation, better water management) were successful in strengthening resilience (Toth et al. 2009; Bruneau et al. 2009). Accordingly, it comes as no surprise that most stakeholders described their vulnerabilities and adaptations to drought (and other water-related concerns, such as flooding) by emphasizing the importance of sound economic and technological instruments. They suggested a need for investing in technological and infrastructure development, and for revising such economic instruments such as crop insurance. When referencing past adaptations, they often talked about how they built water reservoirs and distribution systems, established irrigation projects, and later improved irrigation water use efficiency with new technologies. Many also mentioned how they shifted their operations to produce different crops. Many stakeholders complained about domestic and/or international policies and government actions (or lack thereof) that placed them at a disadvantage. For example, government responses to world market fluctuations, trade barriers, and subsidies were mentioned as key factors that made it difficult, if not impossible, for farmers to compete in the world market.

Autonomy and Individualistic Values

Some stakeholders first mentioned, emphatically, the importance of the individual and local community. This focus on autonomy affected the types of economic activity and adaptive capacity. Respondents often mentioned how they, as individuals and as local communities, dealt with drought by self-resilience and without government aid or help from others outside of their region. Indeed, particularly in Alberta, a number of communities emphasized the importance of keeping government out of their business as much as possible. These responses are indicative of the values of freedom of choice/action and individual autonomy.

Society and Communal Values

Communal adaptive responses were also observed, particularly in relation to times of crisis such as the 2001–2 drought. Respondents told how they drew on resources such as their neighbours to salvage irrigation crops during times of water scarcity (i.e., when water supplies could not meet irrigation demands). In this extended drought, respondents drew on communal commitments and assessed their vulnerabilities and adaptive strategies that best fit the community while balancing individual needs where possible. In Alberta, where initial profiles would normally be positioned squarely in the autonomy and individualistic values quadrant, for instance, some community members in the irrigation districts decided to share water resources despite having priority rights under the first-in-time, first-in-right (FITFIR) system of allocation and adopted "water market characteristics" to transfer water rights (Nicole and Klein 2006). Priority rights holders sometimes shared their water resources by producing a crop on one person's land and leaving the other's land unseeded (in fallow), and later sharing whatever profits accrued. In fact, this adaptive response was cited by a number of institutional respondents to highlight the importance of recognizing community relationships as a resource on which managers should be drawing (Morito 2008a). So, while individual autonomy was strongly valued, shared communal decisions could also be highly valued. The importance of community, neighbourliness, and other non-economic factors was strongly emphasized: "The importance of trust was repeatedly emphasized and a careful approach to building and nurturing trust was advised" (Alberta Environment, initial presentation at Athabasca River basin meetings, Sherwood Park, 15–16 April 2008).

Different stakeholders' points of view are illustrated in the following comments:

> Some of them were right thinking … in a very, very low flow year, unless they got an amazing amount of rain by chance, 60% [of the water allocation needed to meet the irrigated crop water requirement] ain't going to give you a crop. And it's probably going to cost you more in energy costs and whatever to put 60% onto your fields and get no return. So 60% is probably losing money. So those people just chose not to even play … to sign up for the 60% … we aren't going to get crops if we do this so let's use our 60% on your land … I will share in helping you work your land and everything, we'll just fallow mine for a year and we'll put the two 60% on your property and we'll share what we get for a crop. That was huge in terms of adaptation. And they did it. No bureaucrat, no politician, nobody planned it. The tools were put in place. (Morito 2008b)

Or, again:

> Guys in the early 80s basically invented water sharing … [it] … wasn't legal at the time, we made ways of making it work … as long everyone agreed and no one was injured by it. No complaint, no problem. Anyway, by doing that, [we] proved it could work … you had peer pressure amongst the community. It's not going to work if you have to do it over a broad area where people don't know each other. (Morito 2008c)

And:

> There's a variety of personalities who work for the government, some are more successful at making these things work than others. … It's a trust issue more than anything else. … We used to say you only work with three people at a time… Then you gotta fine-tune the systems so that everyone

knows what's going on, and then we can make it work. (Morito 2008c)

That respondents changed their emphasis from "individual choice" to focus on "best communal choice" indicates stakeholders are flexible and will adjust values to suit a particular situation or need. In this Alberta case, the adaptive response appears to run somewhat contrary to expectations, namely, that water scarcity did not generate conflict but rather elicited co-operation.

When given the opportunity to tell their stories about how they responded to drought in the past, many community members discussed other impacts that drought and "unfair economic practices" had on them. They spoke about how unfortunate it was that their children had no future in farming and had left the community to train as professionals or to seek non-farming work. As proud as they were of their children, they lamented their leaving, since it marked the beginning of a loss of a highly valued way of life and heritage. They also talked about how they feared the prospect of losing their schools and churches, as declining populations made the maintenance of these institutions financially unfeasible. Some even emphasized that their school and other community infrastructure were quite important for their sense of identity and community solidarity, since these are the places where most community activities take place. Further, being able to support a hockey team or school team to compete against the neighbouring communities was important to many. These kinds of factors are recognized as "social capital"[2] because they are non-financial, non-commodity resources on which people can draw for a variety of purposes. In our study, we define social capital as local collective social resources and the ability and capacity to work together as a collective or as a community to strengthen resilience and adaptive capacity. Social capital refers to the intangible resources members of a community or society can draw on to accomplish something (e.g., trust, familial/community support). These factors also relate to the presence of a "moral economy"[3] that places value or worth on something that is normally not recognized by market economies (e.g., valuing quality of life, social relations, commitments to sustain a healthy environment). We define the term moral economy as a system of exchange based on moral values and expectations, which enables effective communication and ordered social relations; the moral economy relates to normative orientations that people bring to their social interactions (e.g., equity, fairness, respect).

During the 2001–2 drought, the creative water-sharing and water-market relationships were founded on strong social relationships, pragmatism, and trust. The respondents gave and appreciated aid from both peers and community. They valued and received strength and support from local social/community networks. These elements work alongside—and sometimes in spite of—the commodity-based economy. These observations are a reminder that people often place great value on their lifestyle, their relationships, the integrity of their communities, and the actual "places" in which they live. People express value for many intangible factors and not only on quantifiable economic characteristics or economic wealth.

Place: Local Identity, Ecology, and Place-based Values[4]

The emphasis on "communal values" was sometimes demonstrated as a strong sense of geographic connectedness and identity. This differed from the more anthropocentric view of community in the "social/communal/sharing" quadrant. The values associated with a connectedness to a unique "place" (i.e., the environment in which one lives) emphasize biocentric characteristics with specific human cultural and ecological/geographic identity. The experience of losing a way of life is connected to the place and ecological systems in which the respondents live and on which they depend for survival. Their remorse over children leaving the family farm is tied to their place-based values. The land is not treated merely as an economic commodity to be exploited but as a place in which they are responsible for land stewardship.

The connectedness with place and local identity was also evident in responses by the Blood Tribe (Kainai) First Nation stakeholders. When discussing climate vulnerabilities and adaptation, respondents did not focus on economic vulnerability but rather on the lack of social capital; the Kainai valued trust and the ability to draw on a sense of belonging to a vibrant and respected culture and place. They made it clear that their sense of identity (who they are) and a sense of empowerment (political significance) were crucial to their adaptive capacity. The problems of the community were explained by references to their residential school experience, the imposed band council system, and the past banning of traditional practices by the federal government, all of which have become quite familiar to Canadians. Gangs and drug abuse were cited as symptoms of this problem. Their vulnerability to the impacts of climate change (which in their case actually had much more to do with flooding than drought) had, in many respondents' minds, first to do with social and cultural erosion, before economics. Indeed, many economic instruments have already been used to help the community adapt to contemporary economic exigencies, but they have failed, often because there was little motivation by community members to use them or learn how to use them. While the lack of education was stated as one reason for this, many respondents cited the history of their relationship with the Canadian government and the social challenges previously identified as causes of their vulnerability. They referred to the paternalistic practices toward First Nations people throughout Canadian history, which have deeply eroded their capacity to draw on social capital and the moral economy, and which in turn have alienated their communities from the commodity-based economy. Social capital and the moral economy are therefore seen as important elements to equip people with the capacity to think through problems, communicate effectively, work together, and subsequently move to organize, coordinate, and then respond to various stressors. These perspectives demonstrate how the Kainai value their personal identity, their history, and their political and social situation. This is consistent with Magzul's (2013) and Rojas, Magzul, et al.'s (2009) findings, which describe how understanding First Nations' values is critical for resolving conflicts and implementing effective adaptations.[5]

Responses by the Blood Tribe members imply that they see their value system as having been undermined, violated, and ignored. Here, we must rely on some readers' familiarity with Aboriginal cultural values to make the following summary claim. The Kainai's cultural heritage is based on a close connection to the land and the obligations the Creator set for them to act as keepers of the land (again, a strong emphasis on "place"). Their connectedness to the land is also reflected in the connection people have to one another within their culture. Community is primary, and the connectedness to the land is fundamental. This connectedness, according to our analytic framework, also places the Kainai much more within the community and biocentric sectors.

The strong value of place and cultural identity was also illustrated in the Kainai's sister community, the Peigan (Pikani). Where stakeholders value "place," conflict may arise with stakeholders who value "individualistic autonomy." More traditional members of the Pikani had come into conflict with proponents of the Oldman River Dam in the 1990s over control of water distribution and proposed expansion of irrigation and the agricultural economy. The sacredness of certain cottonwood-inhabited riparian zones was seen as threatened by the flooding of a reservoir zone once the dam was constructed. The Pikani were not convinced (and did not accept) that the economic value of development was sufficient justification for the project (Rojas, Magzul, et al. 2009). Indeed, building the dam would violate or impair deeply held spiritual and non-negotiable values (Magzul 2013; Rojas, Magzul et al. 2009).

Discussion: Stakeholder Values as Identified by IACC Research

The values analysis data gleaned from this research offer insights into differing and overlapping perspectives among stakeholders. When contrasting values exist (opposing quadrants), there is a risk of disagreement or conflict. If only similar values exist (one quadrant), imbalanced decisions may occur. Values mapping provides stakeholders and policy makers with a greater awareness of differences, conflicts, and similarities, which can lead to more balanced decision making and conflict resolution.

An agreed-upon values mapping process can increase mutual understanding and agreement, particularly when expressed through institutional and policy instruments. Stakeholders and decision makers can use this process to develop mutual understanding of differences and seek consensus, with the ultimate goal of creating better adaptation decisions (i.e., planning and implementation actions will improve by incorporating broader interests). People will identify with what matters to them collectively and rally around balanced values and locally relevant adaptive responses.

In contrast, when the diverse, broader stakeholders' values are not recognized, the resulting policies, decisions, and actions will likely create conflict. Ignoring a group's value systems can devalue their moral economy and marginalize that community. Adaptive responses that do not factor in relevant stakeholder values are therefore not likely to be effectively implemented, may undermine a community's adaptive capacity, and may increase vulnerability.

Increasing adaptive capacity and decreasing vulnerability, then, depends crucially on understanding and protecting stakeholder value systems (individuals, communities, institutions). Clearly, for the Blood Tribe, recognizing and incorporating more communal/biocentric values and traditional indigenous knowledge is necessary (though not sufficient) to reduce its members' vulnerability to drought (Rojas, Magzul, et al. 2009; Rojas, Reyes, et al. 2009). Even for other SSRB communities, a movement from individualistic to communal/place-based values is critical during times of drought stress.

Individual, Group, and Institutional Values

Examples of the interconnectedness of individual, group, and institutional values were evident in both the historical literature and the IACC research data. Gray's (1967) *Men Against the Desert* documents how Canada's federal government applied unique place-based agricultural research to address the economic, social, and environmental crisis caused by the multi-year droughts of the 1920s–30s. Agriculture and Agri-Food Canada's Dominion Experimental Farms research was integrated with the creation and efforts of a new institution in 1935: "The federal government established the Prairie Farm Rehabilitation Administration (PFRA) during the greatest environmental and economic crisis in twentieth-century Canada" (Marchildon 2009b).

Gray makes evident the fact that government research was focused on soil and water conservation and water development (i.e., sustainable agri-environmental practices). While this research explicitly targeted the physical harms produced by drought (e.g., the need to find better methods to reduce soil drifting from wind erosion or better water management methods to minimize impacts caused by water scarcity), the institutional efforts had strong social and communal aspects. Research experiments and adaptation activities were conducted with the rural people and were intimately linked with rural communities and the rural populations on the farms they served. Gray's observations show that the institutional values recognized the importance of both individual and communal values. One could argue that recognition of both social capital and the moral economy were implicit in the activities of the PFRA, as the institutions of the day (local, provincial, federal) were working hand-in-hand with the local rural farming communities and farm groups. The agricultural scientists and engineers worked directly with farmers to determine best land-use practices and best crops suited to prairie climate and to find improved soil and water management techniques. This communal effort was driven by a common need to find sustainable farming practices that could ensure greater economic security and vibrant rural communities.

During the course of the IACC research, those departments/ministries most responsible for water management were Alberta Environment and the Saskatchewan Watershed Authority in each respective province. Both provinces have been moving toward a more consultative process with stakeholders, a shift from the top-down regulatory approaches of the past. One high-level government respondent (Morito 2008d) emphasized that this is consistent with a worldwide shift initially established in the 1992 Dublin Principles, now commonly known as "integrated water resource management" (IWRM). IWRM is a process that attempts to involve the interests of all stakeholders when making water management decisions. It emphasizes social and economic values while committing to environmental principles and citizen engagement (World Meteorology Organization 1992; Global Water Partnership n.d.; IRC 2009). Similar integrated approaches are now formalized in Alberta's Water for Life strategy (Alberta Environment 2008, 2003) and Saskatchewan's *25 Year Saskatchewan Water Security Plan* (Saskatchewan Water Security Agency 2012[6]). Collaboration and engagement with citizens and all orders of government (including First Nations) are key factors in the longer-term strategy. Provincial and federal government institutions are also working together on interdisciplinary planning approaches to address interjurisdictional and multi-stakeholder concerns related to climate and water (Corkal et al. 2011, 2007; Diaz et al. 2009; Hurlbert, Corkal, et al. 2009; Hurlbert, Diaz, et al. 2009). Local watershed groups were created and are now developing more holistic watershed plans and advising governments of local needs and interests; their efforts are clearly founded in the "place-based" quadrant and consider economic, social, and environmental factors.

The IACC research data also demonstrate that institutional relationships are interconnected with local individuals and watershed groups. Several provincial government managers in Alberta (Morito 2008a) indicated how they or their colleagues worked with the communities by drawing on friendships and familial ties to engage stakeholders in informal discussions to initiate adaptive responses to drought. These managers appealed to people's senses of honour and neighbourliness to comply with regulations.

Agriculture and the "Voice of the Environment"

The IACC research data indicated that diverse stakeholders expressed an interest in environmental sustainability. Interestingly, in the Alberta case where water is essentially fully allocated, agricultural producers and environmental groups appeared to be more proactively engaged in watershed planning, in essence trying to find consensus. In contrast, in Saskatchewan where water was not fully allocated, environmental and agricultural industry groups appeared to take stronger opposing views, leaving little room for consensus.

A number of respondents from government institutions emphasized that farmers and ranchers are not exploiters of the land (this viewpoint is sometimes identified by those who criticize the sector or its activities that pose risks to the environment). Respondents noted that farmers and ranchers are connected to, and depend on, a healthy natural ecosystem for their livelihood and quality of life. They see the land as their home; their way of agricultural production is a matter of lifestyle. They do not merely depend on the land for economic survival. Rather, their relationship to the land is critical to their identity and forms a kind of agricultural tradition and culture. Respondents told how some dryland farmers refused to become irrigation farmers. Some even felt that those family members who either had made the switch or advocated a switch to irrigation were traitors who were destroying a long-established and hard-won tradition of dryland (rain-fed) agriculture. To be fair, the decision to become an "irrigator" is replete with risks due to, for example, market conditions, investment costs, environmental/climate uncertainty, and long timelines to see a return on their investment (see Chapter 6 by Warren on irrigation in southwest Saskatchewan in this volume). However, the point remains that some respondents emphasized loyalty to long-standing traditional practices of dryland farming as a way of life. Again, in times of stress, more communal and biocentric values become important. As with the more traditional Peigan, who could not compromise their spiritual values to accept the flooding of their sacred cottonwoods, the agricultural sector may also resist certain adaptations that conflict with identity and heritage values, even if a compelling economic rationale exists.

While the agricultural sector is an "economic activity," the production of safe food is also seen as a "managed ecosystem" that tries to balance economic and ecological benefits (Swinton 2008). Farmers themselves see value in protecting water supplies and are adopting beneficial land and water management practices (sometimes as ecological goods and services), often with support from government programs (Corkal and Adkins 2008; Corkal et al. 2004). In effect, these initiatives are recognizing a diversity of values systems at play.

The Role of "Boundary Organizations" in Values Analysis

Bridging differing or conflicting values systems and competing interests will often require boundary organizations working with stakeholders to balance social, economic, and environmental values (Batie 2008; Clark and Holliday 2006). Boundary organizations are non-partisan and work with dual accountabilities, linking policy, science, and user-driven local knowledge to strengthen adaptive capacity.

The PFRA was historically an organization that had the elements of a boundary organization. The severe droughts of the 1920s–30s caused extensive social, economic, and ecological hardship on the Canadian Prairies (Marchildon 2009a, 2009b; Marchildon et al. 2008; Gray 1967). The PFRA's mandate was to identify and promote soil and water conservation techniques, sustainable agricultural practices and land use, and improved water management approaches suited to the unique semi-arid characteristics of the climate and geography of the Canadian Prairies. In essence, the PFRA was promoting agricultural sustainability (market values) that were more suited to the regional, social, and place-based needs of the unique climate and geography of the Prairies (i.e., balancing communal and place-based values with market values).[7]

Many stakeholder groups interviewed by the IACC researchers appreciated the historical role and actions of PFRA, which were applied at a local scale. The stakeholders expressed criticism for approaches that did not take into account the local people or local communities; they criticized "top-down" measures imposed by "far-away" agencies. Stakeholders desire institutions that co-create knowledge and adaptive responses with local efforts (i.e., a dual accountability).

The watershed groups in Alberta and Saskatchewan are modern-day boundary organizations. These groups work to bridge science, policy, and various institutions and programs to help meet the needs of local stakeholders with suitable adaptive responses in their specific geographic locations.

Insights from the IACC Values Analysis

The IACC research provides insights into the importance of values analysis. While there is some commensurability and convergence between the value profiles of the various stakeholder groups and the directions governments are now taking with respect to more holistic water management, all stakeholders require greater effort to truly understand and integrate the diversity of values systems. Interview data indicate that government agencies have not yet begun a concerted effort to understand the role of social capital and the moral economy. Furthermore, the vast balance of research and policy development efforts is currently targeted at physical sciences and economics investigations. There is a need for more integrated natural and social sciences research. For example, lessons learned from Canada's Dust Bowl experience can help us understand the relationship between climate and people (McLeman et al. 2013). Similar insights from Australia emphasize the need for integrative stakeholder-government research combined with adaptive governance approaches to reduce drought risk (Nelson et al. 2008).

Very little research is underway to help institutions and stakeholders better understand and more effectively address divergent or conflicting values systems. As a number of Alberta Environment respondents noted, only brief forays into the valuable role of social sciences and humanities research have been undertaken. Institutions generally agree that it is useful to improve knowledge of the moral economy and social capital, but they lack an understanding of how to apply or integrate this knowledge with physical sciences and economics research. A challenge also exists in applying such integrative knowledge at the local scale, where adaptive change is most likely to happen. This is one of the fundamental conclusions of this chapter:

A comprehensive approach to investigate climate change impacts and adaptation requires a concerted effort to understand relationships between social and physical sciences, and needs to factor in the role of social capital, the moral economy and place-based interests. Such efforts are needed to balance social, economic and environmental factors, and are necessary for stakeholders and government institutions to develop and implement adequate adaptive responses.

Current watershed planning efforts are steps toward holistic planning, but the efforts of watershed groups are largely "advisory" in nature and "at-arms-length" to government. The efforts of these groups for financial self-sufficiency and integration with formal institutions face many significant challenges that risk the sustainability of these groups (Hurlbert, Corkal, et al. 2009; Hurlbert, Diaz, et al. 2009). Even genuinely inclusive processes cannot substitute for concerted research and leadership into the functioning of social capital, the moral economy, and place-based values by government agencies themselves, which more than any other group, sector, or institution are recognized to be ultimately responsible for protecting the public good and preserving Canada's environment for present and future generations. In a 2003 study published by Natural Resources Canada, water stakeholders (government representatives and water users) were asked questions about water management, water apportionment, and environmental values (Bruce et al. 2003). The study found a high level of agreement and support for managing water as a community resource, with due considerations for basin-wide interests and water allocations for environmental protection "since the environment cannot defend itself" (Bruce et al. 2003: 133–38). These informants identify a critical role for government in recognizing and addressing diverse values (social, economic, and environmental).

Power differentials among stakeholders may result in those endowed with less power becoming more exposed (lost access to water) or having less capacity (lost economic opportunities). Stakeholders with little power are likely to be more vulnerable to the impacts of climate change. To avoid power asymmetry among stakeholders, conditions must be established to ensure they perceive each other's concerns and interests as legitimate, regardless of differences in values and interests (Rojas, Magzul, et al. 2009;

Rojas, Reyes, et al. 2009). Using values-analysis profiles can be an effective tool in achieving mutual understanding. Engaging the broad spectrum of stakeholder interests will advance more effective watershed management (e.g., equitable access to water) and reduce stakeholder vulnerability.

This chapter has investigated the potential to use values analysis as a means of addressing existing water conflicts or simply as a means of aiding holistic water management. Figure 2 identifies a simple methodology to construct a stakeholder values analysis, and provides insight on how this approach might be implemented as a decision support tool for use by stakeholders who are dealing with water conflicts, divergent interests, or complex water resource management challenges.

The Intergovernmental Panel on Climate Change (IPCC 2007) identifies potential climate change impacts on systems and sectors, affecting ecosystems, coastlines, water, food production, industry, settlements and society, and human health. In light of the potential social upheavals that drought and other climate change impacts may bring, it would appear ever more pressing to undertake research into the role of stakeholders' values to include elements of social capital, the moral economy, and place-based interests. It will be even more important to find ways to integrate that research with the ongoing physical sciences and economics research on the impacts of climate change. Improving knowledge of values systems and social sciences, combined with natural sciences research, will be critical to resolve conflict and bridge local knowledge with policy makers and programs. Such approaches are also likely to lead to the development of new adaptive governance approaches to address drought and water scarcity (Nelson et al. 2008).

Another important conclusion we draw is that the history of Canadian governance has largely been predicated on the assumptions of a liberal democratic society—that is, the assumption that the individual is primary and that he or she is defined principally as a consumer/producer. But as our results demonstrate, individuals also take much of their identity, meaning, and sense of belonging from the community and the place in which they live. In the Blood Tribe case, the ability to develop and sustain an economic system depends crucially on having a robust moral economy and a system of social capital on which individuals can draw. This is also evident in the agricultural community, as demonstrated by the social upheaval caused by severe multi-year droughts in the 1920s and 1930s. A

Values vary between diverse stakeholders. Contrasting values may lead to opposition or conflict, whereas overly similar group values may lead to imbalanced decisions. Identifying the range of stakeholder values is useful to increase awareness, address conflict, and seek balanced, holistic adaptations. The IACC research mapped water management stakeholder values into four quadrants identified in "Figure 1, The Flower of Values". The motivations of stakeholders were driven by considerations for the "Market" (the economy), "Society" (social/communal), "Autonomy" (individual freedom), and "Place" (local identity and ecology). Generally, the most strongly held or firmly rooted values are depicted the furthest distance from the center of the graphic.

Developing a values map is useful as "a values analysis decision support tool." Values mapping requires input from all vested stakeholders to clearly identify the range of values and motivations for stakeholders (including institutions and policy makers). The values data then need to be categorized, compared, and contrasted in an organized manner, such as the Flower of Values quadrants (Figure 1). In turn, stakeholders can begin to articulate what is at risk for them. Where stakeholder values contrast or are in opposition, further dialogue is required to identify possible common ground.

As a decision support tool, this methodology can be helpful to better understand similarities, differences, and motivations. Such knowledge can help create consensus, manage disagreements, and develop mutual planning approaches. In short, the development of a values mapping graphic is a methodology that allows stakeholders to identify values profiles and differing and potentially conflicting values systems. The knowledge of different values will aid in the resolution of differences and help encourage stakeholder dialogue to find common ground and suitable adaptive responses.

Figure 2. A conceptual "values analysis" decision support tool (or methodology) for water management

more robust economic, social, and environmental balance demonstrated more effective adaptations and more resilient communities.

The ethical dimension of these conclusions can be defined in terms of the relationship between stakeholders and governments. Stakeholders have a trust-based relationship with their governments, which in turn have a fiduciary obligation to protect their stakeholders. More precisely, stakeholders stand as citizens to the governing bodies that they legitimize and empower to make certain kinds of decisions for them. This is to say that underlying the political relationship is a moral one, as defined by the moral economy and social capital (trust, honesty, sense of belonging, obligations of reciprocity among people and between people and their places). These values are, therefore, integral to the relationship and cannot be viewed as external concerns of governance.

Differences in value orientations (and lack of knowledge of values orientations) can hinder the adoption of successful adaptive practices. How can government institutions better organize and structure themselves to factor contributions from stakeholder groups and citizen groups? To what degree can governments more fully engage local decision makers in ways that allow for accountability and recognize differing values between stakeholders? To what degree can competing interests in water management, water development, economic expansion, individual protection, community sharing, environmental protection, and place-based identity be established as mutual factors that are considered in reconciling conflict or stress and lead to new insights and better adaptations? How will today's decisions help build resilience and strengthen present and future adaptive capacity? What decisions can be made today that are pragmatic and can lead to meaningful action?

To paraphrase one of the respondents, these questions simply emphasize our main point—*values must be considered in the evaluation of climate-induced water stress and society's adaptive responses.* One way to address this is to seek ways to incorporate values analysis in the development of water and climate policies and programs. The conceptual values analysis decision support tool presented in this chapter may be used a methodology for identifying values systems of diverse stakeholders, including institutions and policy makers. Values analysis will be helpful for stakeholders to better understand differing positions, to address real or perceived conflicts, and to implement improved adaptive responses for strengthening local capacity and an overall more resilient society.

Acknowledgments

This chapter is primarily based on one element of a much broader interdisciplinary research project conducted within Canada's South Saskatchewan River basin (SSRB) in Saskatchewan and Alberta. During the period from 2004 to 2009, natural and social science research investigated vulnerabilities and adaptations to climate-induced water stress in the semi-arid region, which is heavily reliant on agricultural production. The study, Institutional Adaptations to Climate Change, was conducted both in the SSRB watershed in Canada and the Rio Elqui watershed in Chile, with $2.43 million in funding provided by the Social Sciences and Humanities Research Council of Canada. More information on the research may be found at http://www.parc.ca/mcri/.

NOTES

1 The term "values analysis" derives in part from the concept of "values mapping" (Cragg 1997).

2 Bourdieu (1986) states "social capital" is "a durable network of more or less institutionalized relationships of mutual acquaintance and recognition—or in other words, to membership in a group." For example, earned trust is an intangible resource that people draw on when attempting to satisfy an activity that requires the co-operation of others. Familial obligation is another example that people draw on, particularly during times of need or stress. Portes (1998) states that social capital is "the ability of actors to secure benefits by virtue of membership in social networks or other social structures." Putman (1995) defines social capital as "features of social organization such as networks, norms, and social trust that facilitate coordination and co-operation for mutual benefits." Some individuals and communities in southwestern Saskatchewan and the Special Areas of Alberta demonstrated stronger social capital assets than others and were recognized as having greater capacity to work together to address drought and climate risks. See Warren and Diaz (2012) and Magzul (2013).

3 "Moral economy" relates to the normative orientation that people bring to their social interactions (e.g., debts owed due to past sacrifices of others, exchanges of trust between people and groups, inclusion and recognition within a group, goodwill or malice, sense of fairness and justice, responses of appreciation or resentment toward those who "deserve" such responses, honour and trustworthiness, respectability). The main sources of the concept of moral economy are derived from Perry (1909), Thompson (1971), Scott (1976), Adger (1998, 2001), and Morito (2012). The moral economy places value on characteristics such as social relations, quality of life, fairness and equity, a healthy environment, or other such characteristics not normally considered by conventional principles of market economies. The term moral economy has been used to counter tendencies to use reductive explanatory frameworks in history and other disciplines (Scott 1976).

4 Cantin (2010) describes "place-based approaches" as policies and programs that address complex socio-economic issues in a collaborative manner (i.e., with the contributions of multiple stakeholders) and by targeting activities and interventions at a specific geographic scale. This quadrant in the Flower of Values (see Figure 1) identifies stakeholder values for their "local place."

5 The document *The Oldman River Dam Conflict: Adaptation and Institutional Learning* describes in part how conflict and power differentials are factors that can actually impair a community's capacity to participate in consultations and reduce the potential of adapting to climate stress. (Rojas, Magzul, et al. 2009; Magzul 2013)

6 The Saskatchewan Water Security Agency was created in 2012; it was formerly the Saskatchewan Watershed Authority, which existed from 2002–12).

7 The Prairie Farm Rehabilitation Administration (PFRA) was created in 1935 and was a branch of Agriculture and Agri-Food Canada (AAFC). The PFRA evolved into a national agency in 2009 named the Agri-Environment Services Branch (AESB); it existed until 2012. In July 2012, AAFC's AESB and Research Branch were merged together to form one branch named the Science and Technology Branch.

References

Adger, W.N. 1998. "Observing Institutional Adaptation to Global Environmental Change: Theory and Case Study from Vietnam." Global Environmental Change Working Paper 98-21, Centre for Social and Economic Research on the Global Environment, University of East Anglia and University College London.

———. 2001. "Scales of Governance and Environmental Justice for Adaptation and Mitigation of Climate Change." *Journal of International Development* 13: 921–31.

———. 2006. "Vulnerability." *Global Environmental Change* 16: 268–81.

Alberta Environment. 2003. *Water for Life: Alberta's Strategy for Sustainability.* http://www.waterforlife.alberta.ca/. Accessed 27 September 2013.

———. 2008. *Water for Life: A Renewal.* http://www.waterforlife.alberta.ca/. Accessed 27 September 2013.

Banks, T., and E. Cochrane. 2005. *Water in the West: Under Pressure.* Fourth Interim Report of the Standing Senate Committee on Energy, the Environment and Natural Resources. Ottawa: Parliament of Canada. http://www.parl.gc.ca/Content/SEN/Committee/381/enrg/rep/rep13nov05-e.htm. Accessed 30 September 2013.

Batie, S.S. 2008. "Wicked Problems and Applied Economics." *American Journal of Agricultural Economics* 90, no. 5: 1176–91.

Bourdieu, P. 1986. "The Forms of Capital." Pp. 241–58 in J. Richardson (ed.), *Handbook of Theory and Research for the Sociology of Education*. New York: Greenwood Press.

Bruce, J.P., H. Martin, P. Colucci, G. McBean, J. McDougall, D. Shrubsole, J. Whalley, R. Halliday, M. Alden, L. Mortsch, and B. Mills. 2003. *Climate Change Impacts on Boundary and Transboundary Water Management*. CCAF Project A458/402. Ottawa: Natural Resources Canada. http://www.researchgate.net/publication/273561792_Climate_Change_Impacts_on_Boundary_and_Transboundary_Water_Management. Accessed 2 December 2015.

Bruneau, J., D.R. Corkal, E. Pietroniro, B. Toth, and G. van der Kamp. 2009. "Human Activities and Water Use in the South Saskatchewan River Basin." Pp. 129–52 in G. Marchildon (ed.), *A Dry Oasis: Institutional Adaptations to Climate on the Canadian Plains*. Regina: Canadian Plains Research Center Press.

Cantin, B. 2010. "Integrated Place-Based Approaches for Sustainable Development." *Horizons: Sustainable Places* 10, no. 4. Ottawa: Policy Research Initiative.

Clark, W., and L. Holliday (rapporteurs). 2006. *Linking Knowledge with Action for Sustainable Development: The Role of Program Management*. Workshop Summary, Roundtable on Science and Technology for Sustainability Policy and Global Affairs. National Research Council of the National Academies. Washington, DC: National Academies Press.

Corkal, D.R., and P.E. Adkins. 2008. "Canadian Agriculture and Water." In *Proceedings of the 13th IWRA World Water Congress*. Montpellier, France, 1–4 September 2008.

Corkal, D.R., P.E. Adkins, and B. Inch. 2007. "The Case of Canada—Institutions and Water in the South Saskatchewan River Basin." Working paper from Institutional Adaptation to Climate Change research project, University of Regina. http://www.parc.ca/mcri/paper_browse.php. Accessed 27 September 2013.

Corkal, D.R, H. Diaz, and D. Sauchyn. 2011. "Changing Roles in Canadian Water Management: A Case Study of Agriculture and Water in Canada's South Saskatchewan River Basin." *International Journal of Water Resources Development* 27, no. 4: 647–64.

Corkal, D.R., W.C. Schutzman, and C.R. Hilliard. 2004. "Rural Water Safety from the Source to the On-farm Tap." *Journal of Toxicology and Environmental Health, Part A* 67: 1619–42.

Cragg, W. 1997. "Value Mapping Workshop." Toronto, Ontario, 15–17 October 1997.

Diaz, H., M. Hurlbert, J. Warren, and D.R. Corkal. 2009. *Saskatchewan Water Governance Assessment Final Report*. Institutional Adaptation to Climate Change Research Project. Regina: University of Regina. http://www.parc.ca/mcri/pdfs/papers/gov01.pdf. Accessed 27 September 2013.

Global Water Partnership. n.d. Global Water Partnership website: http://www.gwp.org. http://www.gwpforum.org/servlet/PSP?iNodeID=1345. Accessed 8 October 2009.

Gray, J.H. 1967. *Men Against the Desert*. Winnipeg: Burton Lysecki Books.

Hurlbert, M., D.R. Corkal, and H. Diaz. 2009. "Government and Civil Society: Adaptive Water Management in the South Saskatchewan River Basin." Pp. 181–207 in G. Marchildon (ed.), *A Dry Oasis: Institutional Adaptations to Climate on the Canadian Plains*. Regina: Canadian Plains Research Center Press.

Hurlbert, M., H. Diaz, D.R. Corkal, and J. Warren. 2009. "Climate Change and Water Governance in Saskatchewan, Canada." *International Journal of Climate Change Strategies and Management* 1, no. 2: 118–32.

IPCC (Intergovernmental Panel on Climate Change). 2007. "Climate Change 2007 Synthesis Report." IPCC Plenary XXVII, Valencia, Spain, 12–17 November 2007.

IRC (International Reference Centre for Community Water Supply). 2009. International Reference Centre for Community Water Supply (IRC) website: http://www.ircwash.org/home; Source cited: http://www.irc.nl/page/10433. Accessed 8 October 2009.

Magzul, L. 2013. "Vulnerability and Adaptation to Climate Change in Indigenous Communities in Canada and Guatemala: The Role of Social Capital." PhD dissertation, University of British Columbia.

Marchildon, G.P. (ed.). 2009a. *A Dry Oasis: Institutional Adaptation to Climate on the Canadian Plains*. Regina: Canadian Plains Research Center Press.

———. 2009b. "The Prairie Farm Rehabilitation Administration: Climate Crisis and Federal-Provincial Relations during the Great Depression." *Canadian Historical Review* 90, no. 2: 275–301.

Marchildon, G., S. Kulshreshtha, S.E. Wheaton, and D. Sauchyn. 2008. "Drought and Institutional Adaptation in the Great Plains of Alberta and Saskatchewan: 1914–1939." *Natural Hazards* 45, no. 3: 391–411.

McLeman, R.A., J. Dupre, L. Berrang Ford, J. Ford, K. Gajewski, and G. Marchildon. 2013. "What We Learned from the Dust Bowl: Lessons in Science, Policy, and Adaptation." *Popular Environment*. doi: 10.1007/s11111-013-0190-z.

Morito, B. 2005. *Value and Ethical Analysis in Vulnerability to Climate Change: Establishing an Analytic Framework for Identifying, Classifying and Evaluating Vulnerability Issues*. Institutional Adaptation to Climate Change Research Project. Regina: University of Regina http://www.parc.ca/mcri/pdfs/papers/iacc017.pdf.

———. 2008. *Institutional Values and Adaptive Capacity to Climate Change*. Institutional Adaptation to Climate Change Research Project. Regina: University of Regina. http://www.parc.ca/mcri/pdfs/papers/iacc085.pdf.

———. 2008a. Ibid, citing respondents ATH7, ALB9, ALB10.

———. 2008b. Ibid, citing respondent ALB10.

———. 2008c. Ibid, citing respondent ALB9.

———. 2008d. Ibid, citing respondent ALB2.

———. 2012. *An Ethic of Mutual Respect: The Covenant Chain and Aboriginal-Crown Relations*. Vancouver: UBC Press.

Morito, B., and A. Thachuk, 2008. *Values Analysis of Ethnographic Stakeholder Responses: Report on Cabri-Stewart Valley and the Blood Reserve for the IACC*

Project, Section 1A. Institutional Adaptation to Climate Change Research Project. Regina: University of Regina: http://www.parc.ca/mcri/pdfs/papers/iacc065.pdf.

Nelson, R., M. Howden, and M. Stafford Smith. 2008. "Using Adaptive Governance to Rethink the Way Science Supports Australian Drought Policy." *Environmental Science & Policy* 11, no. 7: 588–601.

Nicole, L.A., and K.K. Klein. 2006. "Water Market Characteristics: Results from a Survey of Southern Alberta Irrigators." *Canadian Water Resources Journal* 31, no. 2: 91–104.

Patiño, L., and D. Gauthier, 2009. "A Participatory Mapping Approach to Climate Change in the Canadian Prairies." Pp. 79–92 in G. Marchildon (ed.), *A Dry Oasis: Institutional Adaptations to Climate on the Canadian Plains*. Regina: Canadian Plains Research Center Press.

Perry, R.B. 1909. *The Moral Economy*. New York: Charles Scribner's Sons.

Portes, A. 1998. "Social Capital: Its Origins and Applications in Modern Sociology." *Annual Review of Sociology* 24: 1–25.

Putnam, R.D. 1995. "Bowling Alone: America's Declining Social Capital." *Journal of Democracy* 6: 65–78.

Rojas. A. 2000. "Land Food and Community I. Course Manual, Agricultural and Environmental Ethics." Course: AGSC 250, Faculty of Agricultural Sciences, University of British Columbia.

Rojas, A., L. Magzul, G.P. Marchildon, and B. Reyes. 2009. "The Oldman River Dam Case Study: Conflict, Adaptation and Institutional Learning." *Prairie Forum* 34, no. 1: 235–60. Also published as pp. 235-60 in G. Marchildon (ed.), *A Dry Oasis: Institutional Adaptations to Climate on the Canadian Plains*. Regina: Canadian Plains Research Center Press.

Rojas, A., B. Reyes, A. Magzul, E. Schwartz, R. Bórquez, and D. Jara. 2009. *Living Waters: What Commitment Is Needed from Institutions in the Era of Climate Change? Support Manual for an Adaptive Resolution to Environmental Conflicts*. Published in the same book in Spanish as: *Aguas de la vida: manual de apoyo*. Regina: Canadian Plains Research Center Press. http://www.parc.ca/mcri/books.php. Accessed 2 October 2013.

Rojas, A., and L. Richer. 2005. "Successful Institutional Adaptations to Climate Change Impacts Posed on Water Resources." IACC Project Working Paper No. 18, University of British Columbia. http://www.parc.ca/mcri/.

Saskatchewan Watershed Authority. 2002. *Annual Report 2002*. Moose Jaw: Saskatchewan Watershed Authority. http://www.swa.ca/Publications/Documents/SwaAnnualReport2002.pdf. Accessed 8 October 2009.

Saskatchewan Water Security Agency. 2012. *25 Year Saskatchewan Water Security Plan*. Moose Jaw, Saskatchewan Water Security Agency. https://www.wsask.ca/About-WSA/25-Year-Water-Security-Plan/. Accessed 27 September 2013.

Sauchyn, D., H. Diaz, and S. Kulshreshtha (eds.). 2010. *The New Normal: The Canadian Prairies in a Changing Climate*. Regina: Canadian Plains Research Center Press.

Scott, J. 1976. *The Moral Economy of the Peasant*. New Haven: Yale University Press.
Swinton, S.M. 2008. "Reimagining Farms as Managed Ecosystems." *Choices* 23, no. 2. American Agricultural Economics Association.
Thompson, E.P. 1971. "The Moral Economy of the English Crowd in the Eighteenth Century." *Past and Present* 50: 76–136.
Toth, B., D.R. Corkal, D. Sauchyn, G. van der Kamp, and E. Pietroniro. 2009. "The Natural Characteristics of the South Saskatchewan River Basin." Pp. 95–127 in G. Marchildon (ed.), *A Dry Oasis: Institutional Adaptations to Climate on the Canadian Plains*. Regina: Canadian Plains Research Center Press.
Warren, J., and H. Diaz. 2012. *Defying Palliser: Stories of Resilience from the Driest Regions of the Canadian Prairies*. Regina: Canadian Plains Research Center Press.
Wheaton, E., S. Kulshreshtha, V. Wittrock, and G. Koshida. 2008. "Dry Times: Hard Lessons from the Canadian Drought of 2001 and 2002." *The Canadian Geographer* 52, no. 2: 241–62.
Wheaton, E., V. Wittrock, S. Kulshreshtha, G. Koshida, C. Grant, A. Chipanshi, and B. Bonsal. 2005. *Lessons Learned from the Canadian Drought Years of 2001 and 2002: Synthesis Report*. Saskatoon: Saskatchewan Research Council.
World Meteorology Organization. 1992. *The Dublin Statement on Water and Sustainable Development*. Geneva: World Meteorology Organization. http://www.wmo.int/pages/prog/hwrp/documents/english/icwedece.html. Accessed 8 October 2009.

CHAPTER 12

BRIDGING KNOWLEDGE SYSTEMS FOR DROUGHT PREPAREDNESS: A CASE STUDY FROM THE SWIFT CURRENT CREEK WATERSHED (CANADA)

Jeremy Pittman, Darrell R. Corkal, Monica Hadarits, Tom Harrison, Margot Hurlbert, and Arlene Unvoas

Introduction

Every year droughts have significant impacts around the globe. These impacts cascade through social-ecological systems, meaning that even localized droughts can have global significance in today's highly interconnected world. Despite the visibility of its effects, drought remains one of the most enigmatic disasters or climate-related disturbances, eluding even a broadly accepted definition.

As with most extreme events, it is typically better to address drought risk proactively, through preparedness planning, rather than solely reacting to drought events. Wilhite (2005, 1996) has demonstrated the benefit of drought preparedness in a number of contexts. Benefits from preparedness are derived from reduced stress on the system, improved ability to make decisions during crises, and lower costs associated with proactive adaptation—all aspects that help reduce the vulnerability of society in general and the rural population in particular. However, preparedness is not a panacea, and it must be accompanied by a suite of reactive adaptation strategies to be effective.

This chapter explores how deliberative, watershed-scale drought preparedness planning fits within broader adaptation strategies and programs in a case study of the Swift Current Creek watershed in Canada. The chapter begins with an overview of the conceptual framework that guided the research and follows with a detailed description of the case study. It then presents the methods used to explore the case and subsequently highlights the main results. Finally, the results are discussed in light of their implications for our understanding of multi-stakeholder, deliberative processes for drought preparedness, and conclusions are presented on the value of working with multiple and diverse stakeholders to bridge knowledge for drought preparedness.

Conceptual Framework

Human adaptation to climate is defined as "the process of adjustment to actual or expected climate and its effects, in order to moderate harm or exploit beneficial opportunities" (IPCC 2012: 5). We see preparedness as a specific type of proactive adaptation, where actors anticipate options and become "ready to respond … and manage … consequences through measures taken prior to an event" (MREM 2011: 4). Preparedness is somewhat synonymous with what Smit et al. (2000) refer to as anticipatory and planned adaptation in that preparedness is deliberately undertaken prior to a potentially problematic climate event. Preparedness occurs in the context of uncertainty, meaning that actors must prepare with incomplete knowledge of the severity, magnitude, timing, and frequency of future events.

Berkes (2009) has shown how processes that engage knowledge from different sources (e.g., scientists, agricultural producers, different sector and industry groups, environmental groups, communities and social groups) can help navigate uncertainty. Here, uncertainty is conceptualized as an irreducible property of social-ecological systems. Social-ecological systems are inherently linked, co-dependent, and co-evolutionary systems comprised of social and ecological dimensions (Berkes and Folke 1998). The rationale for drawing on diverse knowledge to confront such uncertainty in social-ecological systems is to broaden active participation and the breadth of information used in decision making.

Bridging is the process of bringing different knowledge systems together to address problems that are relevant to different groups (Bohensky and Maru 2011). Bridging brings knowledge systems together in ways that maintain the integrity of each system (Reid et al. 2006), and knowledge is translated between actors without coercion (Sundberg 2007). Knowledge-bridging processes can be facilitated using boundary objects (Cash et al. 2003), which are objects that can take many forms (e.g., maps, models, concepts) and allow for knowledge communication and translation between actors with different understandings, interpretations, and interests associated with common problems (Brand and Jax 2007; Star and Griesemer 1989). Boundary objects must be flexible and adaptable to distinct contexts and situations. In this case study, a number of boundary objects were used to bridge knowledge for drought preparedness in the Swift Current Creek watershed.

Overview of the Swift Current Creek Watershed

The Swift Current Creek watershed is located in southwestern Saskatchewan, which is a relatively dry region of the Canadian Prairies (Figure 1). As part of the Palliser Triangle, the watershed has experienced recurring severe droughts over the last century. Some of the most notable droughts occurred in the 1930s, 1960s, 1980s, and 2000s, each having significant implications for agricultural production in the watershed (SRC 2011).

The watershed encompasses a total drainage area of 5,592 km^2. It begins near the Cypress Hills in Saskatchewan and continues to the creek's outlet on the South Saskatchewan River near Stewart Valley (Figure 1). The watershed contains mostly agricultural land and a number of rural communities. There are 5 urban municipalities (UMs) in the watershed, Swift Current and Shaunavon being the largest, and 12 rural municipalities (RMs).

The Swift Current Creek is supplied by snowmelt runoff and a number of groundwater springs. It flows about 160 km from its headwaters, contributing water into the South Saskatchewan River, which ultimately

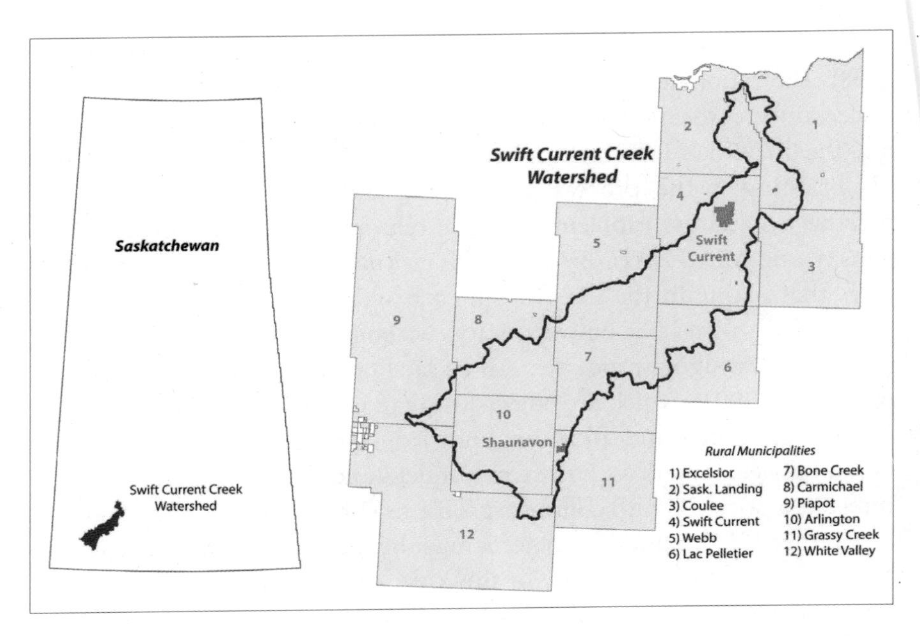

Figure 1. Map of the Swift Current Creek watershed, Saskatchewan

drains into Hudson Bay. The creek provides several services within the Swift Current Creek watershed, such as water for agricultural production (irrigation and livestock), municipal drinking water supplies, and recreation. Developed in 1943, the Duncairn Dam and Reid Lake Reservoir provide some drought protection within the watershed. This infrastructure stores 105,000 dam^3 of water at its full supply level, which supports 7,000 ha of irrigated agricultural land and provides a dependable water supply for the city of Swift Current.

The Swift Current Creek Watershed Stewards (SCCWS), a not-for-profit corporation officially created in 2001, has sought to maintain or improve watershed health since it was organized (Table 1). In 2007, the SCCWS partnered with the Saskatchewan Watershed Authority (now the Water Security Agency), a provincial Crown corporation mandated to manage water in Saskatchewan, to implement a source water protection planning process. This process was part of a broader provincial initiative called the Long Term Safe Drinking Water Strategy, one component of

Table 1. Timeline of milestones and successes for the Swift Current Creek Watershed Stewards (SCCWS)

<table>
<tr><th>Year(s)</th><th>Description</th><th>Key successes</th></tr>
<tr><td>1998</td><td>The City of Swift Current voices concerns over increased water treatment costs at its water treatment plant. Agriculture and Agri-Food Canada's Prairie Farm Rehabilitation Administration is tasked with testing water quality; it is determined that there is no obvious decrease in water quality.</td><td rowspan="8">• Completed a four-year watershed monitoring project
• Worked with many producers to promote beneficial management practices that protect water supplies within the watershed
• Established an effective working relationship with urban and rural municipalities, as well as diverse stakeholders within the watershed
• Created awareness of an invasive species (salt cedar) that was entering the watershed from the United States; in response to a SCCWS flyer on the issue, a stakeholder identified the first salt cedar plant in Saskatchewan and measures began to control this invasive species
• Developed strong working relationships with various government agencies (local, provincial, federal)
• Participated in the Southwest Public Safety Region pilot project to help prepare for emergencies
• Worked with partners on academic research (e.g., Vulnerability and Adaptation to Climate Extremes in the Americas project) to better understand climate risks and adaptation options in the watershed</td></tr>
<tr><td>1999</td><td>An accidental release of raw effluent from the city of Swift Current's lagoons flows into Swift Current Creek. A group of concerned stakeholders, representing the interests of various federal, provincial, and municipal agencies, gathers to discuss various watershed issues, such as effluent releases into the creek and sewage flowing into Lac Pelletier.</td></tr>
<tr><td>2001</td><td>The City of Swift Current is fined $25,000 for the effluent release, with the stipulation that the fine be used to form a creek stewardship group. The City agrees to pay $5,000 per year for five years. The stewardship group is officially formed and becomes incorporated as the SCCWS.</td></tr>
<tr><td>2002</td><td>The SCCWS applies for and receives funding to hire a watershed coordinator. As the group meets, a decision is made to educate water users and other stakeholders within the watershed about water quality and quantity issues and impacts.</td></tr>
<tr><td>2006</td><td>The SCCWS is invited to apply for Agri-Environmental Group Plan funding under the Federal-Provincial Agriculture Policy Framework. SCCWS receives funding to increase awareness of agri-environmental risks in the watershed and begins planning to address these risks.</td></tr>
<tr><td>2007</td><td>The Saskatchewan Watershed Authority asks the SCCWS to develop a source water protection plan.</td></tr>
<tr><td>2009</td><td>SCCWS completes a source water protection plan.</td></tr>
<tr><td>2010–15</td><td>The SCCWS (13 members) continues to operate as a non-profit watershed group and seek opportunities and funding to enhance the watershed's environment, economy, and social systems.</td></tr>
</table>

Table 2. Recommendations and actions related to drought preparedness from the *Swift Current Creek Watershed Protection Plan*

Recommendations	Actions
Research the impact of climate change on water supply, including the variability in flow regimes in the creek, and develop mitigation strategies	Develop adaptation strategies to deal with natural climate variability and cyclical flow regimes in Swift Current Creek
	Determine/estimate extreme cyclical variations and how best to manage them
	Use historical events to better understand and quantify future events
Research and implement measures for drought preparedness, including organization of a drought preparedness workshop	Develop water supply availability information, including surface water and groundwater; identify communities at risk; and organize a drought preparedness workshop

Source: SCCWS 2009.

which was aimed at producing community-based source water protection plans in a number of Saskatchewan's watersheds. These plans were to be produced collaboratively with multiple watershed stakeholders.

In 2009, the *Swift Current Creek Watershed Protection Plan* was completed (SCCWS 2009). The plan contained 62 recommended action items aimed at improving sustainability in the Swift Current Creek watershed and identified different ways that stakeholders and organizations across different governance levels (i.e., local, regional, provincial, federal) could work together to achieve the desired outcomes. The result was two main recommendations, and four subsequent actions, directly related to drought preparedness (Table 2). These recommendations and actions included different elements of stakeholder engagement, adaptation planning, and hydro-climate analysis, and were to be implemented by the SCCWS, the Saskatchewan Watershed Authority, the Saskatchewan Ministry of Agriculture, and Agriculture and Agri-Food Canada.

Methods for Drought Preparedness Planning

In early 2010, a drought preparedness planning project was initiated in the Swift Current Creek watershed to address the recommendations and action items in the *Swift Current Creek Watershed Protection Plan*. The project used a participatory action research (PAR) approach to bridge knowledge systems during development of the plan. PAR is a combination of participatory research, which is research that explicitly includes and engages stakeholders throughout the research process (Cargo and Mercer 2008), and critical action research, which is research undertaken with the intent of producing beneficial outcomes for stakeholders (Kemmis and McTaggart 2000). Rather than a method in itself, PAR is an approach to research that guides the use of a broad range of methods. As such, the specific methods used in PAR can span both qualitative and quantitative inquiry (Cargo and Mercer 2008), as well as draw from both the natural and social sciences (Ravera et al. 2011). PAR is a proven successful technique for bridging knowledge between different groups throughout a research project (Whitfield and Reed 2012).

This project was to be implemented collaboratively between the SC-CWS and the Saskatchewan Watershed Authority, but it required the involvement of other stakeholder groups, government organizations, and research groups. These additional organizations were either engaged directly in the planning process or contributed specific pieces of work or studies that informed the planning process. Other organizations engaged in the planning process included the Southwest Enterprise Region, the Saskatchewan Ministry of Agriculture, the Saskatchewan Ministry of Municipal Relations (formerly Municipal Affairs), Agriculture and Agri-Food Canada through its Agri-Environment Services Branch, the Prairie Adaptation Research Collaborative (PARC), and the Saskatchewan Research Council (SRC). These different groups were engaged during workshops, and many completed complementary studies that were used throughout the planning exercise (see the Acknowledgments). Most specifically, PARC conducted hydro-climate variability assessments and projections (Barrow 2011; St. Jacques et al. 2011; PARC 2010), and SRC completed extreme events characterizations (SRC 2011) that were used in the planning process.

The inclusion of these diverse groups and stakeholders required that knowledge be bridged across a number of boundaries. At the local level, there were participants involved with different modes of agricultural production, including both dryland and irrigation producers of crops, forage, and livestock. Other local-level participants included UMs and RMs, as well as additional community groups (e.g., Southwest Enterprise Region). At the provincial and federal levels, several different agencies were involved, each with different expertise (see above). Knowledge held by these agencies ranged from scientific knowledge regarding hydrology and agrology to more pragmatic knowledge regarding agricultural extension or program and policy development. In addition, climatological knowledge and expertise were provided by PARC and SRC.

The knowledge-bridging process involved two main components: a participatory vulnerability assessment and a participatory adaptation planning exercise. The vulnerability assessment aimed to understand how and why past droughts had been problematic and identify the variety of different adaptation options used to deal with past droughts. In addition, potential vulnerabilities and adaptation options for future droughts were explored. The vulnerability assessment provided the foundation for the adaptation planning exercise, which aimed to identify strategies that could increase preparedness to future droughts in light of existing and potential vulnerabilities. Three main boundary objects were used to bridge knowledge from diverse sources throughout the project: maps, timelines, and scenarios. The utility of these boundary objects has been demonstrated elsewhere (Ravera et al. 2011; Kok et al. 2007), but their applicability in the context of drought preparedness planning in Saskatchewan watersheds was untested prior to completion of this case study.

Outcomes of the Bridging Process

The knowledge-bridging process resulted in a number of outcomes. Most importantly, it facilitated the development of in-depth insights into current and future drought vulnerability within the watershed and provided the foundations for adaptation planning. These insights and outcomes are discussed below in relation to the boundary objects and bridging processes that facilitated the research.

Participatory Mapping and Timeline Construction

The participatory mapping and timeline exercises, using both maps and timelines as boundary objects, allowed participants to discuss the spatial and historical elements of drought vulnerability in the watershed. These exercises provided insights into the long history of the watershed's agricultural sector in dealing with drought (Table 3) and the lessons learned throughout the course of this history. For example, participants discussed how the tillage practices of the 1930s had increased agricultural vulnerability to drought and how significant progress toward soil conservation had been made in the watershed since then (see Chapter 5 by Warren on min till in this volume). Additionally, participants identified a number of beneficial policy and practice cycles, such as water development projects that constructed farm dugouts in the 1980s or the promotion of shallow-buried pipelines in the late 1990s, which significantly reduced agricultural sensitivity and increased preparedness for droughts.

The mapping exercise identified different locations in the watershed that were more or less sensitive to drought and excessive moisture. Also, the locations of key events were recorded in ways that complemented the timeline activity. The mapping exercise provided some interesting insights into different biophysical vulnerabilities in the watershed as well, such as how certain fish populations can become trapped in deep pools along the Swift Current Creek when streamflow is low. Arguably, the mapping activity was the most successful of all the exercises, largely because it gave participants an opportunity to visualize issues and sparked valuable engagement between the different stakeholders.

Participatory Scenarios

Scenarios were an additional boundary object used during the planning process (Table 4). The scenarios were developed based on findings from complementary studies (see the Acknowledgments) and focus group discussions aimed at understanding vulnerabilities. These scenarios were framed as "what if" questions and were developed to represent a range of possibilities related to dry and wet conditions in the watershed. More specifically, these scenarios explored the vulnerabilities and adaptation options under extreme events of different intensity, duration, and frequencies. They also stimulated discussion of existing and potential vulnerabilities and adaptation options.

Table 3. Timeline of important events related to drought preparedness in the Swift Current Creek watershed

Period/Year	Description
1930	Plow and thrasher era; no straw; soil pulverized
1950s	Irrigation development; flooding of flat land; alkali issues
1951	Duncairn Dam almost washed away by flood
1952	The community of Eastend almost washed away by flood
1950s–60s	Widespread drought; trees and shelterbelts planted to catch snow and reduce wind erosion
1970	Heavy snowfall; many calves lost
1976	Cattle walking over corrals because of high snow levels
1978	May – five-day blizzard
1982	May 25 – 1.5 feet of snow; blizzard
1988	Very dry; PFRA dugout program expanded and many dugouts built during this severely dry year
1991	Very wet; two to three inches of rain in spring
1996	Wet winter snow
1997	Large flooding in spring due to rapid thaw; Gravelbourg almost flooded out
1999	Introduction of PFRA shallow pipelines for livestock
2000	Rained approximately 13 inches within 14 hours in Vanguard area; water diverted into Old Wives basin
2001	Widespread drought
2002	Minimal moisture until July; rained hard in August
2005	Improvements in watering techniques to exclude livestock access: fencing of dugouts and using solar-powered and remote watering systems
2007	Duncairn Dam spillway taxed with inflow from a large snowmelt runoff and a rapid spring thaw
2008	Very few sloughs in spring
2010	A record dry winter and spring, followed by an excessively wet summer; beginning in July, dugouts fill, watercourses flow, and soil and land become waterlogged in areas of high rainfall
2010	The town of Maple Creek and surrounding area receive record flooding following intense short-duration rains. A portion of the Trans-Canada Highway infrastructure is washed out. Junction Dam, immediately upstream of the highway, survives the flood, largely because spillway capacity was increased in 2008 to safeguard the dam for larger flows and flood events.

Note: PFRA = Prairie Farm Rehabilitation Administration.

A number of interesting insights were gained from the participatory scenario process.

First, participants highlighted the need for long-term programming to reduce sensitivities and increase adaptive capacities to extreme events, rather than short-term programs or ad hoc responses aimed at coping with events already occurring. This discussion emerged during Scenario A, somewhat in response to excess moisture conditions being experienced during the time of the workshops.

Second, participants noted the challenges associated with adaptation to drought in the watershed and stated that the successes and failures of past adaptation strategies would have significant implications for future drought vulnerabilities. This discussion largely emerged during Scenario B. For example, the irrigation development in the watershed during the 1950s had not necessarily produced the benefits that were intended, such as the production of irrigated, high-value crops in the watershed. The irrigation infrastructure does provide important access to water for crops and forage in times of drought, but these crops and forage are typically of low economic value. At the time of the workshops, much of this infrastructure was publicly owned and required significant maintenance and investment to remain operational. The broad public benefit of this investment had been brought into question, along with the monetary value to the local economy actually added by the irrigation system. As such, the federal government was in the process of divesting the irrigation infrastructure to local groups (see Chapter 6 by Warren on irrigation in this volume). Participants noted how some sub-projects within the irrigation system would probably be sustained under local operation, but many were at risk of being decommissioned. This provided an interesting element to the scenario discussions in that irrigation expansion was not a major theme. Rather, program and policy strategies that promoted small-scale infrastructure investments (e.g., shallow buried pipelines for livestock watering) and improved agronomic practices (e.g., soil conservation) were favoured.

Third, participants viewed increased inter-annual hydro-climate variability as less problematic than longer-term drought or increased frequency of excessive moisture events (Scenario C in Table 4). As such, adaptation options recommended for increased inter-annual variability were similar to those already implemented in the watershed. Participants

Table 4. Participatory assessment of vulnerability and adaptation under the different scenarios

Scenario	Vulnerability	Adaptation
Scenario A: What would happen if a wet year like 2010 happened twice in five years?	• Problems with hay quality • Stressful: long haying season • Rural areas more affected than urban areas • Ranchers may be better able to handle the rain—grass acts as a buffer; hay quality decreases and quantity increases • Timing of rain depends on impact and vulnerability • Silage companies difficult to hire • Cities would require increased budgets for repairs and snow removal (e.g., heaving sidewalks). • Large-scale economic problems • Timing a major concern • Poor-quality crops	• Change calving cycle • Stockpile feed • Ranchers may need to test hay and buy protein blocks. • Use more inoculants (i.e., $5/bale, which is less expensive than silage) • 8+ tonnes hay per acre – could afford to do silage • Start seeding earlier • Learn how to grow rice • Producers need more control of marketing (CWB, options for cattle) • Keep off-farm jobs • Need programs aimed at long-term solutions rather than short-term fixes • Resentment results when some rural municipalities receive relief but others do not.

Scenario B: What would happen if a long-term drought (lasting longer than previously experienced) occurred?	• Price increase could influence stocking rates. • Irrigators – rely on water • Ranchers may not have enough water to maintain grazing on native pasture. • Ranchers may sell some of their herd, which forces them to risk decreasing the quality of their herds' genetics. • Cattle will not walk one mile for water so must haul water • Overgrazing	• Ideally, stockpile two years' supply of hay before the droughts occur • Defer grazing and use grazing management system • Haul cattle to other areas of Saskatchewan and Manitoba to graze • Construct more water storage (farm dugouts) • Increase reservoir storage and dam size • Sell part of herd • Use a variety of feeds • AAFC divestiture of irrigation projects will mean new adaptive responses are required by local operators. • Stakeholders view southwestern Saskatchewan as well-adapted and fairly drought-proof (e.g., pipelines, irrigation projects, dugouts, soil conservation).
Scenario C: What would happen if drought and excessive moisture events switched back and forth from wet to dry years very quickly?	• Longer duration is more devastating than high frequency. • Vulnerability depends on the stage of your career. • Markets dictate what happens.	• Plan for normal precipitation • Have two-year supplies of water and hay available • Use frugal management • Maintain cropping practices – large-scale change does not occur • Need improved lead-time on climate forecasts

Source: SCCWS 2009.
Note: CWS = Canadian Wheat Board; AAFC = Agriculture and Agri-Food Canada.

did highlight that vulnerability would largely depend on the response of international commodity markets to this variability and noted that frugal management of financial and environmental resources would be required. In addition, participants noted how vulnerability to increased variability largely depends on agricultural producers' stage of career, with established producers less vulnerable than younger producers, since they typically have less debt.

Adaptation Planning and Prioritizing Actions

The adaptation planning and prioritization workshop followed the participatory mapping, timelines, and scenario exercises, and aimed to bridge diverse stakeholder knowledge in the co-production of a drought preparedness plan. During this workshop, participants were presented with synthesized findings related to vulnerability and adaptation from the first workshop and subsequently asked to develop adaptation strategies that could help address these vulnerabilities. In addition, participants were presented with information from studies by SRC that characterized extreme climate events to facilitate the planning and knowledge-bridging activities.

The exercises resulted in the development of adaptation strategies aimed at the municipal and agricultural sectors (Table 5). Strategies varied from those focused mostly on infrastructure (e.g., build redundancy into municipal water supply systems) to those focused on capacity-building approaches (e.g., provide training for municipal staff on emergency management). Many of the strategies related to modifying existing practices (e.g., define drought triggers for different levels of response), developing better climate information systems (e.g., increase number of climate observation stations), and then integrating these systems with decision making (e.g., base relief programs partly on reliable climate science).

Discussion: Opportunities and Challenges

Although many of the strategies listed in Table 5 are justifiable and have potential net benefits, several opportunities and challenges have been associated with implementation. This project bridged knowledge from diverse stakeholders while preparing the plan and built a core group of collaborators for implementing drought preparedness projects in the

Table 5. Adaptation strategies and priorities

Theme	Adaptation strategy	Priority
Municipal	Provide training for staff on emergency management	High
	Conduct water supply planning	High
	Define drought triggers for different levels of response	High
	Take a watershed approach to municipal emergency response planning	High
	Identify high-risk areas for landowners and city	High
	Develop framework for implementing water use restrictions	Medium
	Promote coordination between municipalities	Medium
	Develop action plans for different types of drought (hydrological, meteorological, mechanical)	Medium
	Promote water conservation programs (e.g., low-flush toilets)	Medium
	Stockpile resources, such as water pipelines	Medium
	Develop agreements for sharing equipment and expertise across municipalities during emergencies	Medium
	Build redundancy into municipal water supply systems	Low
	Match water quality to water use requirements	Low
Agricultural	Improve access to and availability of climate/weather forecasting	High
	Expand producer crop and weather reporting network	High
	More hydrometric stations for real-time data	High
	Increase number of climate observation stations	High
	Develop effective monitoring and information systems	High
	Promote cross-organizational knowledge	High
	Improve integration of seasonal forecasts into crop planning	Medium
	Develop long-term preparedness and adaptation programs	Medium
	Define drought triggers for support from provincial and federal governments	Medium
	Base relief programs in part on reliable climate science	Medium
	Develop crisis line for drought management prior to drought	Medium

watershed. This coordination is exemplified in the ongoing collaboration between the SCCWS, the Saskatchewan Watershed Authority, and Agriculture and Agri-Food Canada on PARC's Vulnerability and Adaptation to Climate Extremes in the Americas (VACEA) project (2011–2016). VACEA is funded jointly by the International Development Research Centre and Canada's Tri-Council. On VACEA, key actors have been able to maintain their collaborative relationships to advance drought preparedness in the watershed, despite having to take advantage of a different funding source.

The drought preparedness initiatives also had many synergies with different projects already underway by the SCCWS. These projects include their watershed monitoring and invasive species programs, which track and report on watershed health issues and invasive species prevalence. More specifically, the drought preparedness work had synergies with the SCCWS's salt cedar monitoring and removal program, since salt cedar can have negative impacts (e.g., over-salinization) on existing water and soil resources. The negative impacts of salt cedar can amplify agricultural sensitivity to drought.

As noted earlier, several challenges are associated with implementation. For example, there is often a lack of clear responsibility for implementing different projects, which can paralyze the governance network. In some cases, local actors, such as the SCCWS, are left to implement projects on their own, even if they do not have a legislated mandate to do so. This problem is particularly apparent for addressing the salt cedar issue in the watershed, but it is also relevant for implementing many of the strategies in Table 5, such as promoting coordination between municipalities. Without formalized funding sources or programs, it is very difficult and often simply not possible to implement any course of action.

There are also several barriers to collaboration in the watershed. These include a pervasive rural-urban divide, which is relevant in many areas throughout the province (Partridge and Olfert 2009; Hoggart 1990), and also a fear in many municipalities that increased collaboration leads to forced amalgamation. Some of the participatory planning exercises and tools possibly helped address these barriers to some degree, since many of the strategies identified by participants in Table 5 relate to improved collaboration between municipalities, but the benefits of the activities are not necessarily long-lived and are at risk of easily being forgotten. Since the completion of this planning project in 2011, attempts to improve municipal collaboration regarding drought and excessive moisture have had limited success, and stakeholders have not been able to make real progress in developing more specific action plans.

Conclusions

This case study provides several preliminary insights regarding collaborative drought preparedness in Saskatchewan.

First, it demonstrates the key role of local watershed stewardship organizations in preparing for drought. These organizations are able to provide multi-stakeholder, deliberative forums for bridging different perspectives and values regarding the direction of drought adaptation. In addition, watershed groups are able to nurture a forum for collaboration with a broad range of non-government and government actors across different levels (i.e., local, municipal, regional, provincial, and federal). Accordingly, watershed groups help the diversity of stakeholders take advantage of opportunities arising from different funding sources and program frames. However, watershed stewardship organizations only have an informal role in drought preparedness and are not empowered by any formal legislation in the Saskatchewan context, which enables their flexibility but can constrain their ability to act or influence water management decisions.

Second, this case study demonstrates the value of different boundary objects, such as participatory mapping, timelines, and scenario assessments, for engaging with different knowledge systems in deliberative processes. The general utility of these tools has been demonstrated elsewhere (e.g., Ravera et al. 2011; Kok et al. 2007), but this case confirms that they can be useful and practical when working on drought preparedness in Saskatchewan's watersheds. These boundary objects facilitated the development of an innovative drought preparedness plan, which, although preliminary, provided some guidance toward drought preparedness for key actors in the watershed.

Finally, this case reiterates that knowledge-bridging activities during planning are only the first piece of the puzzle in building drought preparedness. The role of the SCCWS has been crucial as a bridging organization to bring stakeholders together to begin preparedness planning. The research work of the case study, and related funding, were catalysts that helped begin participatory planning by diverse stakeholders to consider developing preparedness plans. But plans can only be effective if they are implemented, monitored, and adjusted to ensure the desired results are achieved. Without clearly defined roles for the diverse stakeholders, or sustained commitment by all actors (including all levels of government), preparedness plans will suffer from an implementation gap and fail to realize their potential. Also, changing policy priorities, programs, and funding sources will limit actors' ability to implement plans. The lack of long-term, secure funding means even the sustainability of the watershed groups themselves is not assured. This case study suggests that continued collaboration between a core group of actors with varying interests and expertise can help improve capacity to adjust to changing priorities while maintaining general goals toward drought preparedness and sustainability. It must be emphasized that there is great value in participatory planning with a diversity of stakeholders. Once this planning process is initiated, a real challenge occurs when stakeholders need to move beyond planning into adaptive action.

Acknowledgments

Funding for the preparedness project was received from Natural Resources Canada as part of its Prairie Regional Adaptation Collaborative program (PRAC). As such, many of the ongoing PRAC research projects were used to inform the planning process. The Prairie Adaptation Research Collaboration (PARC) at the University of Regina led the PRAC, with the Saskatchewan Watershed Authority as the PRAC lead on the Drought and Excessive Moisture Preparedness theme for Saskatchewan. PARC contributed significant research expertise on climate change to the project, including on hydro-climate variability assessment and paleoclimate reconstruction. The community-based vulnerability assessment was completed by the University of Regina in collaboration with the Saskatchewan Watershed Authority, as part of both PRAC and a complementary Social Sciences Humanities and Research Council of Canada–funded project (the Rural Community Adaptation to Drought project). The Saskatchewan Research Council completed a characterization of extreme events in the watershed, and Dr. Steven Quiring from Texas A&M University completed an evaluation of climate extremes monitoring. The planning process was also aligned with Agriculture and Agri-Food Canada's Drought Preparedness Partnership, which, although not officially part of PRAC, provided a provincial-level assessment of drought preparedness in Saskatchewan.

References

Barrow, E. 2011. *Preliminary Probabilistic Analyses of Drought Indices in the Prairies*. Regina: Prairie Adaptation Research Collaborative. http://www.parc.ca/rac/fileManagement/upload/Prelimnary%20Probalistic%20Analysis%20of%20Drought%20Indices%20in%20the%20Prairies_June%202011.doc.pdf. Accessed October 2013.

Berkes, F. 2009. "Evolution of Co-management: Role of Knowledge Generation, Bridging Organizations and Social Learning." *Journal of Environmental Management* 90: 1692–1702.

Berkes, F., and C. Folke. 1998. *Linking Social and Ecological Systems: Management Practices and Social Mechanisms for Building Resilience*. Cambridge: Cambridge University Press.

Bohensky, E.L., and Y. Maru. 2011. "Indigenous Knowledge, Science, and Resilience: What Have We Learned from a Decade of International Literature on 'Integration'?" *Ecology and Society* 16, no. 4. doi: 10.5751/ES-04342-160406. http://www.ecologyandsociety.org/vol16/iss4/art6/.

Brand, F.S., and K. Jax. 2007. "Focusing the Meaning(s) of Resilience: Resilience as a Descriptive Concept and a Boundary Object." *Ecology and Society* 12, no. 1.

Cargo, M., and S.L. Mercer. 2008. "The Value and Challenges of Participatory Research: Strengthening Its Practice." *Annual Review of Public Health* 29, no. 1 (April): 325–50. doi: 10.1146/annurev.publhealth.29.091307.083824.

Cash, D.W., W.C. Clark, F. Alcock, N.M. Dickson, N. Eckley, D.H. Guston, J. Jäger, and R.B. Mitchell. 2003. "Knowledge Systems for Sustainable Development." *Proceedings of the National Academy of Sciences* 100, no. 14: 8086–91.

IPCC (Intergovernmental Panel on Climate Change). 2012. *Managing the Risks of Extreme Events and Disasters to Advance Climate Change Adaptation: Special Report of the Intergovernmental Panel on Climate Change*. Cambridge: Cambridge University Press.

Hoggart, K. 1990. "Let's Do Away with Rural." *Journal of Rural Studies* 6, no. 3: 245–57.

Kemmis, S., and R. McTaggart. 2000. "Participatory Action Research: Communicative Action in the Public Sphere." Pp. 559–603 in N. Denzin and Y. Lincoln (eds.), *Handbook of Qualitative Research*. Thousand Oaks, CA: Sage Publications.

Kok, K., R. Biggs, and M. Zurek. 2007. "Methods for Developing Multiscale Participatory Scenarios: Insights from Southern Africa and Europe." *Ecology and Society* 13, no. 1.

MREM (Ministers Responsible for Emergency Management). 2011. *An Emergency Management Framework for Canada*. 2nd edition. Ottawa, ON: Public Safety Canada.

PARC (Prairie Adaptation Research Collaborative). 2010. *Hydroclimatic Variability: South Saskatchewan River Basin*. Regina: PARC. http://www.parc.ca/rac/fileManagement/upload/Hydroclimatic%20Variability_South%20Saskatchewan_River_Basin_%20March10.pdf. Accessed October 2013.

Partridge, M.D., and M.R. Olfert. 2009. "Dissension in the Countryside: Bridging the Rural-urban Divide with a New Rural Policy." Pp. 169–210 in M. Gopinath and H. Kim, *Globalization and the Rural-Urban Divide*. Seoul: Seoul National University Press.

Ravera, F., K. Hubacek, M. Reed, and D. Tarrasón. 2011. "Learning from Experiences in Adaptive Action Research: A Critical Comparison of Two Case Studies Applying Participatory Scenario Development and Modelling Approaches." *Environmental Policy and Governance* 21, no. 6 (November): 433–53. doi: 10.1002/eet.585.

Reid, W., F. Berkes, T. Wilbanks, and D. Capistrano. 2006. *Bridging Scales and Knowledge Systems: Concepts and Applications in Ecosystem Assessment*. Washington, DC: Island Press.

SCCWS (Swift Current Creek Watershed Stewards). 2009. *Swift Current Creek Watershed Protection Plan*. Swift Current, SK: SCCWS. https://www.wsask.ca/Global/Water%20Info/Watershed%20Planning/SwiftCurrentCreekWatershedProtectionPlanFinal.pdf. Accessed October 2013.

Smit, B., I. Burton, R. Klein, and J. Wandel. 2000. "An Anatomy of Adaptation to Climate Change." *Climatic Change* 45: 223–51.

SRC (Saskatchewan Research Council). 2011. "Drought and Excessive Moisture—Saskatchewan's Nemesis: Characterizations for the Swift Current Creek, North Saskatchewan River, Assiniboine River and Upper Souris River Watersheds." Saskatoon, SK: SRC. http://www.parc.ca/rac/fileManagement/upload/2TSk%20Nemesis11%20SWA%20DEM%20Report%20updated.pdf. Accessed October 2013.

St. Jacques, J., Y.A. Huang, Y. Zhao, S.L. Lapp, and D.J. Sauchyn. 2011. *The Effects of Atmosphere-Ocean Climate Oscillations on and Trends in Saskatchewan River Discharges*. Regina: PARC.. http://www.parc.ca/rac/fileManagement/upload/The_Effects%20of%20Atmosphere_Ocean_Climate_Oscillations_on_and_Trends_in_Saskatchewan_River_Discharges.pdf. Accessed October 2013.

Star, S.L., and J.R. Griesemer. 1989. "Institutional Ecology, 'Translations' and Boundary Objects: Amateurs and Professionals in Berkeley's Museum of Vertebrate Zoology, 1907–39." *Social Studies of Science* 19, no. 3 (August 1): 387–420. doi: 10.1177/030631289019003001.

Sundberg, M. 2007. "Parameterizations as Boundary Objects on the Climate Arena." *Social Studies of Science* 37, no. 3 (June 1): 473–88. doi: 10.1177/0306312706075330.

Whitfield, S., and M.S. Reed. 2012. "Participatory Environmental Assessment in Drylands: Introducing a New Approach." *Journal of Arid Environments* 77 (February): 1–10. doi: 10.1016/j.jaridenv.2011.09.015.

Wilhite, D.A. 1996. "A Methodology for Drought Preparedness." *Natural Hazards* 13, no. 3: 229–52.
——— (ed.). 2005. *Drought and Water Crisis: Science, Technology, and Management Issues*. Boca Raton, FL: CRC Press.

PART 6

LEARNING FROM OTHER EXPERIENCES

CHAPTER 13

DROUGHT RISKS AND OPPORTUNITIES IN THE CHILEAN GRAPE AND WINE INDUSTRY: A CASE STUDY OF THE MAULE REGION

Monica Hadarits, Paula Santibáñez, and Jeremy Pittman

Introduction

This chapter focuses on the vulnerability of an agricultural system in Chile and offers potential lessons learned that could apply to Canadian agriculture. Although many differences exist between Canada and Chile, there are similarities at a regional scale between the Canadian Prairies and the Maule region in Chile. Water supplies for irrigation in both countries, for example, are mostly derived from snowmelt in the mountains (the Rockies and the Andes, respectively). Similarities also exist in governance structures in that a private-sector marketing system exists in both countries, whereby producers market their own products. In Canada, this open market, for grain in particular, is a result of very recent policy changes. In the past, producers marketed some grains, wheat, and barley collaboratively on the global market; now they have the option of marketing their product independently. This marketing change, along with other projected changes (e.g., climate, social), may create new risks and opportunities for Canadian producers. The viticulture sector in Chile's Maule region may offer some lessons based on the experiences of Chilean producers.

Climate change poses challenges and opportunities for the agriculture sector, including viticulture (Hadarits et al. 2010; Belliveau et al. 2006; White et al. 2006). Viticulture is particularly sensitive to climate change because small fluctuations in temperature and rainfall can significantly influence wine quality and quantity (Gladstones 2011). In addition, wine grapes (*Vitis vinifera*) are perennial plants, representing a long-term investment for producers of at least several decades, over which the climate is projected to change beyond the optimal range of growing conditions in many regions (Hannah et al. 2013; Jones et al. 2005). The wine industry is growing rapidly in Chile—the number of hectares planted in vinifera grapes almost doubled from 1991 to 2011 (ODEPA 2013). During this same time period, wine production increased by 500%. Wine exports are important to Chile's economy, contributing over US$1.4 billion in 2012 (ODEPA 2013; Vinos de Chile 2012). However, a recent study by Hannah et al. (2013) concluded that mean climatic suitability for viticulture in Chile may decrease by up to 25% and available water discharge may decrease 20%–30% by 2050. Future projected decreases in precipitation will also result in an increasing need for irrigation (Hannah et al. 2013). These projected changes have serious economic and cultural implications, especially when coupled with changes in social and economic conditions (e.g., labour laws, consumer preferences, fluctuations in global markets).

This chapter describes drought-related vulnerabilities for the wine industry in Chile using a case study of the Maule region. It begins with a discussion of the conceptual framework and rationale guiding the work, followed by a description of the study site. It then documents the main findings, discusses some potential lessons learned that may be applicable to Canada, and concludes with a summary of the chapter's main points.

Conceptual Framework and Rationale

Climate Change and Viticulture

The wine industry has observed changes in vine development and fruit maturation in recent years; for example, budbreak, flowering, and fruit maturity have occurred earlier in the growing season in Germany, France, and California (Mira de Orduña 2010: 1844). There is growing concern about the viability of the industry in some well-established wine-producing regions, and as a result, there is a growing body of scholarship investigating the implications of climate change on viticulture and viniculture (Jones and Goodrich 2008; Webb et al. 2008; White et al. 2006; Jones et al. 2005). Holland and Smit (2010) suggest this scholarship falls into four broad categories: i) climate change impacts on wine quality; ii) climate change impacts on grapevine phenology and yield; iii) viticultural suitability and terroir in a changing climate; and iv) the adaptive capacity of the wine industry to climate change.

Much attention has been given to the first three categories, where most of the work has focused on modelling future climate change and assessing the impacts of these changes using phenological and physiological models (Stock et al. 2005). Some research has complemented this work by modelling and estimating the economic impacts on the industry (Webb et al. 2008). Although many studies recognize the need to understand the capacity of the wine industry to adapt to climate change, few studies have explicitly addressed the role of human adaptation in this context (Hadarits et al. 2010; Holland and Smit 2010; Belliveau et al. 2006).

Vulnerability Assessments in Agriculture

Vulnerability assessments have been used successfully to understand how an agricultural system experiences and manages climate and non-climatic risks and opportunities. These assessments have provided invaluable insights from the perspective of producers into current risks and opportunities for their operations, the range of adaptive strategies they draw from, the forces affecting their adaptive capacity, and how climate change may affect them in the future (Hadarits et al. 2010; Young et al. 2010; Reid et al. 2007; Belliveau et al. 2006).

This research adopted a community-based vulnerability approach, where vulnerability is conceptualized as a function of a system's exposure-sensitivity and adaptive capacity (Smit and Wandel 2006). For a more detailed description of these concepts, please refer to Chapter 1. The empirical application of this approach requires the actors within the system itself (e.g., grape growers, wine producers) to identify the relevant exposure-sensitivities and adaptive capacity contributing to their vulnerabilities (Smit and Wandel 2006). Actors are typically engaged through participatory methods such as interviews and focus groups. Moreover, the assessment of current exposure-sensitivities and adaptive capacity provides a lens through which future vulnerabilities to climate change can be understood (Ford and Smit 2004). Qualitative information about current vulnerability can be combined with quantitative output from climate and agricultural production models (e.g., Hannah et al. 2013; Lereboullet et al. 2013; Jones et al. 2005) to provide a more holistic view of future vulnerability.

Integrating the modelling work (described above) with adaptation research is a new approach in the climate change and viticulture field to understand climate change impacts and the adaptive capacity of the wine industry to deal with these impacts in a more holistic manner (e.g., Lereboullet et al. 2013). This chapter integrates the modelling approach with a community-based vulnerability assessment.

Description of the Maule Region, Chile

The Maule region is located in central Chile and spans an area of more than 30,000 km^2 (Figure 1). Maule is the largest wine-producing region in the country, containing the most hectares planted of any region in the country. It also accounts for half of the country's wine production, most of which is exported. Most of the region's soils are loam and loamy clay; near the coast the soils are less fertile than the central valley and eastern foothills. As such, most wine grapes are grown in the central valley.

Approximately 1 million people live in Maule, of which 5,000 are involved with growing wine grapes. Grape-growing operations range in scale from large multinational corporations to very small producers. Vineyards exhibit highly varied degrees of capital investment and agronomic expertise, and range in size from 6 ha to over 2,000 ha. Many growers have invested in wineries, either independently or through co-operatives.

With over 55,000 ha of vineyards planted in 35 *Vitis vinifera* varieties, including Chardonnay, Sauvignon Blanc, Cabernet Sauvignon, Merlot, Carmenère, and Syrah, the region produces almost 400 million litres of wine per year (SAG 2012a, 2012b). Tender fruits are also commonly grown in Maule, including cherries, plums, kiwis, apples, table grapes, blueberries, and raspberries. Many wine grape growers also engage in other

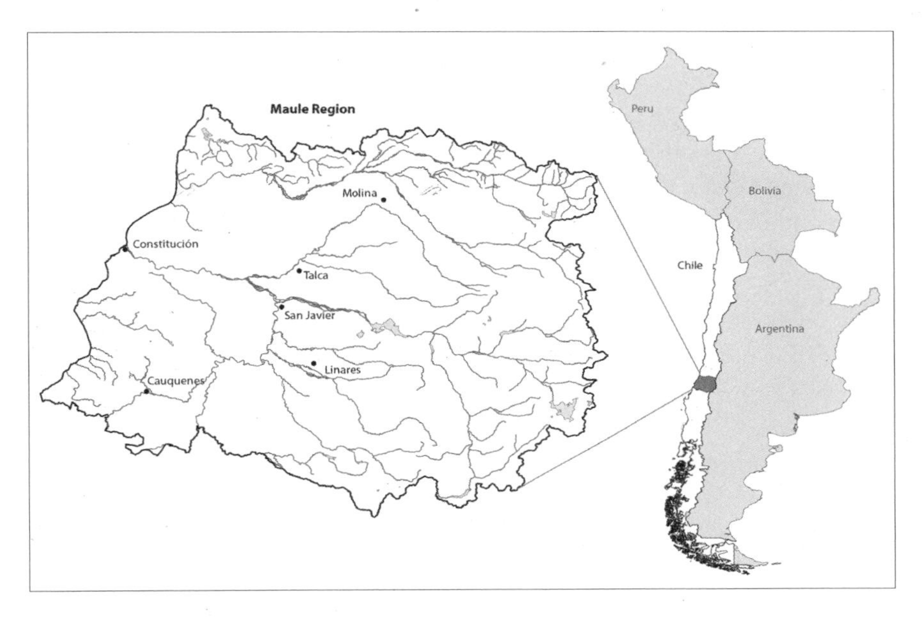

Figure 1. Map of the Maule region, Chile

tender fruit production. Besides viticulture and viniculture, silviculture is also an important economic driver in the region.

The Mediterranean climate in the valley is characterized by heavy winter rains and a long dry period beginning in spring (November) and ending in summer (March), creating ideal growing conditions for wine grapes (Vinos de Chile 2010). The dry period facilitates excellent grape maturation, and since rain during harvest is rare, quality remains relatively consistent from year to year. The sharp contrast between maximum and minimum daily temperatures supports preferred vine development and fruit maturation (Vinos de Chile 2010).

Many of the vineyards in Maule are irrigated by either flood or drip systems. Their primary source of water is derived from snow and glacier melt in the Andes Mountains, which feed the Maule, Lontuè, and Teno Rivers. Water is supplied via canals to agricultural producers (Díaz 2007). In Chile, water rights are held separately from property rights. Under the 1981 Water Code, water rights can be obtained from the government, but once rights are fully allocated, transfers take place through the market (Corkal et al. 2006). Although rights are formally specified according to an allocated volume (e.g., litres per second), in practice rights tend to be expressed as a portion of flow or shares of canals (Bauer 1997). In many regions where water resources are scarce, water rights have a high economic value and can therefore be very expensive to purchase (Gómez-Lobo and Paredes 2001).

Methods

Interviews

A multi-method approach was adopted for this work. Seven semi-structured key informant interviews were conducted in Maule between April and August 2008 to provide context for the research. Key informants were purposefully selected based on their experience and knowledge of the wine industry and included oenologists and governance representatives. Building on the key informant interviews, 46 in-depth semi-structured interviews with grape growers and wine producers were conducted. Interviewees were selected using a purposive, snowball sampling technique. Three key collaborators provided short lists of potential interviewees, all of whom were contacted for interviews, and each person interviewed was asked to provide additional contacts. The interview guide was structured around the vulnerability approach, with exposure-sensitivity and adaptive capacity as the main themes. Interviewees were asked categorical questions describing the characteristics of their operation. They were also asked open-ended questions about recent and current risks and opportunities for their operations, management strategies to reduce risks and capitalize on opportunities (current vulnerability), and about potential future vulnerabilities (see Hadarits et al. 2010). The interviews were complemented by secondary sources to provide additional context and verify the information provided by interviewees. In total, 13 grape growers, 31 grape and wine producers, and 2 wine producers were interviewed. This cross-section of individuals involved in the wine industry helped to provide insights into the vulnerabilities across different production systems. Summary statistics for the sample are listed in Table 1.

Table 1. Summary statistics for interviewees

	Mean	Median	Mode	Range
Vineyard size (ha)	29.3	107.5	150	5–2,000
Winery size (litres)	3,420,037	1,280,000	1,500,000	6,000–18,000,000
Produce other crops?	Yes: 48%		No: 52%	

Climate Change Scenarios

To assess future exposure-sensitivity, climate change scenarios were generated using weather station data and regional climate change models. The baseline (1980–2010) was established by compiling meteorological data obtained from the Chilean National Meteorological Institute and various public and private organizations. This information was supplemented with data provided by the *Agroclimatic Atlas of Chile* (Santibáñez and Uribe 1993: 66); however, the reference period was updated for this study. This atlas contains cartographic information with a spatial resolution of 1 km. The digital version of this cartographic set is available at the Center on Agriculture and Environment website (AGRIMED, Universidad de Chile; http://www.agrimed.cl). For the future climate scenarios, the PRECIS (Providing Regional Climates for Impacts Studies) dynamic downscaling model was applied to the 2050 climate period and A2 scenario (http://www.ipcc.ch/ipccreports/sres/emission/index.php?idp=94).

SIMulator of PROCedures (SIMPROC) Modelling

The SIMulator of PROCedures (SIMPROC) model was used in this study to assess the climate change scenarios and their impacts on wine grape behaviour. The SIMPROC model is a climatic crop simulator that helps identify important changes in agricultural production (MMA 2010; CONAMA 2008; Santibáñez 2001). The model considers weather variables as well as key variables associated with the production system in question to simulate potential crop yields. Gross photosynthesis, potential dry matter production (Penning de Vries and Van Laar 1982), and maintenance respiration (Van Keulen and Wolf 1986: 479; Ludwig et al. 1965) are all incorporated into the model. There is also a subroutine to simulate the water balance of the soil-plant system, and the user can fix a criterion for irrigation watering and consider the efficiency of water applied. The water deficit is represented through a production function, the growth phase and the process of senescence, when soil water content falls below a critical threshold.

For this study, red and white varieties were evaluated separately, as their optimum growing conditions differ greatly. For example, optimum temperatures for photosynthesis in red wine grape varieties range from 22°C to 30°C (Schneider 1989). High temperatures during bud development stimulate fruitfulness (Baldwin 1964), and optimum temperature for flower primordial induction ranges from 30°C to 35°C (Bruttrose 1970). Buds are more fruitful at high temperatures and light intensities, whereas the optimum range for pollen germination is from 25°C to 30°C (10°C is the minimum, 35°C the maximum) (Santibáñez et al. 1989). For white varieties, the optimum temperature for photosynthesis is between 20°C to 25°C (Schneider 1989), and temperatures above 29°C are detrimental to fruit development and quality. SIMPROC was run for red and white grapes for both the baseline and 2050 under full irrigation and a 20% deficit.

Phenology also modulates crop sensitivities to rising temperatures, frosts, and heat stress. Crop sensitivity may differ from one phase to another. The model contains algorithms to simulate frost damage and the effect of water shortages on production, as well as the loss of leaf area index due to frost occurrence. Frost and water stress sensitivity and temperature thresholds are simulated by a phenological sub-model that assigns each phase a different sensitivity. The model also incorporates the accumulation of degree-days above a base temperature through the relative phenological age variable, which varies from 0 at crop just emerged to 1 at maturity (harvest); this variable represents phenological development.

Results and Discussion

Current Drought-related Vulnerabilities

Drought is a complex issue for the wine industry in Maule. Dry years are extremely problematic for producers—57% of interviewees noted drought,

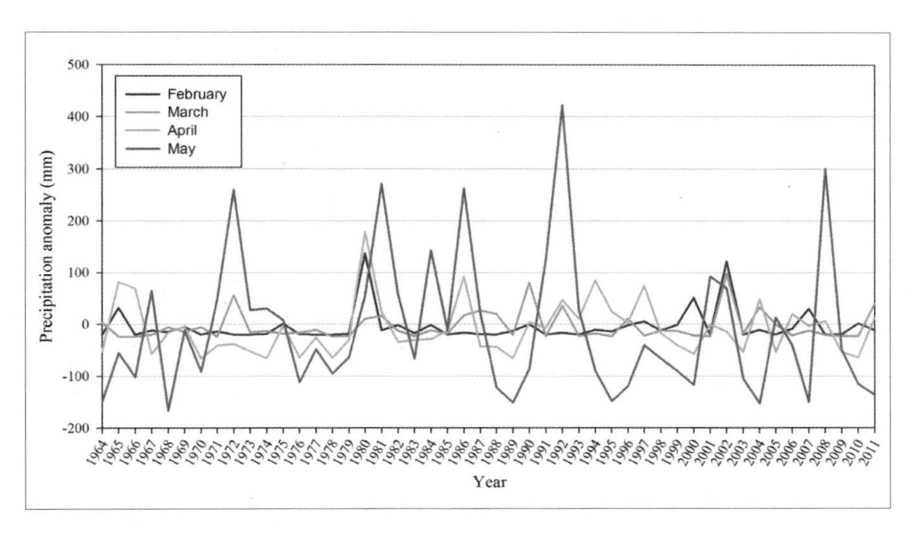

Figure 2. 1964–2010 precipitation anomalies for Parral (1980–2010 baseline)

primarily irrigation water shortages, as contributing to below-average years for their operations. Since most vineyards are irrigated in Maule, adequate winter recharge in the Andes is important to maintain water supplies during the summer (the dry season). In dry years, recharge is often inadequate to satisfy demands (i.e., all water rights allocated on a canal). When water supply declines, producers are unable to irrigate their grapes to their satisfaction and the vines' needs. The vines then experience water stress, which can be advantageous in small amounts but extremely disadvantageous if stress is excessive. Minimal stress is associated with desirable colour and phenolic compound characteristics in wines. Extreme stress, however, is associated with blocked phenolic maturation and reduced production (Lereboullet et al. 2013). Figure 2 shows precipitation anomalies for Parral, located in southern Maule, and highlights the high degree of year-to-year variability growers have to manage.

Many producers reported that production decreases in times of drought because the vines cannot produce the same volume of juices under water stress, and therefore the grapes are smaller (i.e., volume decreases). Since 2001, many grape growers and wine producers experienced up to a 30% decrease in production as a result of drought. Growers have fewer kilograms to market and therefore their economic returns suffer because they are paid by weight. This loss has carry-over effects because less operating money is available for the next growing season (e.g., for inputs, labour). Although affected by decreases in production, producers engaged in high-quality wine production noted that mild drought increases wine quality in some years because the juices become more concentrated. This effect of mild drought on wine quality creates a marketing opportunity for producers. However, excessive drought decreases their production and negatively affects wine quality.

Growers' proximity to the main canal influences their exposure to drought. The canals closer to the main canal receive water before those that are farther away. Growers who receive their water last identified much more severe water shortages than those closer to the main canal. Interviewees attributed this effect to water hoarding, a lack of adherence to rationing rules upstream, and losses to seepage and evaporation, in some cases because people do not maintain their canals.

Grape growers and wine producers have a wide range of adaptive strategies they use in the vineyard to reduce drought risks. Almost all growers monitor conditions very closely—many have installed climate and agronomic monitoring equipment—and assess their vines regularly during the growing season to quickly identify signs of plant stress. In drier parts of the region, they also strategically plant vines in low-lying areas to take advantage of natural drainage, and they harvest before the plants begin to show signs of stress. One wine producer mentioned they harvested 20 days earlier than normal (February instead of March) in one drought year to avoid excessive stress, and this worked well for them. Another producer harvested later than normal to allow grapes to reach the preferred level of maturation, but some of the grapes were dehydrated, and this negatively affected wine quantity and quality.

Access to water and water rights also influences drought vulnerability. Water rights in Maule are scarce and expensive (Gómez-Lobo and Paredes 2001), and some large growers mentioned they purchase additional rights to help them through dry times. This situation has led to an unequal distribution of resources and questions around social equity, as small- and medium-size operations become marginalized because they are unable to afford to participate in the water market (Bauer 2004, 1997). The government has attempted to curtail water-rights hoarding by fining users who do not use their allocation—a small price that large producers are willing to pay for increased water security.

Producers explore alternative sources of water and modify their management strategies in times of drought. A few of the interviewees drilled new groundwater wells or upgraded their existing groundwater pumping capacity. One grower also upgraded their irrigation equipment. Many growers modify their irrigation schedules and ration water; 48% of the interviewees produce other crops and prioritize irrigating their higher-value, more water-sensitive crops in drought years (e.g., they water cherries and kiwis before wine grapes). Wine producers often purchase additional grapes or bulk wine to offset their production losses.

Temperature is commonly identified as the main determinant of vine phenology, or the vine's rate of physiological development from budbreak to flowering, setting, vèraison (change of grape colour), and fruit ripeness (Gladstones 2011: 5). At temperatures above 25°C, net photosynthesis decreases, and at temperatures above 30°C, berry size and weight decrease, and metabolic processes and sugar accumulation may stop (Mira de Orduña 2010: 1845). High summer temperatures, which often accompany drought in Maule, were identified by 20% of interviewees as being problematic. Merlot was identified as particularly sensitive to high temperatures, as exposure results in dehydration, lower yields, and ultimately, reduced financial returns. In addition, when high temperatures are coupled with intense solar radiation, the risk of sunburn increases if growers deleaf and thin their vines too much; this is more of a concern for white wine grapes because it negatively affects quality, specifically colour and taste.

To reduce the risks associated with high temperatures and intense solar radiation, growers reduce de-leafing and thinning, and remove affected bunches at harvest. They also graft different varieties that are not working well for their vineyard. For example, a couple of growers grafted Pinot Noir and Carménère on Cabernet Sauvignon rootstocks in response to market conditions (i.e., better prices) and to experiment with wine grape suitability in their vineyard. Wine producers try to mix out the undesirable flavours and colour, and they also upgrade their winery equipment to better deal with these challenges. For example, one producer invested in cold fermentation tanks to facilitate better aromas in white wines, and another invested in individual cylinders for each wine batch, which resulted in better-quality wines.

Non-drought Related Risks

Although drought creates significant risks and some opportunities for grape growers and wine producers in Maule, several other forces influence their vulnerability. As is the case in the agriculture sector in general, fluctuations in market conditions create economic uncertainty for growers and producers. Many producers export their wine, and therefore the value of the US dollar greatly affects their bottom line. Much of the industry relies on manual labour to complete their vineyard work (i.e., pruning, thinning, harvesting), and labour is becoming increasingly scarce. Vineyards compete with other agricultural producers in the region for labour; since there is widespread high-value production, people can afford to pay their labourers relatively higher wages than grape growers. This situation results in fewer workers being available to complete vineyard tasks or in delays in work, both of which can be detrimental to production. Education can also influence vulnerability. Maule has one of the lowest literacy rates in Chile, which affects access to information, especially regarding government subsidies, grants, or special programs.

The wine industry is relatively new in Chile when compared to wine-producing regions in Europe. The industry has been growing rapidly since the adoption of neo-liberal economic policies in the 1990s. Foreign investment has increased dramatically since then, as has the replacement of lower-quality País grapes with higher-quality, more climate-sensitive French varieties. Many interviewees highlighted the fact that they are learning through practice and experimentation, and are adapting as they go. Growers are also managing a variety of forces that create risks and opportunities for their operations, although their decision making is largely driven by economics.

Future Drought-related Vulnerabilities

Future climate change scenarios project increases in temperature throughout all of Maule in 2050; the maximum temperature in January (the warmest month) is projected to increase between 1°C and 2.5°C, with the most pronounced increases projected in the Andean region (Figure 3). The minimum temperature in July (the coldest month) is also projected to increase between 1°C and 3°C. The Andean region experiences the largest increase in minimum temperatures, and this trend decreases from north to south (Figure 4). Conversely, precipitation is projected to decrease

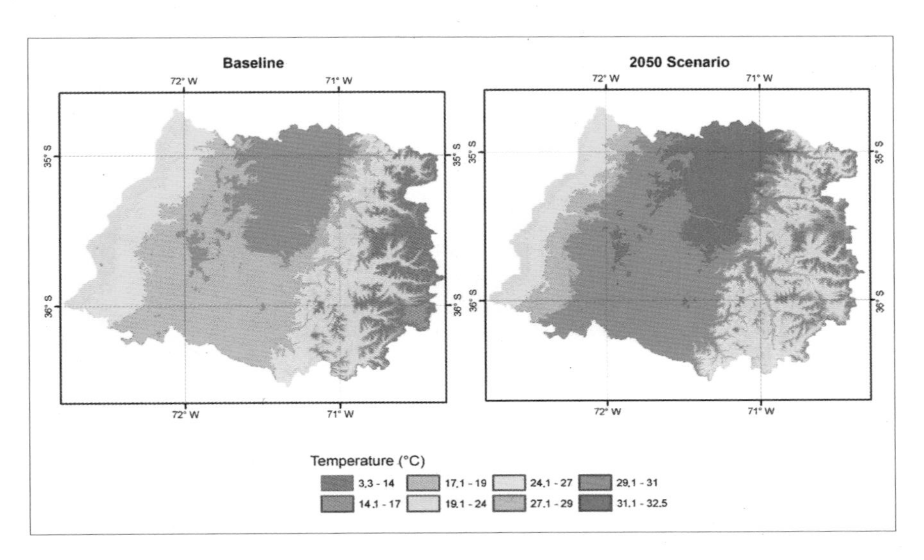

Figure 3. Maximum temperature in January (baseline and 2050) for the Maule region

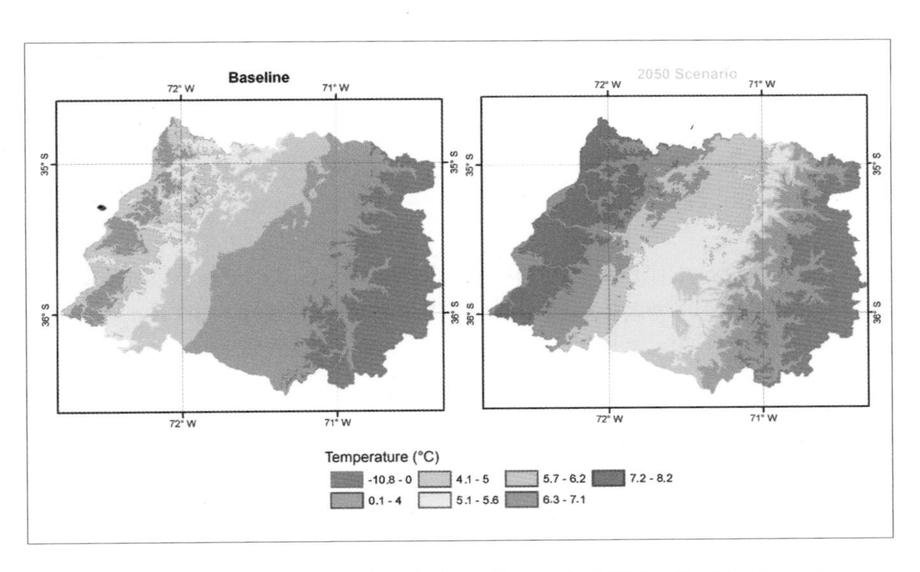

Figure 4. Minimum temperature in July (baseline and 2050) for the Maule region

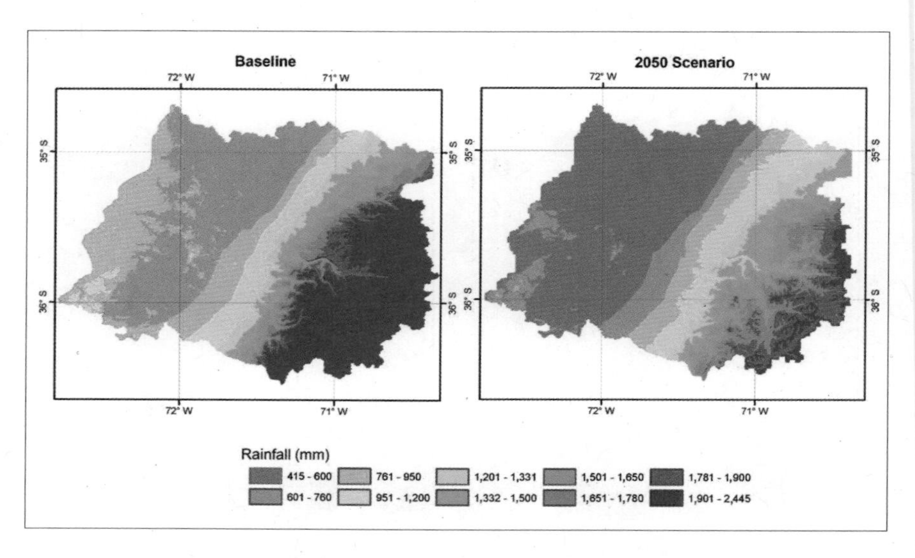

Figure 5. Annual rainfall (baseline and 2050) for the Maule region

(Figure 5). The largest decrease in precipitation is expected on the coast, which could experience up to a 30% water deficit, although the Andean region is also expected to experience a decrease in precipitation (Figure 5). These decreases, coupled with increases in maximum and minimum temperatures, could shift the arid zone in the southern part of the basin by up to 100 km.

The results of the SIMPROC modelling provide insights into grape yields under full irrigation and a 20% water deficit for both the baseline and 2050 for red and white varieties (Figures 6 and 7, respectively). Under full irrigation, red wine grape yields decrease in the northern portion of the central valley, the coast, and the Andean region in 2050 compared with the baseline. Here, optimum growing conditions would shift toward coastal and foothill regions, which are currently too cold for red wine production. However, yields increase by more than one kilogram per hectare per year in the southern portion of the central valley. Under a 20% water deficit, yields decrease on the coast and the northern portion of the central valley in 2050 compared with the baseline, but yields increase in the southern portion of the central valley and in some parts of the Andean region by almost two kilograms per hectare per year. Comparing the

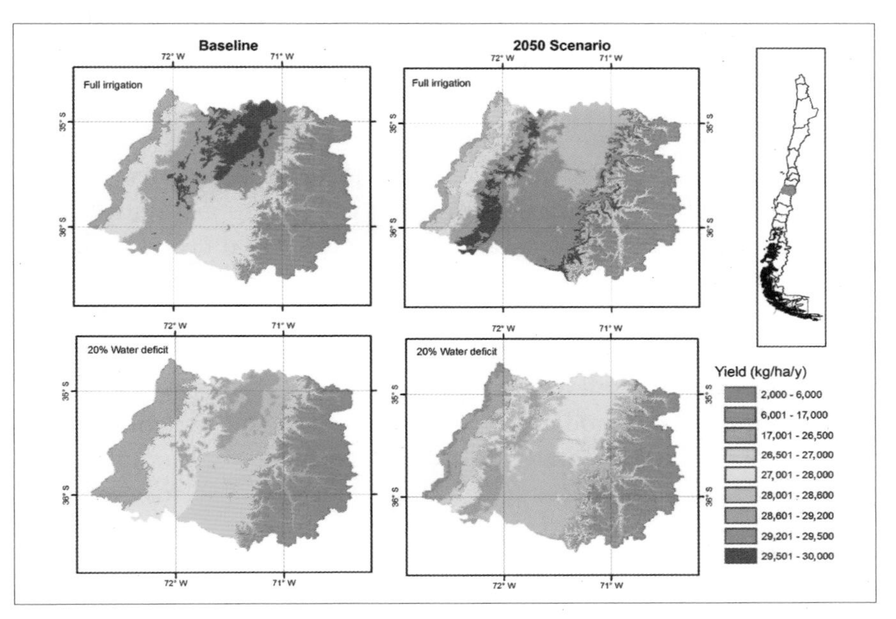

Figure 6. Red wine grape yields under full irrigation and a 20% water deficit (baseline and 2050) for the Maule region

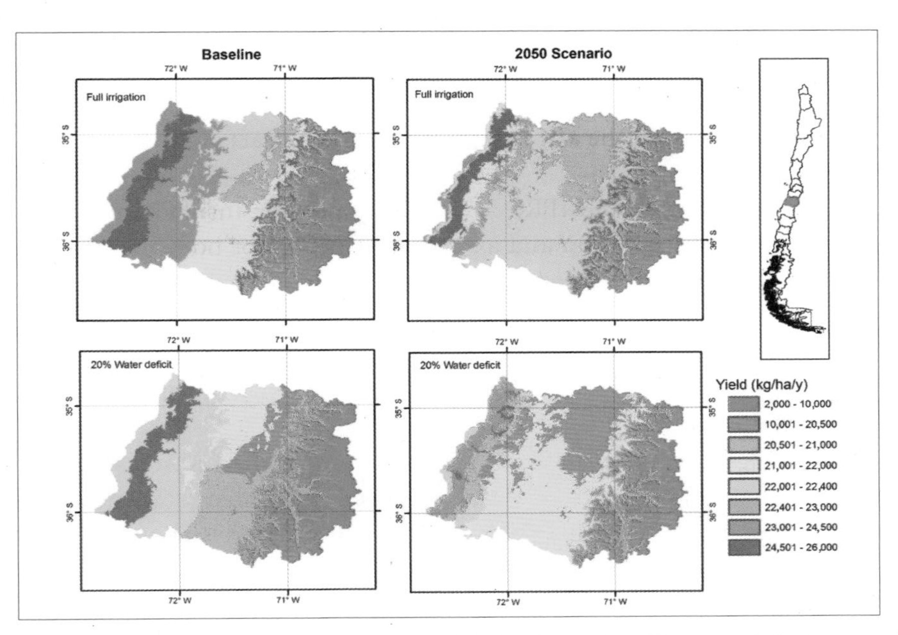

Figure 7. White wine grape yields under full irrigation and a 20% water deficit (baseline and 2050) for the Maule region

full irrigation and 20% water deficit scenarios for 2050, yield decreases throughout the entire region, highlighting the negative impacts of future drought on production.

Under full irrigation for white varieties, yield decreases along the coast and increases in the central valley and the Andean region when 2050 is compared with the baseline, both under full irrigation (Figure 7). For the baseline and 2050 under a 20% water deficit, yield decreases on the coast and the northern portion of the central valley and increases in the southern portion of the central valley and the Andean region. In 2050, yield decreases across most of the region in the 20% water deficit scenario when compared with full irrigation.

To summarize, productivity decreases for red varieties in the north-central valley and parts of the Andean region in the future; this decrease is more pronounced under future water deficits. However, there appear to be opportunities for red varieties in the south-central valley, as future productivity in this area increases in both scenarios. Access to full irrigation is essential for growers, especially in the central valley, to take advantage of the opportunities in 2050 (Figures 6 and 8), as red varieties will require more water (8% for each degree increase in average temperature) due to an increase in evapotranspiration (Figure 8). This underscores the importance of increased efficiency in irrigation water use and reliable water supplies.

For white varieties, productivity decreases along the coast and parts of the north-central valley and increases in the south-central valley, the eastern portion of the north-central valley, and parts of the Andean region (Figure 7). Similar to reds, irrigation requirements will increase for white varieties in the future (Figure 9), and again, access to irrigation water will be essential to maximize opportunities in the future. There is a strip on the coast where irrigation requirements could decrease because the fruit development cycle will be shortened as a result of rising temperatures (Figure 9).

Although there are potential opportunities in the future associated with production increases, growers will need to be able to adapt to the shifts in optimum growing conditions. Vineyards are already planted throughout the region, and growers in the south-central valley may benefit in the future if they have access to water. However, growers in the rest of

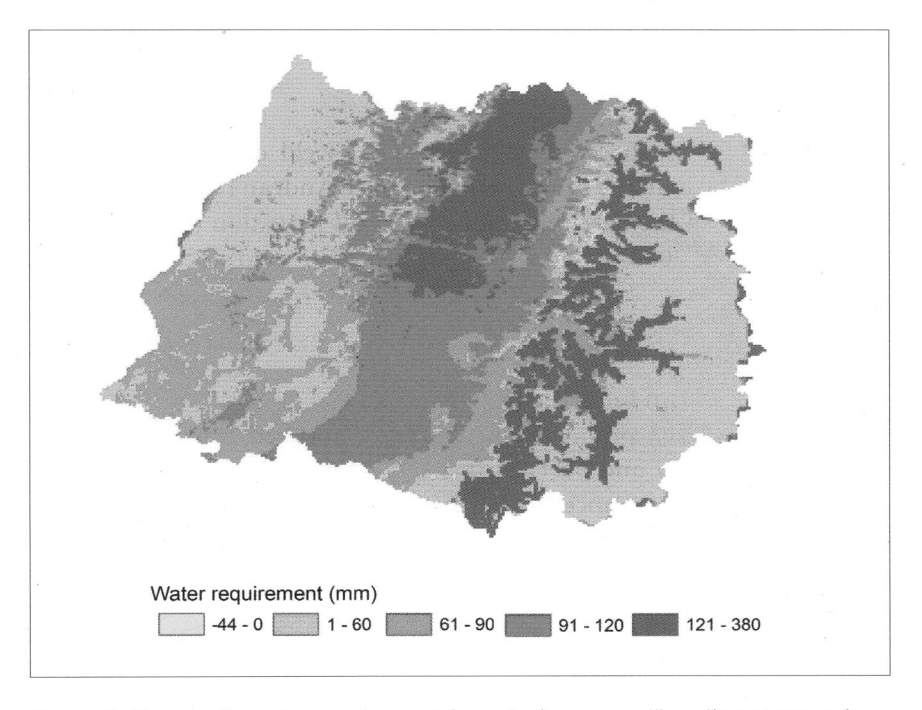

Figure 8. Changes in water requirement for red wine grapes (baseline compared with 2050) for the Maule region

the region may need to make adjustments to accommodate the risks and opportunities projected for them in the future.

Access to capital will influence future adaptive capacity. Large, capital-intensive operations have the ability to invest in water rights, land, and modern equipment; hire well-trained agronomists and winemakers; and take advantage of the projected shifts in optimum growing conditions. Those that both grow grapes and produce wine have more flexibility and are in a better position to adopt a wider range of strategies to reduce drought risks and take advantage of opportunities, as they can make changes not only in their vineyard but also in their winery. Small growers do not have that option if their crop fails or if their quality is reduced.

A few growers were seriously considering acquiring land in new locations to spread their climate risks. Some interviewees mentioned they had purchased or were planning to purchase land in regions located to

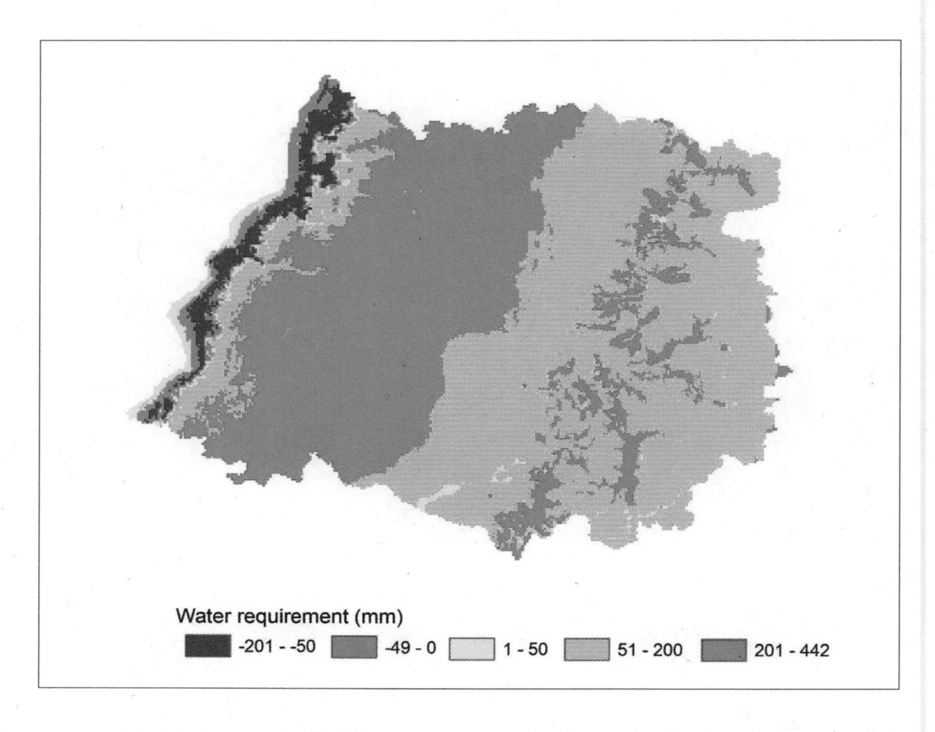

Figure 9. Changes in water requirement for white wine grapes (baseline compared with 2050) for the Maule region

the south of Maule to reduce climate risks as well as explore new terroirs. Many were actively exploring new varieties and experimenting with them to determine the most suitable varieties. They were also adding varieties to their production list to be able to quickly adapt to market demands and maximize their economic returns.

Lessons Learned

Drought has significant impacts on the grape and wine industry in Chile, and climate is an important driver for adaptation; however, economics is always at the forefront of producers' decision making. Profitability is the main concern for producers, resulting partly from the presence of an open market. There are very few government payouts, crop insurance is not widely purchased, and, save for a few small co-operatives, producers market their product independently. Producers spread their economic risk in times of drought and employ adaptation options that help them remain profitable. Many large growers have invested in secondary processing; for example, many growers have established wineries in order to produce bulk wine or fine wine for domestic consumption or export. They also sell grapes, buy grapes, make bulk wine, *and* buy bulk wine in order to remain competitive. Medium and small growers diversify their operations and incorporate high-value crops such as blueberries, cherries, avocados, and olives. Many growers and producers have also worked together to form co-operatives to collectively market their product (grapes and/or wine), but few co-operatives have succeeded in Maule. These are a few examples of how Chilean producers spread their economic risks in times of drought. Some of these adaptations may transfer to the Canadian Prairies and provide guidance from a different context on how producers successfully navigate drought while trying to maintain profitability in a setting where they must market their product independently.

The water market has been used as a tool by agricultural producers to manage drought vulnerability. But despite water being used very efficiently, it is an expensive commodity, which has influenced who is able to participate in the market. For example, there are increasing concerns about hoarding of water rights and conflicts and social equity (Bauer 2004,1997). These types of issues need to be considered when adopting this type of water market system—another important lesson from Chile that may be applicable to Canada in the future.

The SIMPROC modelling work described in this chapter is an innovative approach to understanding the interactions between agro-climatic trends and changes in climate. The model also identifies the subsequent implications of these interactions for crop yields in Chile. This type of modelling could help provide insights into the interactions and implications that exist for the Canadian context and support the agriculture sector with future adaptation planning efforts; for example, it could help identify new crop diversification options and help guide crop science research interests.

Conclusions

Over the past 30 years, Chile has garnered international attention for its production of high-quality wine at very affordable prices. Since then, growers have been capitalizing on the opportunity to engage in high-value agricultural production and have been transitioning their operations to wine grapes and other tender fruit production. They have widely adopted irrigation technology and are continuously learning about the art of viticulture and viniculture as they experiment from year to year. This shift, in turn, has changed their vulnerability to drought, and the broad lessons learned can be transferred to other contexts (e.g., Canada).

Exposure-sensitivity to drought in Maule can be adverse or beneficial for the wine industry, depending on a variety of factors, including the drought's timing, duration, and intensity, as well as the production system characteristics and its adaptive capacity. A small amount of water stress can be beneficial for quality, but it decreases production. As such, moderate water stress provides benefits for some wine producers, but results in income reductions for most grape growers in the region. This situation creates an interesting dynamic in the industry as well as differential vulnerabilities, with wine producers accruing benefits from drought at times and grape growers being negatively affected.

The SIMPROC modelling work suggests there will be changes for both growers and producers in Maule. Productivity (yields) is projected to decrease in the northern and western portions of the central valley and increase in the south, which may create more risks for growers in the north and west, but opportunities for those in the south. However, if production decreases are accompanied by higher-quality production, it may actually create an opportunity for some wine producers. The modelling work also indicates that there will be a decrease in future annual rainfall—potentially affecting irrigation supplies—and that crops will require more water in the future, largely due to increases in temperature. Therefore, droughts may become more frequent, and in order for the industry to succeed, access to sufficient irrigation water will be critical. These future conditions add another level of complexity for growers who have begun to feel confident growing wine grapes and producers who have found their niche.

Over the past few decades, growers and producers have developed a wide range of adaptive strategies they use in times of drought. Capital investments may be necessary to accommodate future changes in optimum growing conditions, as new regions may become more or less suitable for certain varieties of wine grapes. Some growers have even begun purchasing lands in the south in anticipation of future changes. This adaptability and foresight will be beneficial in the future should the projected changes become reality. However, some growers and producers are not prepared for, nor even thinking about, the future. As a result, they may face greater challenges under future droughts, especially when coupled with a variety of external forces (e.g., lack of education and access to capital) that will influence their ability to adapt to drought.

Acknowledgments

The authors thank the project participants for their time and hospitality. Many thanks to Barry Smit (University of Guelph), Harry Polo Diaz (University of Regina), and Fernando Santibáñez (Universidad de Chile) for facilitating the work and providing constructive feedback. Thank you to the three reviewers for their comments and suggestions, to Darrell Corkal for his thoughtful insights, and to Kaitlin Strobbe for editing and formatting the final draft. This research would not have been possible without support from the Social Sciences and Humanities Research Council of Canada (Canada Research Chairs program), the Institutional Adaptations to Climate Change Major Collaborative Research Initiatives project, the Arthur D. Latornell Graduate Scholarship, and the American Society for Enology and Viticulture Scholarship.

References

Baldwin, J.G. 1964. "The Relation between Weather and Fruitfulness of the Sultana Vine." *Australian Journal of Agricultural Research* 15: 920–28.

Bauer, C.J. 2004. *Siren Song: Chilean Water Law as a Model for International Reform.* Washington, DC: Resources of the Future.

———. 1997. "Bringing Water Markets Down to Earth: The Political Economy of Water Rights in Chile, 1976–95." *World Development* 25: 639–56.

Belliveau, S., B. Smit, and B. Bradshaw. 2006. "Multiple Exposures and Dynamic Vulnerability: Evidence from the Grape and Wine Industry in the Okanagan Valley, British Columbia, Canada." *Global Environmental Change* 16: 364–78.

Bruttrose, M.S. 1970. "Fruitfulness in Grapevines: Development of Leaf Primordial in Buds in Relation to Bud Fruitfulness." *Botanical Gazette* 131: 78–83.

CONAMA (Comisión Nacional del Medio Ambiente). 2008. *Analysis of Vulnerability of Agriculture and Forestry Sector, Water and Soil Resources of Chile under Climate Change Scenarios. Chapter IV Analysis of Forestry and Agricultural Sector Vulnerability against Climate Change Scenarios.* http://www.sinia.cl/1292/articles-46115_capitulol_informe_final.pdf. Accessed 14 March 2013.

Corkal, D.R., H. Diaz, and D. Gauthier. 2006. "Governance and Adaptation to Climate Change: The Cases of Chile and Canada." Institutional Adaptations to Climate Change Working Paper, Institutional Adaptations to Climate Change project, Regina, Saskatchewan.

Díaz, J.O. 2007. *Family Farm Agriculture: Factors Limiting Its Competitivity and Policy Suggestions.* Talca: University of Talca, Department of Agricultural Economics.

Ford, J., and B. Smit. 2004. "A Framework for Assessing the Vulnerability of Communities in the Canadian Arctic to Risk Associated with Climate Change." *Arctic* 57: 389–400.

Gladstones, J. 2011. *Wine, Terroir and Climate Change.* Kent Town: Wakefield Press.

Gómez-Lobo, A., and R.M. Paredes. 2001. "Reflexiones sobre el proyecto de modificación del Código de Aguas." *Estudios Públicos* 82: 83–104.

Hadarits, M, B. Smit, and H. Diaz. 2010. "Adaptation in Viticulture: A Case Study of Producers in the Maule Region of Chile." *Journal of Wine Research* 21: 167–78.

Hannah, L., P.R. Roehrdanz, M. Ikegami, A.V. Shepard, M.R. Shaw, G. Tabor, L. Zhi, P.A. Marquet, and R.J. Hijmans. 2013. "Climate Change, Wine, and Conservation." *Proceedings of the National Academy of Sciences* 110: 6907–12.

Holland, T., and B. Smit. 2010. "Climate Change and the Wine Industry: Current Research Themes and New Directions." *Journal of Wine Research* 21: 125–36.

Jones, G.V., and G.B. Goodrich. 2008. "Influence of Climate Variability on Wine Regions in Western USA and on Wine Quality in the Napa Valley." *Climate Research* 35: 241–54.

Jones, G.V., M.A. White, O.R. Cooper, and K. Storchman. 2005. "Climate Change and Global Wine Quality." *Climatic Change* 73: 319–43.

Lereboullet, A.-L., G. Beltrando, and D.K. Bardsley. 2013. "Socio-ecological Adaptation to Climate Change: A Comparative Case Study from the Mediterranean Wine Industry in France and Australia." *Agriculture, Ecosystems and Environment* 164: 273–85.

Ludwig, L.J., T. Saeki, and L.T. Evans. 1965. "Photosynthesis in Artificial Communities of Cotton Plants in Relation to Leaf Area." *Australian Journal of Biological Sciences* 18: 1103–18.

Mira de Orduña, R. 2010. "Climate Change Associated Effects on Grape and Wine Quality and Production." *Food Research International* 43: 1844–55.

MMA (Ministerio del Medio Ambiente). 2010. *Portfolio of Proposals for the Program of Agroforestry Sector Adaptation to Climate Change in Chile.* Santiago: Ministry of Environment, Department of Agricultural Research and Planning, Ministry of Agriculture. http://www.sinia.cl/1292/w3-article-50188.html. Accessed 12 March 2013.

ODEPA (Oficina de Estudios y Políticas Agrarias). 2013. "Estadísticas."ODEPA website. http://apps.odepa.cl/menu/MacroRubros.action;jsessionid=0915FC15FFED D893CF88D3A33224DBED?rubro=agricola&reporte=. Accessed 13 March 2013.

Penning de Vries, F.W., and H.H. Van Laar. 1982. *Simulation of Plant Growth and Crop Production*. Wageningen, The Netherlands: Centre for Agricultural Publishing and Documentation.

Reid, S., B. Smit, W. Caldwell, and S. Belliveau. 2007. "Variability and Adaptation to Climate Risks in Ontario Agriculture." *Mitigation and Adaptation Strategies for Global Change* 12: 609–37.

SAG (Servicio Agricola y Ganadero). 2012a. "Reporte Vinos con Demoninacion de Origen". Santiago: Ministerio de Agricultura. http://svyv.sag.gob.cl/dec_exa/reportes1a.asp. Accessed 14 March 2013.

———. 2012b. "*Catastro vitícola nacional*." Santiago: División Protección Agrícola – Subdepartamento Viñas y Vinos. http://www.sag.cl/ambitos-de-accion/catastro-viticola. Accessed 14 March 2013.

Santibáñez, F. 2001. "The Modeling of the Growth, Development and Production of Corn on Ecophysiological Bases by SIMPROC Model." *Argentina Journal of Agrometeorology* 1: 7–16.

Santibáñez, F., F. Díaz, C. Gaete, S. Daneri, and D. Daneri. 1989. "Agroclimatology and Zoning Chilean Wine Region: Bases for the Designation of Origin of Wines." *Technical Bulletin of the Faculty of Agricultural and Forestry Sciences* 48: 1–26.

Santibáñez, F., and J.M. Uribe. 1993. *Agroclimatic Atlas of Chile*. Santiago: University of Chile, Faculty of Agricultural and Forestry Sciences, Laboratory of Agroclimatology.

Schneider, C.H. 1989. "Introduction à l'écophysiologie viticole. Application aux systemes de conduit." *Bulletin OIV* 701–702: 498–515.

Smit, B., and J. Wandel. 2006. "Adaptation, Adaptive Capacity and Vulnerability." *Global Environmental Change* 16: 282–92.

Stock, M., F.-W. Gerstengarbe, T. Kartschall, and P.C. Werner. 2005. "Reliability of Climate Change Impact Assessments for Viticulture." *Acta Horticulturae* 689: 29–39.

Van Keulen, H., and J. Wolf (eds.). 1986. *Modelling of Agricultural Production: Weather, Soils and Crops*. Wageningen, The Netherlands: Pudoc.

Vinos de Chile. 2010. "Vinos de Chile 2010." Santiago: Vinos de Chile A.G. http://www.vinosdechile2010.cl/indust/index3.htm. Accessed 13 March 2013.

———. 2012. "Información y estadísticas." Santiago: Vinos de Chile. http://www.vinosdechile.cl/contenidos/informacion/informacion/. Accessed 13 March 2013.

Webb, L.B., P.H. Whetton, and E.W.R. Barlow. 2008. "Climate Change and Winegrape Quality in Australia." *Climate Research* 36: 99–111.

White, M.A., N.S. Diffenbaugh, G.V. Jones, J.S. Pal, and F. Giorgi. 2006. "Extreme Heat Reduces and Shifts United States Premium Wine Production in the 21st Century." *Proceedings of the National Academy of Sciences* 103: 11217–11222.

Young, G., H. Zavala, J. Wandel, B. Smit, S. Salas, E. Jimenez, M. Flebig, R. Espinoza, H. Diaz, and J. Cepeda. 2010. "Vulnerability and Adaptation in a Dryland Community of the Elqui Valley, Chile." *Climatic Change* 98: 245–76.

CHAPTER 14

DROUGHT IN THE OASIS OF CENTRAL WESTERN ARGENTINA

Elma Montaña and José Armando Boninsegna

Introduction

This chapter discusses droughts and episodes of water scarcity in the context of the Mendoza River basin, an area in central-western Argentina where a dynamic agriculture emerges in an arid complex Mediterranean climate. The Mendoza River basin is similar to many dryland territorial configurations on both sides of the central Andes or to the Palliser Triangle in the Canadian Prairies, where "green oases" emerge as a result of human-built irrigation systems. As in many semi-arid and arid regions of the Americas, the sensitivity of the regional economy and population to climate variability and the new threats of global warming lead to questions of how to reduce vulnerability of agricultural producers, integrate climate change into their activities, and foster the best possible adaptive strategies for facing inevitable climate change consequences.

The chapter assumes that climate and water-related issues should be understood in terms of coupled natural and social systems. In these terms, the presence and the impacts of droughts should be discussed from a perspective that integrates both the natural and social scientific views. The first section of this chapter deals with the natural scientific perspective; it discusses the climatological conditions that characterize the basin and their impact on regional water scarcities. Following the conceptual approach discussed in Chapter 1, the second section of this chapter deals with the social dimension, focusing on the vulnerabilities of the basin and paying special attention to the social and economic structures of the basin in setting up variable conditions of vulnerability for different producers. The third section focuses on the adaptive capacity of these rural producers, linking this capacity to the social and economic structures. Finally, policy implications for managing future droughts are discussed.

Several natural and social studies, which are the main inputs to this chapter, have been carried out in the region. The natural studies have focused on hydrological cycles, their relationship to agriculture, and the vulnerability of the region to water scarcities, which are the main constraints to economic growth and expansion. Rainfall, runoff variability, and their relationship with large-scale circulation anomalies and different climate conditions have been discussed in both past (Prieto et al. 2000; Compagnucci and Vargas 1998; Rutllant and Fuenzalida 1991; Cobos and Boninsegna 1983) and more recent studies (Gonzalez and Vera 2010; Viale and Norte 2009; Vargas and Naumann 2008; Masiokas et al. 2006; Boninsegna and Delgado 2002). Past droughts have been analyzed by Villalba et al. (2012), and Christie et al. (2011), and Le Quesne et al. (2009) used tree ring series to reconstruct the Palmer Drought Severity Index back to year 1346, providing an insight into the central Andes drought recurrence. Climate change impacts on the Cordillera have been addressed by Bradley et al. (2006), Nuñez (2006), Urrutia and Vuille (2009), Nuñez and Solman (2006), and Vera et al. (2006). Glacier evolution has been the subject of several studies (Le Quesne et al. 2009; Bottero 2002; Luckman and Villalba 2001; Leiva 1999). An estimation of the future streamflow of the San Juan and Mendoza Rivers was made by Boninsegna and Villalba (2006a, 2006b).

In the region, risk and vulnerability studies have focused mainly on water and water scarcity–related issues, as well as on the potential impacts of climate change. The historical perspective of the relationship between water and society can be found in Marre (2011), Montaña (2011, 2008a, 2007) and Montaña et al. (2005) while studies concerning the possible role of global change in altering risk patterns appear in Scott et al. (2012) and Salas et al. (2012). Vulnerability has been addressed by Masiokas et al. (2013), Montaña (2012a, 2012b) and Diaz et al. (2011). New ways of thinking about conservancy ethics, society, and adaptation applied to the central Andes region are found in Montaña (2012a), Montaña and Diaz (2012) and Diaz et al. (2011). As indicated earlier, the chapter integrates this diversity of studies.

Climate Variability and Droughts in Mendoza

The central western Andes region of Argentina, where the province of Mendoza and the Mendoza River are found, is complex in terms of its orography, climate, and socio-economic development. The topography is characterized by the steep north-south barrier of the Andes, which strongly modulates the climate between the western (Chilean) and eastern (Argentinian) sides of the mountains. In the eastern slope of the Andes, located in central-western Argentina, Mendoza is part of what is called the "dry pampas," where most of the scarce yearly precipitation occurs during the hot summer months. During the cold winters, snow accumulates in the higher mountains and occasionally in the valley, where frost episodes are common. As in the case of the Canadian Prairie provinces, the winter snow melts during the spring due to the rise of temperature, increasing river runoff, which peaks in the summer months of December and January. This runoff provides water for human consumption, agricultural activities, and hydroelectric production. This hydrological cycle, which is conditioned by climate, is crucial to sustain human activities in the region. In this context, the Andes have been defined as a "natural water tower" capable of collecting, storing, and distributing water from rain and melting snow (Viviroli et al. 2007, Vitale 1941).

Most of the irrigated agriculture in the Mendoza region is both intensive and diversified. Famed worldwide for its viticulture, Argentina produces approximately 1.5 billion litres of wine per year. Other agriculture-based industries—such as olive oil, and canned and preserved food production—are also highly developed. Tourism, while a relatively new industry, is becoming an important source of revenue for the region. A more marginal agriculture, which has limited access to irrigation and is highly dependent on summer precipitation, extends throughout the eastern and northern part of the province. Only 3.6% of the provincial territory is irrigated, an area identified as "the oasis," which provides the conditions for agriculture and urban development. This concentration of activity clearly highlights the region's potential for producing goods but also its increasing vulnerability to climate variability.

In recent years, there has been increasing concern about the implications of climate change for the region. Recent projections indicate a probable decrease in snow in the mountains and a rise in temperature during the present century, two factors that could seriously increase the water deficit and compromise the survival of the oasis. Reducing (as much as possible) the uncertainties in the long-term forecast of hydrological supply is essential to designing adaptive measures whose implementation will require long-term efforts.

The amount of snowfall and its accumulation, the variation of temperature, and the influence of some climate forcing systems, such as the El Niño–Southern Oscillation (ENSO), regulate the quantity of water available in the mountains. The water regime is highly dependent on the amount of snow that falls during winter and accumulates in the high basins. Temperature regulates the occurrence and rate of the snowmelt and runoff volumes to the extent that the seasonal temperature cycle produces variations in the height of the 0°C isotherm. The position of this line allows for estimating the surface on which the accumulation and/or melting of the snow will occur. The melting of the accumulated snow produces runoff, with the largest volumes produced during the spring and summer months (Figure 1).

The largest inter-annual climate variability driver in the tropical and subtropical regions of South America is the ENSO. Montecinos and Aceituno (2003) noted that during El Niño there is a tendency for the occurrence of above-average rainfall between 30° and 35° S in winter and from 35° to 38° S in late spring. Precipitation anomalies are opposite during La Niña episodes. Increased blockages in the southeastern Pacific during El Niño events produce westerly winds at lower latitudes—a key element that explains winter humidity conditions in central Chile.

At larger scales, the Pacific Decadal Oscillation (PDO) and long-term trends in the Antarctic Oscillation have influenced the climate in the Andes. Wetter conditions are present during the positive phase of the PDO in the subtropical belt of South America. Consistent with these observations,

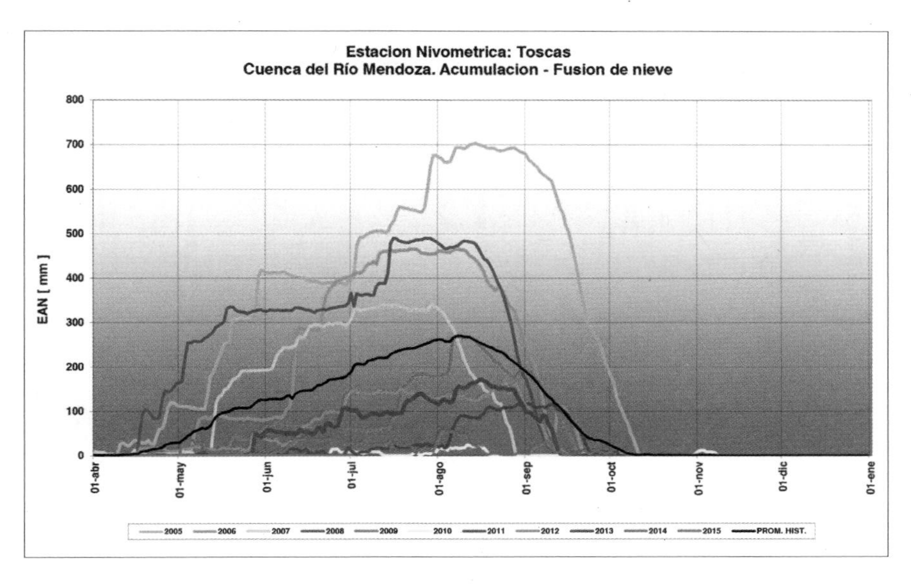

Figure 1. Annual snow accumulation measured at Toscas station (3,100m asl). Note the existence of an accumulation period, from early May until late September when snow starts melting. The snowmelt is quite fast at the measuring station and is completed by mid-November. Snow persists at higher elevations until the end of summer. Also note the large variability in the timing and quantity of the snowfall. (Source: DGI (General Department of Irrigation). 2012)

Masiokas et al. (2010) have reported an association between periods of low and high runoff in Andean rivers (between 30° and 35° S) and in the negative and positive phases of the PDO, respectively, during the twentieth century.

Figure 2 shows periods of snow and runoff shortages between 1910 and 2010. Assessing these periods as drought onsets is not easy, because of the high variability in snowfall and streamflows.

Long-term climate change assessed by different regional circulation models indicates that there will be an increase in temperature (+2.5°C to +3.0°C A2 scenario), less snowfall and runoff (-10% to -15%), and an increase in summer precipitation in the oasis (+30%) between 2090 and 2100. The model of CONAMA (2006), for example, shows a decrease in precipitation on the upper mountain, which could experience up to a 15%–20% snow deficit, but a steady increase in precipitation (rain) concentrated from October to March in the productive oasis (Figure 3) of the

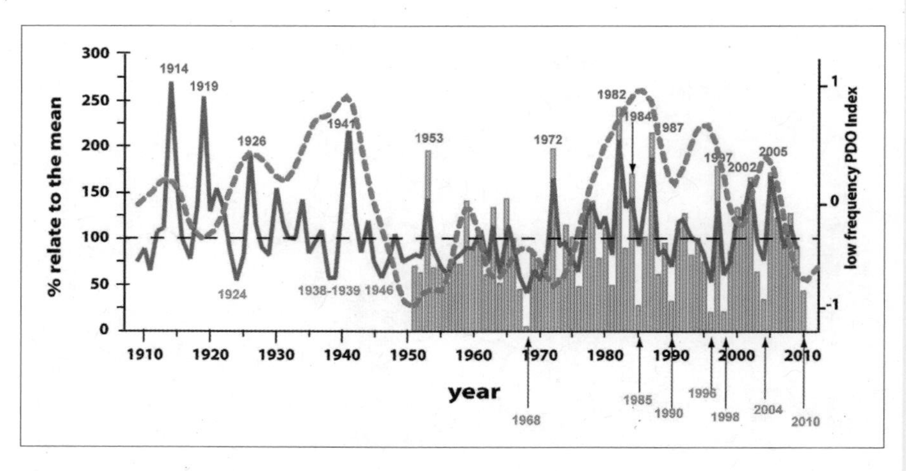

Figure 2. Mendoza River snowfall and streamflow, showing the regional average of snowfall measurements and regional average of streamflow measurements for the period November to February, expressed as percentage of water equivalent, base period 1966–2000. Note the variability in the annual record of snowfall and runoff. Years in blue are El Niño–Southern Oscillation years. The correlation coefficient between the series is $r = 0.945$, $p < 0.001$ (highly significant). Filled bars are mean snow measurements (mean of 10 stations), green line indicates streamflow measurements (mean of 8 stations), and dashed line represents the Pacific Decadal Oscillation (PDO) low-frequency index.
(Modified from Masiokas et al. 2010).

valleys. This summer input of water in the oasis hydrology differentiates Mendoza's oasis from the central Chile situation, where the summers are normally very dry.

It is extremely important to account for all interactions among the variables. For instance, if mountain temperatures rise steadily, snow cover will melt early in the year, provoking a rise in runoff during the early spring months and a drop during the summer, just when agriculture most needs irrigation (Figure 4). The increase in summer precipitation could make the situation even worse, since summer precipitation has been found to be detrimental to vineyard yields due to the increase in hailstorms and fungal diseases (Agosta et al. 2012).

In a viticulture-based agriculture (with the grapes rapidly growing from January to the harvest period in March), this change in the hydrogram produces an agricultural drought situation even with above-average

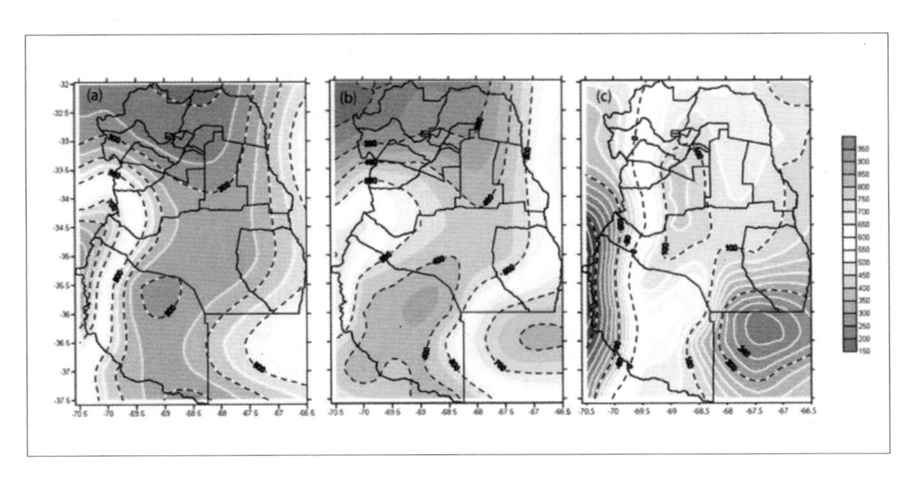

Figure 3. Shown are (a) Mendoza annual precipitation 1961–90, (b) Mendoza annual precipitation 2071–2100 according to PRECIS (Providing Regional Climates for Impacts Studies) circulation model, and (c) difference between future and present estimates. The figure shows a steady increase in precipitation to the east of mountain foothills but a decrease in the higher Cordillera.

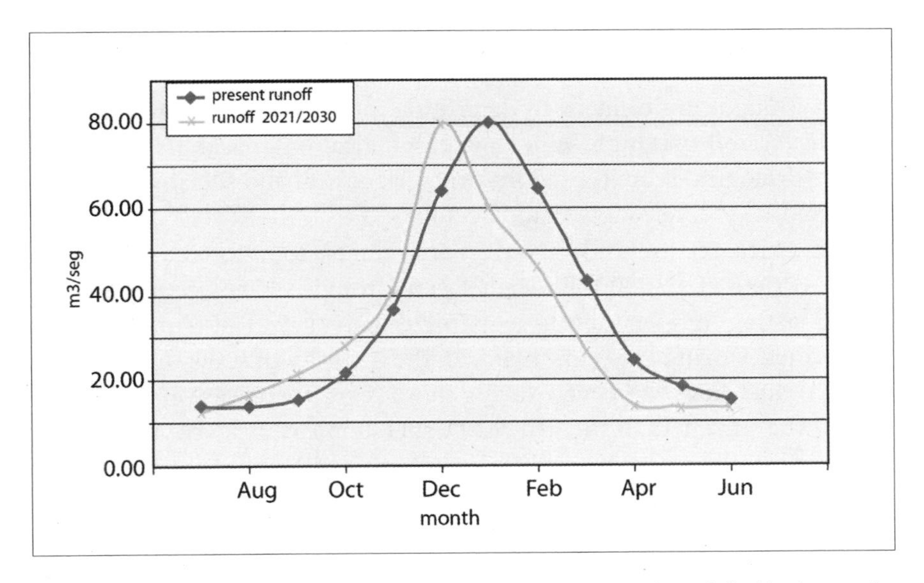

Figure 4. Hydrograph of Rio Vacas (Mendoza River basin) and modelled hydrograph of Rio Vacas (Mendoza River basin) with similar snow cover, but temperature +1.5°C higher than present values

snowfall in the mountains, a situation that calls for adaptive measures such as increasing reservoir capacity.

Drought is a regional natural hazard and should be considered a normal part of climate rather than a departure from normal (Glantz 2003). In these terms, it is difficult to assess whether a drought exists and/or its degree of severity. In the Argentinian and Chilean central Andes, instrumental and paleoclimate records could provide a method to analyze periods of water scarcity, at least from a meteorological point of view. However, it is difficult to define each of those periods as "drought," with all its negative connotations. Periods of drought are not by themselves a disaster; qualifying them as a disaster depends on anthropogenic and environmental impacts. The magnitude of the impacts in turn depends on the timing, duration, and intensity of the phenomenon.

All types of drought originate from a deficit in precipitation (Wilhite and Glantz 1985). If the precipitation scarcity lasts for an extended period, a meteorological drought could occur. But there are also hydrological droughts that are defined as the departure of surface water and groundwater supplies from average conditions. This situation can occur even with average or above-average precipitation.[1] It is when these conditions affect agriculture that an agricultural drought is declared. The start and end of drought are difficult to determine, particularly in regions where climate variability is high. When meteorological water scarcity is coupled with hydrological scarcity, and in turn agricultural and social systems are impacted, a severe drought could occur.

Historical records and narratives are valuable tools for analyzing such events (Prieto et al. 2000). Indeed, severe droughts have occurred in the region lasting long enough to jeopardize the survival of fruit trees and grapevines. During the years 1966–70, the quantity of snow in the Andes was extremely low. Moreover, during those years, there was very little rain during the summers in the valleys. Desperate measures were used to try to cope with the situation, such as draining Cordillera lakes using explosives. This recent historical episode reveals how vulnerable the region is to lasting hazards (Prieto et al. 2010).

However, it is not always easy to measure the cause of a climatic disaster. Several factors, including economic conditions, agricultural mismanagement, hailstorms, frost, plant diseases, and drought affect wine production; untangling the roll of each factor and isolating the effect of drought is particularly difficult. Frederick (2011), for example, pointed out that "the drought occurring between 1967 and 1970 brought agricultural expansion to a temporary halt." On the other hand, after analyzing the contribution of wine production to the provincial gross domestic product (GDP) of Mendoza during the twentieth century, Coria (2014) concludes that the main stressor seems to be several economic crises due to government mismanagement. But coincidentally, the analysis shows a 50% reduction in the added value of wine in the industry in 1969. However, there is no mention of water shortages.

Vulnerability to Drought in Mendoza

Mendoza looks like an idyllic oasis fed by waters from Andean snows. It has a territorial configuration characterized by two opposing landscapes. On the one hand, there are green oases with neat rows of grapevines, tree-bordered roads and streets, and irrigation channels and drains. On the other, there are non-irrigated lands (the "desert") occupied by a scattered population of goat breeders in a "no-man's land," defined as a subordinate space that is empty and void of interest.

This territorial configuration follows a very similar pattern from north to south along the central Andes, both in Chile and Argentina. In these drylands, agriculture is only possible through systematic irrigation, and the oases develop as intensively exploited territories. In the province of Mendoza, for example, oases account for only 3.6% of the total area, but they are home to 98.5% of the province's population and are the centre of most economic activities, among which grape growing and winemaking stand out. Most of the population resides in the city of Mendoza, with a population of about one million people. The second largest city has close to 100,000 inhabitants, followed by a number of small towns in the margins of the oases.

Agriculture is less diversified than in the central Chilean valleys or the Canadian Prairies, as grape growing accounts for about half of the cultivated area of the Mendoza River basin, followed by horticulture (23%) (CNA 2002). Agriculture in the basin is highly integrated with the industrial sector, because 99% of grape production is destined for winemaking. Approximately 23,000 irrigators in the basin account for 89% of surface water use (DGI 2007). In the basin, however, only 45% of farmers irrigate with surface water; 27% irrigate with groundwater only (CNA 2002), and 28% use both surface water and groundwater. Marked differences occur among grape-growing and horticultural productive systems and among wine growers in particular. Among wine growers, there are small producers with traditional vineyards at one end of the spectrum and representatives of the "new viticulture," which is part of a modern global agribusiness, at the other end of the spectrum. In the case of horticulture, small- and medium-scale producers, in some cases of Bolivian origin, produce vegetables for regional consumption, relying on social and family networks to organize their production to successfully develop their agricultural activities.

The agricultural, industrial, and urban sectors are highly dependent on the water resources provided by the Mendoza River, but this dependence affects the region's social and political life, going well beyond a functional dependence on water resources. As Worster (1985) argues, the situation here constitutes a modern "hydraulic society," in which the social tissue is strongly associated with a comprehensive and intensive manipulation of water resources within an order imposed by a hostile environment. The conflicts of the hydraulic society (Montaña 2008a) become palpable precisely when the hostility of these drylands is exacerbated by water scarcities and drought.

Under these conditions, the agricultural communities of the Mendoza River basin are inherently vulnerable to drought. Studies carried out by Montaña (2012a, 2012b) in Mendoza show that drought is the climate exposure most mentioned by agricultural producers in the Mendoza River basin as affecting their operations, followed by hail and frost. References to water scarcity are not only limited to the flow of the river or the water received through irrigations ditches but also involve groundwater, particularly in summer. Structural water scarcities and growing demands add up to recurrent drought crises. When asked about the causes of water scarcity or drought, not all of the explanations provided by farmers are associated with climate, climate change, or hydrological factors. Explanations also include human factors, such as upstream expansion of agriculture and urban sprawl, rightly perceiving the converging natural and human processes that are involved in defining a situation of vulnerability and which give rise to water conflicts that could turn a "natural" phenomenon into a disaster. Farmers are used to dealing with water scarcity and drought, and they consider them a structural problem that will be increasingly problematic in the future. However, they do not seem to be aware of the severity of the droughts that could arise in the context of significant modification of climate conditions, as discussed in the previous section of this chapter.

Drought, as a climate hazard, could overlap with other stressors, such as economic and social crises, generating double exposures (Leichenko and O'Brien 2008). When agricultural producers of Mendoza are asked about the most important problems affecting their productive activity, they identify exposures to economic and social stressors to be just as relevant as climate and water factors. These stressors include economic exposures (national macroeconomics trends, the labour market), social exposures (migration of young rural dwellers and aging of the remaining rural population), and socio-cultural exposures (the Bolivian origin of the horticultural producers and the Aboriginal origin of goat breeders from the desert lands). Even in situations where water scarcity is not the main stressor, it certainly becomes the *coup de grace* that puts small-scale producers, already impacted by the new rules of the globalized agriculture, on the edge of the agricultural system or just expulsed to the urban sector.

Indeed, drought does not impact everyone in the same way. Sensitivity to drought in the oasis of the Mendoza River varies depending on the actors being considered. Consumers of potable water are less sensitive because legislation gives them priority over other uses. It is different for those in the agricultural sector, where every producer would like to be prepared for a drought situation but where only the wealthiest succeed. It is a circular process: the better adapted, the less sensitive (and vice versa). Differences also emerge in relation to the type of water rights and sources accessed by producers, where having access to a well, being in an upstream/downstream location in the basin, and having access to infrastructure, as well as other factors, make a difference. No less relevant is the nature of the productive system; for example, viticulture is less sensitive to drought because grapes tolerate water stress relatively well, whereas horticulture is more sensitive to drought because it requires a reliable water supply.

In the context of these exposures, sensitivity and adaptation are defined by access to factors such as natural, human, social, and institutional capital, as well as economic resources, technology, and infrastructure. As a result of the sensitivity/adaptive capacity equation that involves these factors, vulnerabilities are also unequally distributed in space and in relation to different social groups or actors. Producers who irrigate with only surface water are more vulnerable to reduced river flows than are those with alternative water sources, such as wells, who are better adapted to face droughts. But it is not a question of simply having a well; it is also necessary to have the economic resources required to maintain it, to keep the pump in good condition, and to bear the energy costs associated with using it. Many small-scale producers in the Mendoza River basin, who were better off in the past, had managed to establish wells on their farms. But due to decreasing agricultural profits in the context of the agribusiness model, these farmers are now unable to bear pumping costs. This is an example of exposure to water-related factors aggravated by economic exposures, reinforcing circular patterns of vulnerability and poverty.

During periods of water crisis, irrigators located at the tail end of the systems may receive less water than they are entitled to, so farmers seek to settle at the head of the distribution systems or canals to ensure that they will receive their due. Farmers with better access to water are more likely to succeed in their agricultural activities and, in turn, to have access to better (and more expensive) locations. Oppositely, farmers in the lowest part of the irrigation system are impoverished by poor agricultural performances. Lacking resources, they are unable to move to better locations. Higher temperatures and evapotranspiration reinforce these patterns, concentrating wealthier producers in the upper and cooler areas with higher thermal amplitude and relegating poor producers, who try hard to keep their farms afloat, to the lower and warmer zones. Extreme hydroclimatic events, such as extended droughts, will only consolidate or accelerate the existing tendency toward a spatial and socio-economic segregation of agricultural producers in the oases of central-western Argentina, widening the gap between the dominant players of the local agribusiness and those who are barely able to survive as subsistence producers. In many cases, these small-scale producers have no other alternative than to neglect their farms and seek jobs to generate the necessary income, or even worse, to exit agriculture and migrate to cities, where they join the growing population of urban poor. Thus, scenarios of increasing water deficits suggest an intensification of the current process of socio-spatial segregation: wealthy producers get wealthier uphill, while the smaller producers do increasingly poorly downstream (Montaña 2012a).

Drought Adaptation and Preparedness

Differences in vulnerability among the different players in this hydraulic society may change considerably based on specific adaptive practices. The flows of Río Mendoza are regulated by the Potrerillos Dam, which was built in the upstream section of the basin to both reduce seasonal and inter-annual variability in the river discharge and to compensate for spring and autumn water deficits in crops. However, given the spatial distribution of agricultural lands and the less favourable, scattered location of goat breeders in the drylands downstream from the main users of the Río Mendoza waters, more intense and regulated use of water upstream by the groups with highest social power (i.e., those in the oasis) clearly reduces the possibilities of water "escaping" to the downstream part of the basin where desert communities (and goat breeders) are located. This evident inequity constitutes an interesting case in which a particular set of adaptive practices benefit some sectors and increase the vulnerability of other groups, exposing the complexity of this coupled natural and social system. Although their subordinate social position largely determines the sensitivity of these poor rural communities to various economic and environmental factors, extended droughts will ultimately be the trigger for increased conflicts in this hydraulic society (Masiokas et al. 2013; Montaña 2012b, 2008).

At the farm level, a broad classification of adaptive strategies can differentiate between traditional and "innovative" technologies. There is a clear distinction between the capitalized farmers who apply the latter and the small-scale producers and peasants who are restricted to the former. Every agricultural producer tries to have access to groundwater and make more efficient use of the resource, but not all of them can afford access to the aquifers. Just a small proportion of producers (less than 10%) can pay for permanent irrigation, which differs from many producers in the central Chilean valleys, where modern irrigation is more widespread as a result of a more capitalistic approach to viticulture. Smaller producers and peasants have to settle for more passive forms of adaptation, such as irrigating only the more profitable crops, or simply abandoning their water quantity and quality expectations and their productivity prospects and looking for alternative livelihoods. This is one of the reasons for the increasing impoverishment of small farmers, who become increasingly dependent on off-farm sources of income or sell their lands and migrate. For them, drought is devastating. Among the more highly capitalized farmers, innovative strategies in response to water scarcity involve, among other things, automatic irrigation systems (drip irrigation and others). The capitalized agricultural producers cannot sustain their activity without access to groundwater, as it is inherent to their technology and management style, allowing them to become less dependent on the surface irrigation scheme. This technology enables them to adopt another innovative adaptive strategy: diversifying locations and relocating properties in the foothills upstream to minimize hydro-climatological risks, an action that impacts negatively on aquifer conservation and on the agro-ecological conditions of downstream lands of the basin (Montaña 2012a).

An obvious adaptive measure is to more efficiently use the resource. In terms of developing agriculture with a reduced water demand, there is ample room for improvements through the modernization of irrigation systems. In the case of the Mendoza River basin, where grapes are the main crop, this modernization would have a very positive impact, both for individual producers and for the basin as a whole. But, as has been said, only the medium or large and well-capitalized farmers fully engaged in the "new viticulture" are able to make these investments. Even at low interest rates, loans to reconvert irrigation systems are not suitable for small growers, as their profitability is not sufficient to sustain this level of debt. As already experienced by under-capitalized producers in Chilean agriculture and many small farmers in the Canadian Palliser Triangle, small producers would probably not survive water efficiency measures, given that their adoption would entail economic, social, and political costs difficult to cope with. There is also potential to conserve more of the water currently used by households and industry in the region. A flat rate tariff for drinking water, low efficiency in residential facilities, and obsolete pipelines in the distribution system, together with little control over its use for garden irrigation and recreation, are all factors that explain the high rate of consumption per capita of a growing population and which certainly could be improved.

Local actors claim that more investment in infrastructure is required to deal with water shortages, but the Mendoza River basin has already benefited more from infrastructure works to improve the supply of the resource (dam, reservoir, waterproofing of irrigation channels, irrigation water distribution systems) than from measures aimed at controlling the demand, both rural and urban. And it is the rising water consumption of a growing population and increased economic activities that make the hydroclimatic scenarios where droughts are a central feature more complex.

From the perspective of increasing regional resilience, it is also necessary to understand the different types of drought and their patterns to decide which adaptive strategy to adopt and which type of investment to make. For instance, the region has limited adaptive potential in the long term to respond to hydrological droughts (characterized by decreasing streamflow in the river and dams). A potential adaptive strategy would be to capture and store summer rainwater that is not currently contributing to the river flow, making better use of groundwater and fostering the combined use of surface/groundwater. This strategy needs to occur not just on an individual basis as it is today in the basin, but as a planned collective strategy for managing the hydraulic system as a whole.

It should be noted, however, that the pursuit of efficiency in water use, although a worthy objective in terms of adaptation, is not a new issue. Mendoza, like other basins in Chile and the Canadian Prairies, is a region where rural people have had to historically coexist with water scarcities. Making more efficient use of water is an old goal, inherent to water management in drylands. Moreover, it constitutes a permanent adaptive measure to pursue, with or without climate change. This long adaptive history, however, has not resulted in rural people developing a healthy adaptive capacity to droughts. Rather, it seems that adaptation to increasing drought risk and climate variability will only be taken seriously when an extreme drought is declared. If historical memory is not enough to remind us of the impacts of intense and prolonged droughts in the regions of central-western Argentina, the scientific studies presented above should alert us to the need to take action before it is too late.

In coping with drought, there is always the idea that water limitations can be overcome and the oasis can be expanded. In fact, the "new viticulture" and its modern irrigation based on intensive use of groundwater have pushed the agricultural frontier over the foothills, degrading the agronomic and ecological conditions downstream[2] of the basin. From a social perspective, this degradation affects the small producers in downstream areas, who, already harassed by economic difficulties and a reduced income, become increasingly vulnerable. It is a desertification process that takes place within the oasis itself. Recovering land from the desert in the upstream (or losing it, according to one's point of view) means a desertification of the downstream by moving the oasis upstream (Montaña 2008a). Any surplus of water that might be used to expand the oasis must first be used to recover the old oasis areas that are becoming degraded.

Lessons Learned and Pending Tasks

In the context of intense droughts, climate predictions at local and regional spatial scales are valuable not only for societal benefits but also for planning and managing socio-economic sectors sensitive to climate variations. Several international scientific research centres currently provide global-scale climate predictions. Their ability to make climate predictions at regional and local scales, however, is very limited, not only because of the restricted levels of predictability but also because of the limited ability of current climate models to represent fundamental regional and local physical processes. Climate predictions for the Andes region are particularly challenging in terms of the current models. However, the fact that ENSO and other contributing forces of climate variability can be accurately predicted by the current climate models provides a basis from which to further explore predictability and develop climate prediction tools for the region with a minimum degree of certainty.

Physical indicators and climate indices, such as snowfall, snowfall distribution, mountain temperature, runoff (if possible from all the tributaries), reservoir and lake levels, temperature and precipitation at the oasis, groundwater levels, different uses of water (human consumption, irrigation, hydropower, industry, agriculture, natural ecosystems, cultural uses), and surface cultivation with annual and perennial crops, among other factors, are the main input variables for such models. These physical indicators and climate indices must then be combined with socio-economic variables to predict impacts on communities, assess vulnerabilities, and adopt the most appropriate adaptive strategies.

Based on this case study of Mendoza, it is apparent that preparedness for future drought cannot rely on the short historical memory of local communities. Preparedness planning should also integrate insights from long-term studies. In addition, preparedness plans must incorporate a dynamic model of the oasis that considers both natural and social systems, making sure that the adaptive actions of some groups do not create new vulnerabilities for others, at least not without proper remediation. Moreover, it has become clear that the scope of this undertaking should not be limited to the oasis—the more visible part of the territory—but should also encompass the Cordillera as the main source of water and look downstream to the "invisible spaces" of the desert (Montaña 2005), protecting the cultural diversity associated to the indigenous groups which live there and conserving the ecosystem services that support their style of development.

Advances in better planning and mitigation tools have been made, and these tools are now available worldwide. The main challenge, however, is to transform the social, economic, and political structures that have created an unequal distribution of vulnerability in the basin; without such transformation, drought (a rather normal climate event) becomes a hazard and a disaster for many. It is also fundamental to support governments and decision makers and to empower local people and other social actors to overcome the "short-termism" of the market drivers and narrow economic interests. Once this is achieved, more effective drought preparedness and mitigation plans can be prepared. In the agricultural sector, new and more efficient irrigation systems are needed, reclaimed waters could be better exploited, and training and social organization could contribute to developing more efficient traditional irrigation systems. In the urban sector, new water-conserving technologies need to be explored to more efficiently use water, and urban residents need increased awareness of water limitations. As a case study, Mendoza illustrates that drought takes a number of different forms and that adaptive strategies must be tailored to cope with the particularities of each one within the natural and social context.

NOTES

1 It is the Cordillera's precipitation (especially snow) that feeds the irrigation system. The region can benefit from rain in the foothills and in the plains, but the irrigation system is not designed to capture these resources.

2 . This process has been particularly studied for the River Tunuyán basin, south to the Mendoza River basin. See Chambuleyron (2002).

References

Agosta, E., P. Canziani, and M. Cavagnaro. 2012. "Regional Climate Variability Impacts on the Annual Grape Yield in Mendoza, Argentina." *Journal of Applied Meteorology and Climatology* 51: 993–1009.

Boninsegna, J., y R. Villalba. 2006a. *Los condicionantes geográficos y climáticos. Documento marco sobre la oferta hídrica en los oasis de riego de Mendoza y San Juan*. Primer informe a la Secretaria de Ambiente y Desarrollo Sustentable de la Nación.

———. 2006b. *Los condicionantes geográficos y climáticos. Documento marco sobre la oferta hídrica en los oasis de riego de Mendoza y San Juan*. Segundo informe a la Secretaria de Ambiente y Desarrollo Sustentable de la Nación.

Boninsegna, J., and S. Delgado. 2002. "Atuel River Streamflow Variations from 1575 to Present Reconstructed by Tree Rings: Their Relationships to the Southern Oscillation." Pp. 31–34 in D. Trombotto and R. Villalba (eds.), *IANIGLA, 30 Years of Basic and Applied Research on Environmental Sciences*. Mendoza, Argentina: Zeta Editores.

Bottero, R. 2002. "Inventario de glaciares de Mendoza y San Juan." Pp: 165–69 in D. Trombotto and R. Villalba (eds.), *IANIGLA, 30 Years of Basic and Applied Research on Environmental Sciences*. Mendoza, Argentina: Zeta Editores.

Bradley, R., M. Vuille, H. Diaz, and W. Vergara. 2006. "Threats to Water Supplies in the Tropical Andes." *Science* 312: 1755–56.

Chambuleyron, J. (ed). 2002. *Conflictos ambientales en tierras regadías. Evaluación de impactos en la cuenca del río Tunuyán, Mendoza, Argentina*. Mendoza, Argentina: Universidad Nacional de Cuyo.

Christie, D.A., J. Boninsegna, M. Cleaveland, A. Lara, C. Le Quesne, M.S. Morales, M. Mudelsee, D. Stahle, and R. Villalba. 2011. "Aridity Changes in the Temperate-Mediterranean Transition of the Andes since AD 1346 Reconstructed from Tree-rings." *Climate Dynamics* 36: 1505–1521. doi: 10.1007/s00382-009-0723-4.

CNA (Argentine National Census). 2002. *Censo Nacional Agropecuario*. Buenos Aires: INDEC, Gobierno de la Nación Argentina y DEIE, Ministerio de Economía, Gobierno de Mendoza.

Cobos, D.R., and J. Boninsegna. 1983. "Fluctuations of Some Glaciers in the Upper Atuel River Basin, Mendoza, Argentina." *Quaternary of South América and Antarctic Peninsula* 1: 61–82.

Compagnucci, R.H., and W.M. Vargas. 1998. Inter-annual Variability of the Cuyo Rivers' Streamflow in the Argentinean Andes Mountains and ENSO events." *International Journal of Climatology* 18: 1593–1609.

CONAMA (Corporación Nacional del Medio Ambiente). 2006. *Estudio de la variabilidad climática de Chile para el siglo XXI. Informe Final.* Santiago de Chile, Chile: Departamento de Geofisica de la Universidad de Chile.

Coria, L.A. 2014. "La participación vitivinícola en el Producto Bruto Geográfico de Mendoza en el siglo XX/ The wine participation in the Gross Geographical Product in the 20th Century." *Revista RIVAR, IDEA-USACH,* ISSN 0719-4994, V1 N° 3, septiembre 2014: 69–88.

DGI (General Department of Irrigation). 2007. *Plan Director de Ordenamiento de Recursos Hídricos– Informe Principal. Volumen II: Cuenca del Río Mendoza.* Mendoza: Gobierno de Mendoza

DGI (General Department of Irrigation). 2012. "Boletín Hidrometeorológico Junio 2012." Departamento de Evaluación de Recursos Hídricos. Mendoza: Gobierno de Mendoza.

Diaz, H., R. Gary-Fluhmann, J. McDowell, E. Montaña, B. Reyes, and S. Salas. 2011. "Vulnerability of Andean Communities to Climate Variability and Climate Change." Pp. 209–24 in W. Leal Filho (ed.), *Climate Change and the Sustainable Use of Water Resources.* Berlin: Springer-Verlag.

Frederick, K. 2011. *Water Management and Agriculture Development. A Case Study of the Cuyo Region of Argentina.* New York: Earthscan.

Glantz, M.H. 2003. *Climate Affairs: A Primer.* Washington, DC: Island Press.

Gonzalez, M., and C.S. Vera. 2010. "On the Interannual Wintertime Rainfall Variability in the Southern Andes." *International Journal of Climatology* 30: 643–57.

Leiva, J.C. 1999. "Recent Fluctuations of the Argentinian Glaciers." *Global Planetary Change* 22: 169–77.

Le Quesne, C., C. Acuña, J. Boninsegna, A. Rivera, and J. Barichivich. 2009. "Long-term Glacier Variations in the Central Andes of Argentina and Chile, Inferred from Historical Records and Tree-ring Reconstructed Precipitation. *Palaeogeography, Palaeoclimatology, Palaeoecology* 281: 334–44.

Leichenko, R., and K. O'Brien. 2008. *Double Exposure: Global Environmental Change in an Era of Globalization.* New York: Oxford University Press.

Luckman, B.H., and R. Villalba. 2001. "Assessing the Synchroneity of Glacier Fluctuations in the Western Cordillera of the America during the Last Millennium." Pp. 119–40 in V. Markgraf (ed.), *Inter-Hemispheric Climate Linkages.* New York: Academic Press.

Marre, M. 2011. *El agua no es suficiente.* Mendoza: Editorial Universidad Nacional de Cuyo.

Masiokas, M.H., R. Villalba, B.H. Luckman, C. Le Quesne, and J.C. Aravena. 2006. "Snowpack Variations in the Central Andes of Argentina and Chile,

1951–2005: Large-scale Atmospheric Influences and Implications for Water Resources in the Region." *Climate* 19, no. 24: 6334–52.

Masiokas, M.H., R. Villalba, B. Luckman, and S. Mauget. 2010. "Intra-to Multidecadal Variations of Snowpack and Streamflow Records in the Andes of Chile and Argentina between 30° and 37°S." *Journal of Hydrometeorology* 11: 822–31.

Masiokas, M.H., R. Villalba, B.H. Luckman, E. Montaña, D. Christie, E. Betman, C. Le Quesne, and S. Mauget. 2013. "Recent and Historic Andean Snowpack and Streamflow Variations and Vulnerability to Water Shortages in Central-western Argentina." In R. Pielke (ed.), *Climate Vulnerability, vol. 5: Vulnerability of Water Resources to Climate*. Burlington, VT: Elsevier Science.

Montaña, E. 2007. "Identidad regional y construcción del territorio en Mendoza, Argentina: memorias y olvidos estratégicos." *Bulletin de l'Institut Français d'Etudes Andines* 36, no. 2: 277–97.

———. 2008a. "Las disputas territoriales de una sociedad hídrica. Conflictos en torno al agua en Mendoza, Argentina." *Revista Interamericana de Economía Ecológica, Revibec* 9: 1–17.

———. 2008b. "Central Andes Foothill Farmers Facing Global Environmental Change." International Human Dimensions Programme (IHDP) on Global Environmental Change. *IHDP Update* 2 (October): 36–40.

———. 2011. *Compartir la escasez. Disputas por el agua en Mendoza, Argentina. En Estudios sociales sobre el riego en Argentina*. Primera parte. San Juan: INTA.

———. 2012a. *Escenarios de cambio ambiental global, escenarios de pobreza rural. Una mirada desde el territorio*. Buenos Aires: CLACSO-CROP.

———. 2012b. "Vulnerabilidades pasadas y presentes de los productores agrícolas y ganaderos de la cuenca de Mendoza." Pp. 59–74 en S. Salas, E. Jiménez, E. Montaña, R. Garay-Flühmann, D. Gauthier, y H.P. Diaz, *Vulnerabilidad al cambio climático Desafíos para la adaptación en las cuencas de Elqui y Mendoza*. La Serena: InterAmerican Institute for Global Change Research y Universidad de La Serena.

Montaña, E., and H.P. Diaz. 2012. "Global Environmental Change, Culture and Development: Rethinking the Ethics of Conservation." *The International Journal of Climate Change, Impacts and Responses* 3, no. 3: 31–40.

Montaña, E., L. Torres, E. Abraham, E. Torres, y G. Pastor. 2005. "Los espacios invisibles. Subordinación, marginalidad y exclusión de los territorios no irrigados en las tierras secas de Mendoza, Argentina." *Región y Sociedad* 32 (enero-abril): 3–32.

Montecinos, A., and P. Aceituno. 2003. "Seasonality of the ENSO-related Rainfall Variability in Central Chile and Associated Circulation Anomalies." *Journal of Climate* 16: 281–96.

Nuñez, M. 2006. *Desarrollo de escenarios climáticos en alta resolución para Patagonia y zona cordillerana. Período 2020/2030. Proyecto Desarrollo de Escenarios Climáticos y Estudios de Vulnerabilidad*. Informe N° 3. Secretaria de Ambiente y Desarrollo Sustentable.

Nuñez, M., y S. Solman. 2006. *Desarrollo de escenarios climáticos en alta resolución para Patagonia y zona cordillerana. Período 2020/2030*. Proyecto Desarrollo de Escenarios Climáticos y Estudios de Vulnerabilidad. Informe Nro 2. Secretaria de Ambiente y Desarrollo Sustentable.

Prieto, M.R., H.G. Herrera, T. Castrillejo, and P.I. Dussel. 2000. "Recent Climatic Variations and Water Availability in the Central Andes of Argentina and Chile (1885–1996). The Use of Historical Records to Reconstruct Climate" (in Spanish). *Meteorologica* 25, nos. 1–2: 27–43.

Prieto, M.R., D. Araneo, and R. Villalba. 2010. "The Great Droughts of 1924–25 and 1968–69 in the Argentinean Central Andes: Socio-economic Impacts and Responses." In *2nd International Regional Climate Variations in South America over the late Holocene: A New PAGES Initiative*. Valdivia, Chile: PAGES-IANIGLA-CONICET-Bern University.

Rutllant, J., and H. Fuenzalida. 1991. "Synoptic Aspects of the Central Chile Rainfall Variability Associated with the Southern Oscillation." *International Journal of Climatology* 11: 63–76.

Salas, S., E. Jiménez, E. Montaña, R. Garay-Flühmann, D. Gauthier, and H.P. Diaz. 2012. *Vulnerability to Climate Change. Challenges for Adaptation in the Elqui and the Basins*. La Serena, Chile: InterAmerican Institute for Global Change Research.

Scott, C.A., R.G. Varady, F. Meza, E. Montaña, G.B. de Raga, B. Luckman, and C. Martius. 2012. "Science-Policy Dialogues for Water Security: Addressing Vulnerability and Adaptation to Global Change in the Arid Americas." *Environment* 54, no. 3: 30–42.

Urrutia, R., and M. Vuille. 2009. "Climate Change Projections for the Tropical Andes Using a Regional Climate Model: Temperature and Precipitation Simulations for the End of the 21st Century." *Journal of Geophysical Research* 114 (Atmospheres). doi: 10.1029/2008JD011021.

Vargas, W., and G. Naumann. 2008. "Impacts of Climatic Change and Low Frequency Variability in Reference Series on Daily Maximum and Minimum Temperature in Southern South America." *Regional Environmental Change* 8: 45–57.

Vera, C., G. Silvestri, B. Liebmann, and P. González. 2006. "Climate Change Scenarios for Seasonal Precipitation in South America from IPCC-AR4 Models." *Geophysical Research Letters* 33, no. 13. doi: 10.1029/2006GL025759.

Viale, M., and F. Norte. 2009. "Strong Cross-barrier Flow under Stable Conditions Producing Intense Winter Orographic Precipitation: A Case Study over the Subtropical Central Andes." *Weather and Forecasting* 24: 1009–31.

Villalba, R., A. Lara, M.H. Masiokas, R. Urrutia, B.H. Luckman, G.J. Marshall, I.A. Mundo, D.A. Christie, E.R. Cook, R. Neukom, K. Allen, P. Fenwick, J.A. Boninsegna, A.M. Srur, M.S. Morales, D. Araneo, J.G. Palmer, E. Cuq, J.C. Aravena, A. Holz, and C. Le Quesne. 2012. "Unusual Southern Hemisphere Tree Growth Patterns Induced by Changes in the Southern Annular Mode." *Nature Geoscience* 5: 793–98.

Vitale, G. 1941 (2005). *Hidrología mendocina: contribución a su conocimiento.* Mendoza: Ediciones Culturales.

Viviroli, D., H.H. Dürr, B. Messerli, M. Meybeck, and R. Weingartner. 2007. "Mountains of the World, Water Towers for Humanity: Typology, Mapping, and Global Significance." *Water Resources Research* 43, no. 7. doi: 10.1029/2006WR005653.

Wilhite, D.A., and M.H. Glantz.1985. "Understanding the Drought Phenomenon: The Role of Definitions." *Water International* 10: 111–20.

Worster, D. 1985. *Rivers of Empire. Water, Aridity and Growth of the American West.* New York: Pantheon Books.

PART 7

CONCLUDING REMARKS

CONCLUSION

Margot Hurlbert, Harry Diaz, and Jim Warren

In a recent survey of people's perceptions about the most pressing risks facing society conducted by the World Economic Forum, *water crisis*, of which drought is one aspect, is ranked as the number one societal risk (WEF 2015). The water crisis, however, is a crisis with multiple contributing factors: increasing economic development, population growth, rapid urbanization, climate change, and governance issues to name a few. People cause and contribute to these factors; therefore, people have a large role to play in whether in fact a "crisis" occurs. The greatest environmental risks come in the form of extreme weather (of which drought is an example) and failures to adapt to climate change (WEF 2015). The high levels of awareness identified in the World Economic Forum survey suggest a widely held sense of urgency for discussing the topic of drought in the context of future climate change, social vulnerability, and adaptation.

This book underlines the need for an interdisciplinary approach to understand drought—the most significant natural hazard affecting livelihoods on the Canadian Prairies and other parts of the world. It builds on historical and empirical field studies focused on the social and economic impact of droughts in some of the semi-arid regions of Canada, Argentina, and Chile, and the capacity of local people and institutions to reduce the severity of these impacts. The uniting methodological approach to droughts in this book is that of a vulnerability perspective, wherein droughts are viewed through the lens of vulnerability—vulnerability as a function of both natural conditions and human occupancy, as well as the ability of natural and social systems to adapt. The chapters in this book detail drought impacts that have occurred and are yet to occur. In relation to those changes that have occurred, many agricultural producers and communities have demonstrated a substantial capacity to adapt and cope with these impacts in the context of other stressors. However, many farmers and ranchers have disappeared from the rural scene as a result of changes to the economic and political context during the last 50 years. An important lesson we have learned is that vulnerability is fluid, increasing or reducing the exposure of producers as a result of new social conditions.

A central concern that has informed the need for these studies is the threat of climate change. Most climate change scenarios developed for the three regions covered in this book indicate that future climate variability ranges will exceed those we have experienced in the recent past. These forecasts mean more frequent and severe droughts, a context wherein drought-related risks are potentially significant and cannot be ignored (see Chapter 3 by Wheaton et al. for the Canadian case). Four major "hotspots of vulnerability," summarized below, increase the seriousness of droughts in the context of increasing climate change.

Access to Natural Capital: Increasing Water Scarcities

Water resources constitute a natural capital that is fundamental to many human activities. A reduced availability of this capital is certainly problematic for many human activities, such as agriculture. The certainty of climate change and its impacts, as well as the human incapacity to mitigate them, makes access to this capital a highly risky enterprise in semi-arid regions. The impacts of droughts and the adaptations made in the three regions over recent decades are described in many of the preceding chapters (Chapters 4–8, 11, 13, and 14). In the Canadian case, these adaptations contributed to enhanced drought resilience in many rural communities. Those adaptations include changes in practice, such as the adoption of min till and irrigation described by Warren in Chapters 5 and 6. However, despite the high levels of technical, social, and institutional adaptation that have occurred over the past decades, Prairie agricultural producers and their communities remain vulnerable to drought.

The findings of paleoclimate research (discussed in Chapter 2 by Sauchyn and Kerr), together with projections based on future climate change scenarios (outlined in Chapter 3 by Wheaton et al.), suggest that droughts could become more frequent, more severe, and longer in duration in the Canadian Prairies over the course of the twenty-first century. As these authors point out, "dry times are expected to become much drier, and wet times wetter." Severe droughts are expected to become a more permanent feature in some areas of the Canadian Prairies as well as in the Maule and Mendoza regions of Chile and Argentina (Chapters 13 and 14, respectively). As pointed out by Kulshreshtha et al. in Chapter 4, drought, such as the one of 2001–2, has disastrous implications for Canada, the provinces, the communities, and the economic returns to agricultural producers. This is troubling, given the current drought resilience threshold of two to three years for agricultural production units identified by many research projects and reported in Chapter 9 by Hurlbert. This constitutes an adaptive range that is not sufficient to face the projected longer and more severe droughts predicted in Chapters 2 and 3.

Some of the chapters also indicate the high variability in access to water resources among rural people. By virtue of where agricultural production units are located in regions characterized by different geographical characteristics and microclimates, some producers have greater access to natural capital, such as adequate precipitation and reliable surface water and groundwater resources. Physical location in the water basin is sometimes a key for success, as demonstrated in the cases of Argentina and Chile. Chapter 14 by Montaña and Boninsegna highlights the dramatic situation of small producers in the tail end of the basin, who never seem to receive the necessary amount of water, while rich wine producers located on the other side of the basin seem to have ample access to water resources. Chapter 13, by Hadarits et al., reveals a similar situation in the Chilean region of Maule, where proximity to main irrigation canals influences exposure to drought. However, access to irrigation is not always the solution. As Warren demonstrates in Chapter 6, investment in irrigation infrastructure does not always ensure drought resilience. These problems seem to be more institutional in nature rather than just locational issues.

And, clearly without changes to the institutional context they present, cleavages among producers in accessing natural capital will be multiplied under future climate conditions.

Access to Economic Capital: The Threat of Double Exposure

It is not difficult to argue that agricultural producers' exposure to drought affects their economic capacities. Drought stress contributes to a significant reduction of productive assets, reducing the adaptive capacity of producers to respond to future droughts and other climate events. However, the relationship between droughts and economic processes is not unidirectional. Rather, these two dimensions are interlocked in complex ways and affect people's vulnerability in similarly intricate ways. For communities and individual producers, a reduction in the availability of economic capital due to economic stress imposes a significant constraint on their adaptive capacity in the face of prolonged drought. In this case, the overlap between climate and economic stressors multiplies the negative impacts of each of them.

The predominant source of economic stress in the Canadian case has been the unequal relationship between input costs and commodity prices—traditionally defined as the "cost-price squeeze"—over recent decades, which has limited the amount of capital available to withstand back-to-back crop failures. Until recently, the prices available for the commodities produced by Canadian farmers and ranchers in the Prairies have been poor. Prices for cereal grains experienced a brief peak in 1975, and that price level was not achieved again until the 2000s. In the case of cattle prices, the recent bovine spongiform encephalopathy (BSE) crisis (2003–7) produced a significant decline in producer incomes when important export markets were closed. At the same time, commodity prices have been excessively low, for decades, relative to increasing input costs. Prices for machinery, fuel, fertilizer, labour, and many herbicides and pesticides rose apace for decades, while agricultural product prices remained low by comparison. This strain on economic capital was historically a contributing issue to the disaster of the 1930s drought in Canada, as indicated by Marchildon in Chapter 8. Risks associated with reduced access to economic capital are perceived currently by producers as a significant exposure in a future characterized by climate change, as argued by Kulshreshtha et al. in Chapter 4, due to the increasing advent of free trade and exposure of Canadian producers to world market prices.

The cases from Argentina and Chile show that opportunities to have consolidated access to economic resources follow different paths for small and large producers. While large producers have permanent and secure access to economic assets, small producers have a reduced capacity to secure the same assets due to socio-economic segregation. This segregation is leading to a situation in which small producers have to neglect their productive units and search for non-agricultural jobs that can provide the family with a minimal income, a path that is becoming increasingly familiar to many rural families in Canada. This constant threat of double exposure—to drought and limited economic conditions—is increased by institutional failures.

Institutional Capital: Absent Governments

As identified by Hurlbert in Chapters 9 and 10, and by Marchildon in Chapter 8, institutional capital—those organizational resources and capacities that support decision making and manage risk in relation to drought and climate variability—is essential for community adaptation. In the three regions covered in this book, this capital is available to producers through a variety of local, regional, and national organizations and agencies. Significant changes in the fabric of government programs and policies in later years have not only diminished their capacity to reduce rural vulnerabilities but also have impeded the development of proper climate policies for facing the threat of climate change.

In relation to the Canadian case, Marchildon, in his historical account of the 1930s, covers two institutional adaptations from the 1930s, the Special Areas Board and the creation of the Prairie Farm Rehabilitation Administration (PFRA). Although the former is still in existence, the latter has been disbanded, with many negative consequences for the local community (see Chapter 6 by Warren). Aggravating the loss of the PFRA is the reduction of government staff, including engineers and scientists, in federal government agencies, such as Environment Canada, Natural Resources Canada, and Agriculture and Agri-Food Canada, and over the last few decades at all levels of government. In addition to this loss of personnel,

Fletcher and Knuttila (Chapter 7) point out the loss of significant historical programs that assisted agricultural producers, including the Canadian Wheat Board single-desk system and the Crow Rate, which assisted grain transportation, while Hurlbert (Chapter 9) mentions reduced payouts under AgriStability, an important government program designed to reduce the impacts of declines in producer incomes. In the cases of Argentina and Chile, the existence of a formal institutional framework oriented to ensure the economic viability of large producers to the detriment of small producers is clearly a significant institutional gap in the development of a robust, coordinated, and anticipatory approach to reduce the risks associated with climate events.

In this context, there is a clear need for a more profound analysis of neo-liberalism and its adoption into government policies, programs, and practices. The adoption of this strategy by governments involves minimal government intervention in business and reduced public expenditures, and favours markets and individual responsibility over social welfare spending. Chile has been more radical in its commitment to neo-liberalism, with its own particular impacts on natural resources such as water. Canada and Argentina, on the other hand, have redefined and shaped many of their economic programs and policies according to the fundaments of liberalism but still maintain many of their social programs. However, in both countries, the market and the private sector have taken a central role in economic development, while the government is increasingly focusing its energies and efforts on establishing a proper normative and institutional framework for the development of a liberal economy.

Social Capital: Disempowered Communities

The existence of social capital ensures the availability of collective resources and capacities to deal with a variety of stressors. Local institutions, including informal social networks, contribute significantly to strengthening this social capital in rural society. Work-trading arrangements and labour-sharing activities, such as community brandings and firefighting, strengthen community bonds, and networks for mutual support provide forums for knowledge sharing and for the existence of a social capital fundamental to facing the hazards of climate and other stressors.

However, several trends are foreboding for social capital: the increase in farm size and corresponding decrease in the number of farmers (cited by Fletcher and Knuttila in Chapter 7 and Marchildon in Chapter 8), and the increase in farm debt (cited by Fletcher and Knuttila in Chapter 7). The first trend was also identified by Corkal et al. in Chapter 11 as threatening a highly valued way of life and heritage as reduced population increased the prospect of losing community schools and having churches close. Depopulation would reduce a variety of institutions and practices, including local hockey teams and intercommunity sporting events. Ultimately, local networks of strength and social support (the community safety net) would suffer. In Chapter 7, Fletcher and Knuttila also identify the trend of increasing debt as threatening the support network of farm women. Social capital in Argentina and Chile is following a similar path, where an increasing process of differentiation between modern and traditional agriculture is eroding the fabric of rural communities.

Although the four "hot spots of vulnerability" identified above are cause for serious concern, two developments related to the buffering of social capital show cause for optimism. First, as reported by Hurlbert in Chapter 10, Warren in Chapter 5, and Corkal et al. in Chapter 11, some very promising displays of strong social capital have emerged during times of water scarcity. The most significant example is water sharing by irrigators and communities, which occurred during the 2001–2 drought. Not only did irrigators share, transfer, assign, and optimize water interests regardless of legal rights to optimize returns and benefits during this time, but local communities also entered into voluntary water-sharing arrangements and water-reduction strategies to preserve this important resource.

The next significant development of social capital is recounted by Pittman et al. in Chapter 12. Bottom-up local governance initiatives to respond to drought have occurred with increasing frequency in the Prairie provinces to plan proactively for times of water shortages. New groups of actors are emerging with new roles in navigating drought risks. Knowledge-bridging activities between diverse stakeholders through local watershed stewardship organizations are not without difficulties, but they show great promise and increasing value in building social capital, enhancing resiliency, and reducing vulnerability to anticipated future droughts.

Overall, the chapters in this book increase our understanding of drought and its impacts on natural and social systems. Furthermore, this book outlines adaptations that have been made by people to cope with drought and improve resiliency, increasing our ability to understand the hazard of drought and transform its impact by proactively planning for drought and creating and taking advantage of opportunities. Future research surrounding drought is needed to clearly identify opportunities associated with our increasingly wetter and drier climate and to take advantage of them. More information is needed not only on changes in agricultural production techniques, crops, and geographical area, but also on the specific configurations of social and governance arrangements that can take advantage of opportunities. In addition, this book has demonstrated that studying changes in adaptive capacity over time, as was done here by comparing adaptation to the drought of the 1930s to that of 2001–2, can provide useful information. A time-path study of changes in adaptive capacity and changes in the capitals (economic, technological, human, natural, infrastructure, and institutional) of communities would provide invaluable information for policy planners. This study could also provide information on the cumulative impacts resulting from recurrent drought. Further comparative case studies of the adaptation of irrigated agricultural producers, the adaptation of dryland producers, viticulture and horticulture, or the specific type and duration of drought together with adaptive strategies in different communities would offer insight into contextual and institutional determinants of adaptation.

In the context of comparative studies, it is fundamental to understand the cases of other regions and countries. In the same way that we pay attention to future climate scenarios to identify future vulnerabilities, the social and economic situation of other countries could be analyzed as potential social scenarios for Canada. Given the process of neo-liberalization of Canadian society, the restructuring of the agricultural sector, and the increasing reduction of state services, the cases of Argentina and Chile appear as concerning probable scenarios for Canada's rural people. The chapters on Argentina and Chile provide illustrative insight into what maladaptation might look like as a result of economic, social, and institutional decisions, such as the complete privatization of water in Chile and its impact on human and social capital. These chapters also provide insight into the growing inequity between capitalized irrigators and smaller irrigators in Argentina and the associated spatial inequality.

This book provides insight into the conditions generating challenges for the future regarding droughts and the measures required to reduce the vulnerability of rural communities to them. Meeting these future challenges will require developing a greater understanding of the social forces and conditions that have contributed to enhanced resilience, as well as those which detract from successful adaptation. The most important conclusion of this book is that the problem of drought is a vast and pernicious problem. However, solutions lie within the actions and planning of people, as well as local, municipal, provincial, and national governments. As humans we have perceptions that exceed our immediate needs, which allow us to understand the world around us. This same ability will allow for consistent, principled, and far-sighted action plans needed to combat the problem of drought in the future. This book moves us down that path.

References

WEF (World Economic Forum). 2015. *Global Risks Report 2015*. Geneva: WEF.

INDEX

A

B

C

D

E

F

G

H

I

N

O

P

Q

R

S

W

Z